Fundamentals of Abstract Algebra

Fundamentals of Abstract Algebra is a primary textbook for a one-year first course in Abstract Algebra, but it has much more to offer besides this. The book is full of opportunities for further, deeper reading, including explorations of interesting applications and more advanced topics, such as Galois theory. Replete with exercises and examples, the book is geared towards careful pedagogy and accessibility, and requires only minimal prerequisites. The book includes a primer on some basic mathematical concepts that will be useful for readers to understand, and in this sense, the book is self-contained.

Features
- Self-contained treatments of all topics
- Everything required for a one-year first course in Abstract Algebra, and could also be used as supplementary reading for a second course
- Copious exercises and examples

Mark DeBonis received his PhD in Mathematics from the University of California, Irvine, USA. He began his career as a theoretical mathematician in the field of group theory and model theory, but in later years switched to applied mathematics, in particular to machine learning. He spent some time working for the US Department of Energy at Los Alamos National Lab as well as the US Department of Defense at the Defense Intelligence Agency, both as an applied mathematician of machine learning. He held a position as Associate Professor of Mathematics at Manhattan College in New York City, but later left to pursue research working for the US Department of Energy at Sandia National Laboratory as an applied mathematician of machine learning. His research interests include machine learning, statistics and computational algebra.

Textbooks in Mathematics
Series editors:
Al Boggess, Kenneth H. Rosen

Classical Analysis
An Approach through Problems
Hongwei Chen

Classical Vector Algebra
Vladimir Lepetic

Introduction to Number Theory
Mark Hunacek

Probability and Statistics for Engineering and the Sciences with Modeling using R
William P. Fox and Rodney X. Sturdivant

Computational Optimization: Success in Practice
Vladislav Bukshtynov

Computational Linear Algebra: with Applications and MATLAB® Computations
Robert E. White

Linear Algebra With Machine Learning and Data
Crista Arangala

Discrete Mathematics with Coding
Hugo D. Junghenn

Applied Mathematics for Scientists and Engineers
Youssef N. Raffoul

Graphs and Digraphs, Seventh Edition
Gary Chartrand, Heather Jordon, Vincent Vatter and Ping Zhang

An Introduction to Optimization with Applications in Data Analytics and Machine Learning
Jeffrey Paul Wheeler

Encounters with Chaos and Fractals, Third Edition
Denny Gulick and Jeff Ford

Differential Calculus in Several Variables
A Learning-by-Doing Approach
Marius Ghergu

Taking the "Oof!" out of Proofs
A Primer on Mathematical Proofs
Alexandr Draganov

Vector Calculus
Steven G. Krantz and Harold Parks

Intuitive Axiomatic Set Theory
José Luis García

Fundamentals of Abstract Algebra
Mark J. DeBonis

https://www.routledge.com/Textbooks-in-Mathematics/book-series/CANDHTEXBOOMTH

Fundamentals of Abstract Algebra

Mark J. DeBonis

CRC Press
Taylor & Francis Group
Boca Raton London New York

CRC Press is an imprint of the
Taylor & Francis Group, an **informa** business

A CHAPMAN & HALL BOOK

First edition published 2024
by CRC Press
2385 NW Executive Center Drive, Suite 320, Boca Raton FL 33431

and by CRC Press
4 Park Square, Milton Park, Abingdon, Oxon, OX14 4RN

CRC Press is an imprint of Taylor & Francis Group, LLC

© 2024 Mark J. DeBonis

Library of Congress Cataloging-in-Publication Data

Names: DeBonis, Mark J., author.
Title: Fundamentals of abstract algebra / Mark J. DaBonis.
Description: First edition. | Boca Raton : C&H/CRC Press, 2024. | Series:
Textbooks in mathematics | Includes bibliographical references and
index. | Summary: "Fundamentals of Abstract Algebra is a primary
textbook for a one year first course in Abstract Algebra, but it has
much more to offer besides this. The book is full of opportunities for
further, deeper reading, including explorations of interesting
applications and more advanced topics, such as Galois theory. Replete
with exercises and examples, the book is geared towards careful pedagogy
and accessibility, and requires only minimal prerequisites. The book
includes a primer on some basic mathematical concepts that will be
useful for readers to understand, and in this sense the book is
self-contained"-- Provided by publisher.
Identifiers: LCCN 2023048005 (print) | LCCN 2023048006 (ebook) | ISBN
9781032367019 (hbk) | ISBN 9781032370910 (pbk) | ISBN 9781003335283
(ebk)
Subjects: LCSH: Algebra, Abstract--Textbooks.
Classification: LCC QA162 .D436 2024 (print) | LCC QA162 (ebook) | DDC
512/.02--dc23/eng/20231220
LC record available at https://lccn.loc.gov/2023048005
LC ebook record available at https://lccn.loc.gov/2023048006

ISBN: 978-1-032-36701-9 (hbk)
ISBN: 978-1-032-37091-0 (pbk)
ISBN: 978-1-003-33528-3 (ebk)

DOI: 10.1201/9781003335283

Typeset in Latin Modern font
by KnowledgeWorks Global Ltd.

Publisher's note: This book has been prepared from camera-ready copy provided by the authors

To my parents Clement and Josephine DeBonis.

Contents

Preface

I started this text some time ago, just after my PhD, when I began teaching abstract algebra to university undergraduate mathematics majors. It was a way to organize my lectures and not have to rethink my presentation from scratch each time I taught the course (obviously I am not unique in this idea). However, later I used the text as a way to record interesting topics, examples, techniques and applications and to answer those questions that were often referred to, but often without providing any mathematical proof. This text became my algebraic memoirs and a compendium of basic notions of abstract algebra I collected during my journeys through this field. This journey still continues even as I write this preface. For instance, applications in neural networks that are *group equivariant* and reinforcement learning and a *partially observable Markov decision process* (POMDP) which utilizes group representation. Abstract algebra is finding its way into a myriad of important *real world* (whatever that means) applications.

I remember my first encounter with abstract algebra as an undergraduate mathematics major, a standard course in the major. It was this mathematics course that I instantly fell in love with, and knew even then that this would be the area of study I would pursue as a graduate student in pure (a very pretentious adjective) mathematics. I even remember that day in class when I was overwhelmed with amazement by the abstraction this topic offered. We were using Fraleigh's text, *A First Course in Abstract Algebra*, and the topic was discovering when group cosets formed a well-defined group in and of themselves, and this led to the notion of normal subgroups. I had a terrible instructor for that course and wound up reading the text on my own, and probably learned more than I ever would have even with a good teacher.

This text spans more material than a standard one-year course in abstract algebra. However, the text contains as a subset the standard material for a one-year class. It is self-contained, in the sense that it has all the necessary background for the topics it presents. It is highly recommended that the student has had a course on how to do mathematical proof, however I would be hypocritical if I insisted on this, not having taking the course myself. Some very basic topics are presented in the first chapter that are foundational for what is to follow and is highly recommended, although most can be found in that introduction to mathematical proof course I mentioned. Even if you have seen this material, it would not hurt to cover it twice. The text is divided into two parts – groups (which typically covers one semester), then rings and fields for the second semester. There are some who argue that the two parts should be reversed, however I decided to start with groups, since there are less axioms involved for this algebraic structure and perhaps this makes it more manageable as a first topic in abstract algebra.

Here are my suggestions for a first pass through the text, although, of course, the instructor of this course would know best what to cover. As already stated, Chapter 1 is a good idea. Most of Chapter 2 is also fundamental, except perhaps Section 2.9.1 on semi-direct products. Chapter 3 can be skipped without any disruption in the continuity and reflects that aspect of the text which is more of a compendium of information. I see Chapter 4 as critical which hopefully the instructor can reach by the end of the semester. It is a wonderful application of group theory. Section 4.7 can be omitted. Chapter 5 and 6 can also be skipped on a first pass, although the information therein is important. So far as rings and fields are concerned all of Chapter 7 is essential, as well as Chapter 8 which is typically covered in a first course in abstract algebra. In Chapter 9, one can skip 9.3 (although an interesting application if you are interested in such things). Section 9.5 is more theoretical and may be skipped. That should finish one year of abstract algebra, but for those who want to delve in Galois theory, say in a follow on course, Chapter 10 makes for such a course (perhaps with a review of Chapter 9).

Thank you for considering this text for your adventures in abstract algebra. I hope you discover the beauty in this topic just as I did some time ago.

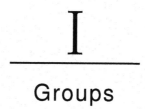

Groups

CHAPTER 1

Background Material

I N THIS CHAPTER, we present a collection of topics which serve as a background for the ideas you will see in this text. In Section 1.1, we present equivalence relations which foreshadow many important structures in abstract algebra such as the cosets of a group, the orbits of a group action, and much more. In Section 1.2, we present basic ideas about functions since all the structures presented in this text will have special functions associated with them. Section 1.3 will highlight some fundamental properties of the integers as a prelude to the more general topic of rings. In Section 1.4, we describe in more detail congruence modulo n which is an equivalence relation with additional properties. The equivalence classes of this relation produce one of the basic examples of a group and of a ring. Although, not explicitly presented, ideas in basic set theory are prevalent in this text and are an important background topic.

1.1 EQUIVALENCE RELATIONS

The notion of a relation on a set is important in many fields of mathematics. We shall see many applications of a particular type of relation (called an equivalence relation) in this text. We start by defining a relation and then narrow things down to an equivalence relation.

Definition 1.1 *A **relation** $\sim$ on a set A is simply any subset of the cartesian product $A \times A$, i.e. a collection of ordered pairs of elements of A.*

If (a, b) is an element of $\sim$, we instead write $a \sim b$ and we say *a relates to b*.

Example 1.1 *Here, we list a number of examples of relations.*

1. *Let $A = \{a, b, c, d\}$ and let $\sim$ be the subset $\{(a, b), (b, b), (c, d)\}$ of $A \times A$. For instance, according to our definition of $\sim$, we have $c \sim d$ or c relates to d.*

2. *Let $A = \mathbb{Z}$ (the set of integers) and $\sim$ be $<$. In other words, (n, m) is in $\sim$ or $n \sim m$ exactly when $n < m$.*

3. *Set $A = \mathcal{P}(\mathbb{Z})$ which represents all the subsets of $\mathbb{Z}$ (called the **power set** of $\mathbb{Z}$). Let $\sim$ be $\subseteq$, i.e. subset. In other words, a subset X relates to another subset Y of $\mathbb{Z}$ exactly when $X \subseteq Y$.*

DOI: 10.1201/9781003335283-1

4. *Take any set A and let $\sim$ be equality, i.e. $a \sim b$ exactly when $a = b$. In other words, $\sim$ is the set $\{(a, a) \; : \; a \in A\}$.*

5. *Let $f : A \to B$ be a function from a set A to another set B. Define a relation on A as follows: $a \sim b$ iff $f(a) = f(b)$.*

6. *Let $A = \mathbb{Z}$ and define $\sim$ as follows: $n \sim m$ iff there exists an integer k such that $m = nk$. One says that n **divides** m and we write $n|m$. For instance, $(3, -15)$ is in $\sim$, since 3 divides -15 because $-15 = 3(-5)$.*

7. *Define a relation on the set $\mathbb{Z}$ as follows: Fix a positive integer n and define $m \sim k$ iff $n|(m - k)$. This relation is called **congruence modulo n**, and in place of $m \sim k$, we typically write $m \equiv_n k$ or $m \equiv k \pmod{n}$. Section 1.4 is dedicated to this important relation.*

8. *Define a relation on $\mathbb{Q}$ as follows: $\frac{a}{b} \sim \frac{c}{d}$ iff $ad = bc$. So for instance $\left(\frac{1}{2}, \frac{-3}{-6}\right)$ is in $\sim$ or $\frac{1}{2} \sim \frac{-3}{-6}$, since $(1)(-6) = (2)(-3)$.*

There are various properties one may wish to investigate with respect to a relation. We list a few below.

Definition 1.2 *Let $\sim$ be a relation on a set A. We say $\sim$ is*

1. **reflexive** *if for all $a \in A$, we have $a \sim a$.*

2. **symmetric** *if for all $a, b \in A$, we have $a \sim b$ implies $b \sim a$.*

3. **transitive** *if for all $a, b, c \in A$, we have $a \sim b$ and $b \sim c$ implies $a \sim c$.*

4. **irreflexive** *if for all $a \in A$, we have $a \not\sim a$.*

5. **anti-symmetric** *if for all $a, b \in A$, we have $a \sim b$ and $b \sim a$ implies $a = b$.*

Some examples of types of relations that are of particular importance in mathematics are the following:

Definition 1.3 *Let $\sim$ be a relation on a set A. We say that $\sim$ is*

1. *a **partial ordering** of A if it is reflexive, anti-symmetric and transitive.*

2. *an **equivalence relation** on A if it is reflexive, symmetric and transitive.*

3. *a **function** on A if whenever $a \sim b$ and $a \sim c$, then it must be that $b = c$.*

Example 1.2 *We see now that $\subseteq$ is a partial ordering on $\mathcal{P}(\mathbb{Z})$ and $\equiv_n$ is an equivalence relation on $\mathbb{Z}$. Note that if we restrict the relation **divides** to a relation on positive integers, then it becomes a partial ordering on positive integers.*

The focus of our discussion for the remainder of this section is equivalence relations. Let's list here the examples introduced already which are equivalence relations. The reader should take the time to prove that they are indeed equivalence relations.

Example 1.3 *Here are some examples of equivalence relations.*

1. *Take any set A and let $\sim$ be equality, i.e. $a \sim b$ exactly when $a = b$.*

2. *Let $f : A \to B$ be a function from a set A to another set B. Define a relation on A as follows: $a \sim b$ iff $f(a) = f(b)$.*

3. *Define a relation on the set $\mathbb{Z}$ as follows: Fix a positive integer n and define $m \sim k$ iff $m \equiv k \pmod{n}$.*

4. *Define a relation on $\mathbb{Q}$ as follows: $\frac{a}{b} \sim \frac{c}{d}$ iff $ad = bc$.*

Example 1.4 *Here are some equivalence relations specific to linear algebra for those familiar with the material.*

1. *Matrix equivalence is an equivalence relation on the set of $m \times n$ matrices; i.e. for two $m \times n$ matrices A and B define $A \sim B$ iff there exist a finite number of elementary row operations which convert A into B.*

2. *Matrix similarity is an equivalence relation $n \times n$ matrices; i.e. for two $n \times n$ matrices A and B define $A \sim B$ iff there is an invertible matrix P such that $B = P^{-1}AP$.*

3. *Isomorphism is an equivalence relation on the set of vector spaces; i.e. two vector spaces V and W relate iff they are isomorphic. For instance, $\mathbb{R}^3$ is equivalent to P_2, since 3-tuples are isomorphic to polynomials of degree 2 or less (because they have the same dimension, namely 3).*

As we have stated already, for the remainder of this section, we will be assuming that $\sim$ is an equivalence relation, and as such, we typically use the notation $\equiv$ in place of $\sim$.

Definition 1.4 *Let $\equiv$ be an equivalence relation on a set A and $a \in A$. The **equivalence class** of a **with respect to** $\equiv$, written*

$$[a]_\equiv \quad equals \quad \{b \in A \ : \ a \equiv b\}.$$

*The element a is sometimes called a **representative** of the class $[a]_\equiv$. The collection of all equivalence classes of A with respect to $\equiv$, in other words $\{[a]_\equiv \ : \ a \in A\}$, is denoted by $A/\equiv$ and is called the **quotient set** of A.*

At times we will simply write $[a]$ in place of $[a]_\equiv$ when the equivalence relation is understood, and we may simply call $[a]$ the **class of** a for brevity. Some other notation for an equivalence class which the reader may encounter here or in other texts is $\overline{a}$ in place of $[a]$.

Example 1.5 *Let's compute some equivalence classes for the examples already presented.*

1. *The equivalence classes for equality on a set A are singleton sets, i.e. $[a] = \{a\}$, since no other element besides a relates to a.*

2. *For the equivalence relation we defined on $\mathbb{Q}$ (see Example 1.3.4), an equivalence class represents all the different ways we can represent a particular fraction. For instance, the equivalence class*

$$\left[\frac{1}{2}\right] = \left\{\frac{1}{2}, \frac{-3}{-6}, \frac{12}{24}, \ldots\right\}.$$

3. *Consider the equivalence relation congruence modulo 3 on $\mathbb{Z}$. There are exactly three distinct equivalence classes. Each class contains integers which when divided by 3 yield the same remainder.*

$$[0] = \{0, \pm 3, \pm 6, \ldots\}$$

$$[1] = \{\ldots, -8, -5, -2, 1, 4, 7, \ldots\}$$

$$[2] = \{\ldots, -7, -4, -1, 2, 5, 8, \ldots\}$$

One can view equivalence relations as a generalization of equality. Each class in a sense contains all the elements of a set which we view as being the same. Just consider the example of the equivalence class of $\frac{1}{2}$. In practice, we view $\frac{1}{2}$ and $\frac{-3}{-6}$ as being the same even though symbolically they look very different. Equivalence classes are simply a formal way of equating things which we wish to view as being equal.

We now prove a result which uncovers the essential properties of an equivalence relation.

Lemma 1.1 *Let $\equiv$ be an equivalence relation on a set A.*

1. *For all $a \in A$, we have $a \in [a]$.*

2. *For all $a, b \in A$, we have $[a] = [b]$ iff $a \equiv b$.*

3. *For all $a, b \in A$, either $[a] = [b]$ or $[a] \cap [b] = \emptyset$.*

Proof 1.1 *The first part follows immediately from the reflexive property. For the second part, assume first that $[a] = [b]$. Now since $a \in [a]$, we have $a \in [b]$ and so by definition and symmetry $a \equiv b$. Now assume that $a \equiv b$. Using transitivity and symmetry, notice that $c \in [a]$ iff $a \equiv c$ iff $c \equiv b$ iff $c \in [b]$ and so $[a] = [b]$. For the last part, either $[a] = [b]$ or $[a] \neq [b]$. In the latter case, we show that $[a]$ and $[b]$ are disjoint, thus proving the result. Indeed, we prove this by proving the contrapositive. Therefore, assuming $[a] \cap [b] \neq \emptyset$, there is some $c \in [a] \cap [b]$. Then $c \in [a]$ and $c \in [b]$ and so $c \equiv a$ and $c \equiv b$. Using symmetry and transitivity, we have $a \equiv b$ and so by the second part $[a] = [b]$.* □

Notice that the second part of the lemma says that any element of a class can represent that class, i.e. if $b \in [a]$, then $[b] = [a]$. The first and third part of the lemma says that equivalence classes divide the set A into a union of disjoint subsets.

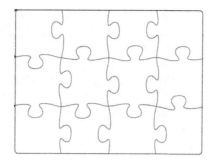

Figure 1.1 A partition is a puzzle.

Example 1.6 *Let's consider what equivalence classes look like in the case of our linear algebraic examples of an equivalence relation.*

1. *Consider the equivalence relation on M_{mn} matrix equivalence. Each equivalence class can be represented by a reduced row-echelon form matrix.*

2. *Consider the equivalence relation similarity on $n \times n$ diagonalizable matrices. Each equivalence class can be represented by a diagonal matrix.*

3. *Consider the equivalence relation isomorphism on finite dimensional vector spaces. Each equivalence class can be represented by $\mathbb{R}^n$, i.e. if $\mathcal{V}$ represents the collection of all finite dimensional vector spaces, then*

$$\mathcal{V}/\equiv \quad equals \quad \{[\mathbb{R}^n] \mid n = 0, 1, 2, \ldots\},$$

if we allow $\mathbb{R}^0$ to be the set $\{0\}$.

Let's formally define this notion of a partition of a set.

Definition 1.5 *Let A be a non-empty set and $\mathcal{P}$ be a family of non-empty subsets of A. We say $\mathcal{P}$ is a **partition of A** or $\mathcal{P}$ **partitions** A if*

1. *For all $a \in A$, there is an $X \in \mathcal{P}$ such that $a \in X$.*

2. *For all $X, Y \in \mathcal{P}$ distinct, we have $X \cap Y = \emptyset$.*

One can think of a partition of a set as a puzzle (see Figure 1.1) where each puzzle piece is an element of the partition, and when you put all the puzzle pieces together, you get set A. According to this formal definition, we see from the lemma that $A/\equiv$ is a partition of A.

Example 1.7 *Consider the earlier example of congruence modulo 3 an equivalence relation on $\mathbb{Z}$. The partition into equivalence classes, namely $\mathbb{Z}/\equiv_3$, consists of three puzzle pieces, namely $[0]$, $[1]$ and $[2]$. These three classes are pairwise disjoint and their union is all of $\mathbb{Z}$.*

EXERCISES

1 For the examples in Example 1.1, list three elements in each relation.

2 State and prove which of the properties in Definition 1.2 has each of the examples in Example 1.1.

3 Let A be the set of integers and $\sim$ be the relation $\leq$. State and prove which of the properties in Definition 1.2 this relation has.

4 Verify that $\subseteq$ is a partial ordering on $\mathcal{P}(\mathbb{Z})$ (you may use your work in Exercise 2).

5 Verify that $\equiv_n$ is an equivalence relation on $\mathbb{Z}$ (you may use your work in Exercise 2).

6 Verify that if we restrict the relation **divides** to a relation on positive integers, then it becomes a partial ordering on positive integers.

7 Verify that the relations listed in Example 1.3 are equivalence relations (you may use your work in Exercise 2).

8 If $f : \mathbb{R} \to \mathbb{R}$ is the function defined by $f(x) = x^2$, describe the equivalence classes of Example 1.3.2.

9 For each of the following relations on $\mathbb{Z}$, decide if it's an equivalence relation. If it is, then verify; otherwise, give a counter example.

   a. $a \sim b$ iff $|a - b| \leq 3$.

   b. $a \sim b$ iff $2|(a + b)$.

   c. $a \sim b$ iff $3|(a + b)$.

   d. $a \sim b$ iff $ab \leq 0$.

   e. $a \sim b$ iff $ab \geq 0$.

10 For each of the following relations on a set A, decide if it's an equivalence relation. If it is, then verify; otherwise, give a counter example.

   a. A is the set of all English words. For $x, y \in A$ define $x \sim y$ iff x has a letter in common with y.

   b. $A = \mathbb{R}^2$ and for $(a, b), (c, d) \in A$ define $(a, b) \sim (c, d)$ iff $a + d = b + c$.

   c. Let $A = \{2, 3, 4, \ldots\}$. For $m, n \in A$ define $m \sim n$ iff there is a prime p such that $p \mid m$ and $p \mid n$.

11 Define the following relation on $\mathbb{R}^2$: $(a, b) \sim (c, d)$ iff $a^2 + b^2 = c^2 + d^2$.

    a. Verify that $\sim$ is an equivalence relation.

    b. Describe geometrically an equivalence class for $\sim$.

12 For each of the following relations, verify that it is an equivalence relation and then describe clearly a typical equivalence class.

    a. A is the set of all differentiable functions in one variable and $f \equiv g$ iff $f'(x) = g'(x)$.

    b. $A = \mathbb{R}^2$ and $(a, b) \equiv (c, d)$ iff $a = c$.

13 Suppose a relation $\sim$ on a set A has the following two properties:

    a. For all $a \in A$, $a \sim a$.

    b. For all $a, b, c \in A$ if $a \sim b$ and $b \sim c$, then $c \sim a$.

Prove that $\sim$ is an equivalence relation on A.

14 Consider a circle divided into four equal sectors.

    a. If we can color each of the four sectors either black or white, how many different colorings are there?

    b. Let A be the set of colorings in part a, and define a relation as follows: Two colorings of the circle are equivalent if you can get from one to the other by rotating the circle by either $0°$, $90°$, $180°$ or $270°$. Prove that this relation is, in fact, an equivalence relation.

    c. List the set of equivalence classes for this equivalence relation.

15 Prove the following statements:

- Given a partition $\mathcal{P}$ of a set A, the relation defined by $a \equiv b$ iff there is an $X \in \mathcal{P}$ such that $a, b \in X$ defines an equivalence relation whose equivalence classes consist precisely of the elements of $\mathcal{P}$.

- Conversely, if one starts with an equivalence relation $\equiv$ on a set A and forms the partition into equivalence classes, and then defines an equivalence relation as we just did in the previous bullet, then we wind up with the same equivalence relation as we began with.

1.2 FUNCTIONS

In this section, we present the facts about functions that we use from time to time in the course of this text. We start with the definition.

Definition 1.6 *Let A and B be sets.*

1. The **cartesian product** of A and B, written

$$A \times B = \{(a, b) \ : \ a \in A \ \text{ and } \ b \in B\}.$$

2. A **relation** R from A to B is any subset of $A \times B$.

3. A **function** (or **map**) f from A to B, written $f : A \to B$, is a relation from A to B with the added properties that for each $a \in A$,

   - there is some $b \in B$ such that $(a, b) \in f$, and
   - if $(a, b), (a, c) \in f$, then $b = c$.

4. Given a function f from A to B, set A is called the **domain** of f and the set B is called the **codomain** of f. The **range** (or **image**) of f, written $f(A) = \{b \in B \ : \ (a, b) \in f\}$.

5. Given a function f from A to B, if $(a, b) \in f$, then element $a \in A$ is called an **input** for f and the element $b \in B$ is called an **output** for f. We employ the notation $f(a) = b$ in this situation.

6. Two functions f and g from A to B are **equal** if they are equal as sets. In other words, for all $a \in A$, we have $f(a) = g(a)$.

Example 1.8 Here, we present some specific examples of functions.

1. Let $A = \{a, b, c\}$, $B = \{1, 2, 3\}$ and define $f : A \to B$ by $f = \{(a, 3), (b, 1), (c, 1)\}$. Then $f(a) = 3$, $f(b) = 1$, $f(c) = 1$ and $R(f) = \{1, 3\}$.

2. Functions can also be defined in terms of a formula. For instance, we can define $f : \mathbb{Z} \to \mathbb{Z}$ by $f(x) = x^2$. In which case

$$f = \{(0, 0), (1, 1), (-1, 1), (2, 4), (-2, 4), \ldots\}.$$

We will move away from the ordered pair representation of a function, i.e. $(a, b) \in f$, and make use of the notation $f(a) = b$ from now on.

Definition 1.7 Let $f : A \to B$ be a function.

1. f is **injective** (or **one-to-one**) if $f(a_1) = f(a_2)$ implies $a_1 = a_2$. In other words, in terms of the contrapositive statement, two different inputs cannot yield the same output.

2. f is **surjective** (or **maps onto** B) if for every $b \in B$, there is an $a \in A$ such that $f(a) = b$. In other words, the image and codomain are equal.

3. f is a **bijection** if it is both injective and surjective.

Example 1.9 These examples help illustrate the definitions just related and also demonstrate that there is no dependency between the two concepts of injective and surjective, i.e. neither property implies the other.

1. Let $f : \mathbb{Z} \to \mathbb{Z}$ by $f(n) = 2n$. This function is one-to-one but does not map onto $\mathbb{Z}$.

2. Let $f : \mathbb{Z} \to \{0, 1\}$ by

$$f(n) = \begin{cases} 0, & \text{if } n \text{ is even} \\ 1, & \text{if } n \text{ is odd} \end{cases}$$

This function is not one-to-one and maps onto $\{0, 1\}$.

3. Let $f : \mathbb{Z} \to \mathbb{Z}$ by

$$f(n) = \begin{cases} n/2, & \text{if } n \text{ is even} \\ n, & \text{if } n \text{ is odd} \end{cases}$$

This function is not one-to-one and maps onto $\mathbb{Z}$.

4. Let $f : \mathbb{R} \to \mathbb{R}$ by $f(x) = 2x - 3$. This function is a bijection.

5. Let $f : \{0, 1\} \to \{0, 1\}$ by $f(0) = 0$ and $f(1) = 0$. This function is neither one-to-one nor maps onto $\{0, 1\}$

Example 1.10 *We introduce some* **important** *examples of functions which we shall see in this text.*

1. *For any set A, the* **identity map** *on A, written $1_A : A \to A$ is defined by $1_A(a) = a$ for all $a \in A$. Certainly this map is a bijection.*

2. *If A is a proper subset of a set B, the* **inclusion map** *written $i : A \to B$ is defined by $i(a) = a$ for all $a \in A$. This map is again one-to-one but certainly cannot map onto B.*

3. *For any sets A and B, the* **projection map** *onto A, written $\pi_A : A \times B \to A$ is defined by $\pi_A(a, b) = a$. This map maps onto A but cannot be one-to-one if $|B| > 1$.*

4. *Consider any function $f : A \to B$. The* **restriction map** *of f to $C \subset A$, written $f \upharpoonright C : C \to B$ is defined by $(f \upharpoonright C)(c) = f(c)$ for all $c \in C$.*

5. *Consider any function $f : A \to B$. An* **extension map** *of f to $C \supset A$, written $\tilde{f} : C \to B$ is any map from C to B having the property that $\tilde{f}(a) = f(a)$ for all $a \in A$.*

   A simple example of this is the following extension of $1_{\mathbb{Z}}$ to $\mathbb{Q}$ defined by

$$\tilde{1}_{\mathbb{Z}} : \mathbb{Q} \to \mathbb{Z} \quad \text{by} \quad \tilde{1}_{\mathbb{Z}}\left(\frac{m}{n}\right) = mn, \text{assuming that } \frac{m}{n} \text{ is in lowest terms.}$$

6. *Let $\equiv$ be an equivalence relation on a set A. The* **quotient map** *(or* **canonical map**) *$\nu : A \to A/\equiv$ is defined by $\nu(a) = [a]$. This map certainly maps onto $A/\equiv$.*

7. *Let A be any set. A function $\sigma : A \to A$ is a* **permutation** *of A if σ is a bijection. The collection of all permutations of A is denoted by $Sym(A)$ and is called the* **symmetric group on** A. *When $A = \{1, 2, \ldots, n\}$, then $Sym(A)$ is typically denoted by S_n and is called the* **symmetric group on** n.

Lemma 1.2 *Let $f : A \to B$ be a function.*

1. *f is a bijection iff for every $b \in B$ there exists a unique $a \in A$ such that $f(a) = b$.*

2. *If A and B are finite sets of the same size, then f is one-to-one iff f maps onto B.*

Proof 1.2 *We leave the first part as a simple exercise. To prove the second part, set $A = \{a_1, \ldots, a_n\}$ and $B = \{b_1, \ldots, b_n\}$. Then $f(A) = \{f(a_1), \ldots, f(a_n)\} \subseteq B$. First assume that f is one-to-one. Then $f(a_1), \ldots, f(a_n)$ are all distinct. Thus, $|f(A)| = n = |B|$ and so $f(A) = B$ proving f maps onto B. We prove the converse by contrapositive. If f is not one-to-one, then $f(a_1), \ldots, f(a_n)$ are not all distinct. Thus, $|f(A)| < n = |B|$ and so $f(A)$ is a proper subset of B proving f does not map onto B.*

Definition 1.8 *Let $f : A \to B$ and $g : B \to C$. The* **composition** *function, written $g \circ f = \{(a, c) \ : \ \exists \, b \in B \text{ such that } (a, b) \in f \text{ and } (b, c) \in g\}$. In other words, $(g \circ f)(a) = g(f(a))$ (note that the domain of g is B so that $g(f(a))$ always makes sense).*

Definition 1.9 *Let $f : A \to B$ be a function. The function $g : B \to A$ is an* **inverse** *of f if $g \circ f = 1_A$ and $f \circ g = 1_B$, i.e. for all $a \in A$ and $b \in B$,*

$$g(f(a)) = a \qquad and \qquad f(g(b)) = b.$$

Example 1.11 *Check for $f, g : \mathbb{R} \to \mathbb{R}$ defined by $f(x) = 2x - 3$ and $g(x) = \frac{x+3}{2}$ that g is an inverse of f.*

In a sense, the inverse g of a function f *undoes* what the function f does. Below are some pertinent results concerning functions and their inverses:

Theorem 1.1 *Let $f : A \longrightarrow B$ be a function from a set A to a set B.*

1. *f has an inverse iff f is one-to-one and maps onto B, i.e. f is a bijection.*

2. *If f has an inverse, then it has exactly one.*

3. *If f has an inverse, then the inverse is also one-to-one and maps onto B, i.e. the inverse is a bijection.*

4. *If f_1 has inverse g_1 and f_2 has inverse g_2, then $f_1 \circ f_2$ has an inverse, namely $g_2 \circ g_1$.*

Proof 1.3 *To prove the first statement, first we assume that f has an inverse g. We show f is one-to-one. For $a_1, a_2 \in A$, if $f(a_1) = f(a_2)$, then $g(f(a_1)) = g(f(a_2))$, i.e. $(g \circ f)(a_1) = (g \circ f)(a_2)$. By definition of inverse, this equation reduces to $a_1 = a_2$, and we have proved that f is one-to-one. To show that f maps onto B, take any $b \in B$. We have to find an $a \in A$ such that $f(a) = b$. The element $a = g(b)$ does the trick, since $f(g(b)) = (f \circ g)(b) = b$.*

Now assume that f is a bijection. Define a function $g : B \longrightarrow A$ as follows: For $b \in B$, by Lemma 1.2, there is a unique $a \in A$ such that $f(a) = b$. We then define $g(b) = a$. Note that g is indeed a function, since the element $a \in A$ is uniquely determined. We now prove that this g is the inverse of f. First, for any $b \in B$, we have $(f \circ g)(b) = f(g(b)) = f(a) = b$. Second, take any $a \in A$. Set $b = f(a)$. Note that by definition of g, we have that $g(b) = a$. Hence, $(g \circ f)(a) = g(f(a)) = g(b) = a$.

To prove the second statement, suppose that g_1 and g_2 are inverses of f. We will show that $g_1 = g_2$ and so f has only one inverse (when it exists). For any $b \in B$, since f is a bijection, by Lemma 1.2, there is a unique $a \in A$ such that $f(a) = b$. Then

$$g_1(b) = g_1(f(a)) = (g_1 \circ f)(a) = a = (g_2 \circ f)(a) = g_2(f(a)) = g_2(b).$$

We leave the remaining proofs as exercises. □

Because of the fact that when an inverse exists there is only one, we can assign it notation without any confusion. The inverse of f will be denoted by f^{-1}. Take note that this is simply notation and should not be taken literally as $1/f$. There is, however, a use of the notation f^{-1} which does not assume that the inverse of f exists.

Definition 1.10 *Let $f : A \to B$ be a function and $C \subseteq B$. The **inverse image** (or **preimage**) of C under f, written*

$$f^{-1}(C) = \{a \in A \ : \ f(a) \in C\}.$$

Example 1.12 *Consider Example 1.9.2. The inverse image of 1 under f, i.e. $f^{-1}(\{1\})$ equals the set of odd numbers.*

The last topic of this section deals with the notion of a **well-defined** map. This topic arises when an input can be represented in more than one way. In particular, we shall look at the situation when inputs are equivalence classes. Basically we want to check when such a map is indeed a function, i.e. If $a = b$ (two representations of the same input), then $f(a) = f(b)$. For otherwise, we would have a single input being sent to two different outputs contradicting the definition of a function. In our particular situation, we might have an equivalence relation $\equiv$ on a set A and $f : A/ \equiv \to B$. We want to make sure f is well-defined, i.e. that f is by definition a function. This entails checking that whenever $[a_1] = [a_2]$ we have $f([a_1]) = f([a_2])$. This property is essential, for instance, when we define our group operation for cosets of a group modulo a normal subgroup. In Section 1.4, we will already see an important instance of well-definition.

Example 1.13 *We will illustrate this concept and its verification with several examples.*

1. *Consider the equivalence relation congruence modulo 3 on $\mathbb{Z}$. Define $f : \mathbb{Z}/\equiv_3 \to \mathbb{Z}$ by $f([n]) = n$. This map is **not** a well-defined function, since for instance $[0] = [3]$ while $f([0]) = 0 \neq 3 = f([3])$.*

2. *Consider the equivalence relation congruence modulo 2 and define $f : \mathbb{Z}/\equiv_2 \to \mathbb{Z}$ by*

$$f([n]) = \begin{cases} 0, & \text{if } n \text{ is even} \\ 1, & \text{if } n \text{ is odd} \end{cases}$$

We claim f is a well-defined function. If $[n] = [m]$, then $n \equiv_2 m$ and so $2|(n-m)$. Therefore, $n - m$ is even and so either n and m are both even or both odd. In either case $f([n]) = f([m])$.

3. *Consider the equivalence relation $\equiv$ we defined earlier on $\mathbb{Q}$ (Example 1.3.4) and consider*

$$f : \mathbb{Q}/\equiv \to \mathbb{Q} \text{ defined by } f([a/b]) = a^2/b^2.$$

We show that f is a well-defined function. If $[a/b] = [c/d]$, then $a/b \equiv c/d$ and so $ad = bc$. Using properties of $\mathbb{Z}$ we have $(ad)^2 = (bc)^2$ and so $a^2 d^2 = b^2 c^2$. Then $a^2/b^2 \equiv c^2/d^2$ which implies $f(a/b) = f(c/d)$.

EXERCISES

1 For each function in Example 1.9, verify the statements made for injective and surjective.

2 For each of the following functions, decide if it is injective, surjective or a bijection:

    a. $f : \mathbb{R}^{>0} \to \mathbb{R}^{>0}$ by $f(x) = \frac{1}{x}$.

    b. $f : \mathbb{R}^* \to \mathbb{R}^*$ by $f(x) = \frac{1}{x^2}$, where $\mathbb{R}^* = \mathbb{R} - \{0\}$

    c. $f : \mathbb{R} \to \mathbb{R}$ by $f(x) = \sin(x^2)$

    d. $f : \mathbb{Z} \times \mathbb{Z} \to \mathbb{Z}$ by $f(m, n) = 3m + n$.

    e. $f : \mathbb{Z} \times \mathbb{Z} \to \mathbb{Z}$ by $f(m, n) = 4m + 2n$.

    f. $f : \mathbb{Z} \times \mathbb{Z} \to \mathbb{Z} \times \mathbb{Z}$ by $f(m, n) = (2m, m + n)$.

    g. $f : \mathbb{R} \times \mathbb{R} \to \mathbb{R} \times \mathbb{R}$ by $f(x, y) = (2x, x + y)$.

3 Carefully explain why the example given in Example 1.10.5 is indeed an extension.

4 Let $f : A \to B$, $g : B \to C$ and $h : C \to D$. Prove the following statements:

    a. Composition is associative, i.e. $(h \circ g) \circ f = h \circ (g \circ f)$.

b. f, g one-to-one implies $g \circ f$ one-to-one.

c. If f maps onto B and g maps onto C, then $g \circ f$ maps onto C.

d. $g \circ f$ one-to-one implies f one-to-one.

e. $g \circ f$ maps onto C implies g maps onto C.

f. Give a counterexample to the following statement: $g \circ f$ one-to-one implies g one-to-one.

g. Give a counterexample to the following statement: $g \circ f$ maps onto C implies f maps onto B.

5 Let $A = \{a, b, c\}$ and $B = \{x, y, z\}$.

a. How many possible functions $f : A \rightarrow B$ are there?

b. How many possible surjective functions $f : A \rightarrow B$ are there?

c. How many possible injective functions $f : A \rightarrow B$ are there?

d. How many possible bijective functions $f : A \rightarrow B$ are there?

6 Consider the functions $f : A \rightarrow B$ and $g : B \rightarrow C$.

a. Prove that if f and g are bijections, then $g \circ f$ has an inverse and $(g \circ f)^{-1} = f^{-1} \circ g^{-1}$.

b. Give a concrete example where $g \circ f$ is a bijection, but f and g are not.

7 Prove Lemma 1.2.1

8 Prove the remaining parts of Lemma 1.1.

9 Compute $f^{-1}(\{1\})$ for Example 1.9.3.

10 Decide whether or not each of the following functions is well-defined:

a. $f : \mathbb{Z}/\equiv_n \rightarrow \mathbb{Z}/\equiv_n$ by $f([a]) = [2a]$.

b. $f : \mathbb{Z}/\equiv_n \rightarrow \mathbb{Z}/\equiv_n$ by $f([a]) = [ma + b]$, for some fixed integers m and b.

c. $f : \mathbb{Z}/\equiv_n \rightarrow \mathbb{Z}/\equiv_n$ by $f([a]) = [a^2]$

d. $f : \mathbb{Z}/\equiv_n \rightarrow \mathbb{Z}/\equiv_n$ by $f([a]) = [a^k]$, for some fixed positive integer k.

e. $f : \mathbb{Z}/\equiv_2 \rightarrow \mathbb{Z}/\equiv_4$ by $f([a]_2) = [a]_4$.

f. $f : \mathbb{Q} \rightarrow \mathbb{Z}$ by $f\left(\frac{m}{n}\right) = m$.

g. $f : \mathbb{Z}/\equiv_4 \rightarrow \mathbb{Z}/\equiv_6$ by $f([x]_4) = [3x]_6$.

h. $f : \mathbb{Q} \rightarrow \mathbb{Q}$ by $f\left(\frac{m}{n}\right) = \left(\frac{2m}{3n}\right)$.

i. $f : \mathbb{Z}/\equiv_4 \rightarrow \mathbb{Z}/\equiv_6$ by $f([x]_4) = [x]_6$.

    j. $f : \mathbb{Z}/\equiv_4 \to \mathbb{Z}/\equiv_3$ by $f([m]_4) = [m^3]_3$.

    k. $g : \mathbb{Q} \to \mathbb{Q}$ by $g\left(\left[\frac{a}{b}\right]\right) = \left[\frac{a+b}{b}\right]$.

11 Consider $f : \mathbb{Z}/\equiv_{35} \to \mathbb{Z}/\equiv_5 \times \mathbb{Z}/\equiv_7$ by $f([x]_{35}) = ([x]_5, [x]_7)$.

    a. Prove f is a well-defined function.

    b. Is f is injective? (justify)

    c. Is f is surjective? (justify)

    d. Is f is bijective? (justify)

    e. Does f have an inverse? (justify)

12 Prove that f is one-to-one iff for any set C and all functions $h : C \to A$ and $k : C \to A$ we have that $f \circ h = f \circ k$ implies $h = k$.

13 Consider a function $f : \mathbb{Z}/\equiv_n \to \mathbb{Z}/\equiv_k$ defined by $f([x]_n) = [mx]_k$ where n, k and m are positive integers. Show that f is well-defined iff $k|mn$.

1.3 BASIC NUMBER THEORY

Some basic concepts in abstract algebra make use of number theory implicitly. In this section, we collect together concepts and results which we will need in later portions of the text. These concepts include the Division Algorithm, the greatest common divisor and results about prime numbers.

    We have already been introduced to the notion of one integer dividing another and this relation on integers is both reflexive and transitive. Furthermore, if we restrict ourselves to positive integers, then *divides* is also anti-symmetric. Here are two further properties of *divides*.

Lemma 1.3 *Let $m, n, d \in \mathbb{Z}$.*

    1. If $m|n$ and $n|m$, then $m = \pm n$.

    2. If $d|m$ and $d|n$, then $d|(am + bn)$ for all $a, b \in \mathbb{Z}$.

Proof 1.4 *We prove the first part and leave the second as an exercise. Since $m|n$ and $n|m$ we have $n = mk$ and $m = nl$ and so $m = mkl$ which can be rewritten as $m(1 - kl) = 0$. Hence, either $m = 0$ (and so $n = 0k = 0$ proving the result) or $1 - kl = 0$. In the latter case, $kl = 1$ which as integers implies either $k, l = 1$ or $k, l = -1$ and so $m = nl = \pm n$.*

    We now prove the Division or Euclidean Algorithm for integers. We leave as an exercise the following result which will be used in the theorem: For all $n \in \mathbb{Z}$ we have $n + |n| \geq 0$.

Theorem 1.2 (Division/Euclidean Algorithm) *Let $n, d \in \mathbb{Z}$ with $d > 0$. There exist unique $q, r \in \mathbb{Z}$ having the property that $n = qd + r$ with $0 \leq r < d$.*

Proof 1.5 *To show existence of q and r consider the set $D = \{n - xd \;:\; x \in \mathbb{Z}$ and $n - xd \geq 0\}$ which is certainly a subset of $\mathbb{N}$, the natural numbers. Now D is non-empty, since for instance $n - (-|n|)d = n + |n| \geq 0$ is in D. Because $D \subseteq \mathbb{N}$, it must have a smallest element, say r, and $r \in D$ implies $r = n - qd$ for some $q \in \mathbb{Z}$. Solving for n we have $n = qd + r$ which is half of what we need to show. We already know $r \geq 0$ so it remains to show that $r < d$. Notice that $r - d = n - qd - d = n - (q + 1)d$ which is of the right form to be in D. However, $r - d < r$ and r is the smallest element of D which implies that $r - d$ must fail to be non-negative, i.e. $r - d < 0$ or $r < d$.*

To show uniqueness of q and r, suppose that $n = qd + r$ and $n = q'd + r'$ with $0 \leq r, r' < d$. Without loss of generality, assume that $q \geq q'$. Suppose, to the contrary, that $q > q'$ and so as integers we would have $q \geq q' + 1$. Then $r' = n - q'd \geq n - (q - 1)d = r + d \geq d$, a contradiction. Hence, $q = q'$ and so $r = n - qd = n - q'd = r'$ thus proving uniqueness.

The integer multiples of $n \in \mathbb{Z}$ will be denoted by $n\mathbb{Z}$. For instance,

$$3\mathbb{Z} = \{0, \pm 3, \pm 6, \pm 9, \ldots\}.$$

In the next lemma, we will prove some useful facts about $n\mathbb{Z}$ which will be used later in the text.

Lemma 1.4 *Consider the set $n\mathbb{Z}$ for some $n \in \mathbb{N}$.*

1. *If $x, y \in n\mathbb{Z}$, then so are $x + y, x - y \in n\mathbb{Z}$. We say that $n\mathbb{Z}$ is closed under addition and subtraction.*

2. *If X is any non-empty subset of the integers closed under addition and subtraction (see part 1), then $X = n\mathbb{Z}$ for some $n \in \mathbb{N}$.*

Proof 1.6 *The first part is easy and left as an exercise. For the second part, should $X = \{0\}$, then $X = 0\mathbb{Z}$ and we are done. Otherwise X has positive elements. Indeed, if we take any non-zero element $m \in X$. Then $0 = m - m \in X$ by assumption and so $-m = 0 - m \in X$ again by assumption. Since m and $-m$ are in X and one of these must be positive we can conclude X has positive elements. Now let n be the smallest positive element in X. We show that $X = n\mathbb{Z}$. First, take $x \in n\mathbb{Z}$ so that $x = nk$. If $k = 0$, then $x = 0$ and as we saw above $0 \in X$. If k is positive, then $x = \underbrace{n + \cdots + n}_{k} \in X$ by assumption (and induction). If k is negative, then as we saw above $-n \in X$ and $x = \underbrace{(-n) + \cdots + (-n)}_{-k} \in X$ by assumption and induction. Hence, $n\mathbb{Z} \subseteq X$. For the reverse inclusion, take any $x \in X$ and write $x = nq + r$ with $0 \leq r < n$ using the Division Algorithm. Since $n\mathbb{Z} \subseteq X$ and X is closed under addition and subtraction, it follows that $r = x - nq \in X$. Since n is the smallest positive integer in X it must be that $r = 0$ and so $x = nq \in n\mathbb{Z}$.*

Definition 1.11 *Let a and b be two non-zero integers. The integer d is the* **greatest common divisor** *of a and b, written $d = gcd(a, b)$ if*

1. $d > 0$

2. $d|a$ and $d|b$ (common divisor) and

3. $e|a$ and $e|b$ implies $e|d$ (greatest).

Example 1.14 *$gcd(-120, 36) = 12$. We note here that the Division or Euclidean algorithm also refers to a systematic way for finding the greatest common divisor of two integers. Its method relies on Theorem 1.2 of the same name.*

Theorem 1.3 *For any non-zero integers a and b the greatest common divisor of a and b exists and is unique.*

Proof 1.7 *To show existence, consider the set $C = \{ax + by : x, y \in \mathbb{Z}\}$. Certainly $C \neq \emptyset$ (since for instance $a = a \cdot 1 + b \cdot 0 \in C$). Furthermore, C is closed under addition and subtraction, since*

$$(ax_1 + by_1) \pm (ax_2 + by_2) = a(x_1 \pm x_2) + b(y_1 \pm y_2) \in C.$$

Therefore, by Lemma 1.4, we know that $C = d\mathbb{Z}$ for some positive integer d. We show now that $d = gcd(a, b)$. We already have $d > 0$. Since $a, b \in C = d\mathbb{Z}$ we can write $a = dk$ and $b = dl$ for some integers k and l which implies $d|a$ and $d|b$. Finally, if $e|a$ and $e|b$, since $d = d \cdot 1 \in d\mathbb{Z} = C$ we can write $d = ax_0 + by_0$ for some $x_0, y_0 \in \mathbb{Z}$. Hence, by Lemma 8.1, $e|(ax_0 + by_0)$, i.e. $e|d$.

To show uniqueness, assume d and d' are both greatest common divisors of a and b. Since $d|a$ and $d|b$ and $d' = gcd(a, b)$ this implies $d|d'$. Reversing roles yields $d'|d$ and so by Lemma 8.1, we have $d = \pm d'$. However $d, d' > 0$ so we can conclude that $d = d'$.

Corollary 1.1 *Let a and b be non-zero integers.*

1. If $d = gcd(a, b)$, then there exist $x_0, y_0 \in \mathbb{Z}$ such that $d = ax_0 + by_0$.

2. $gcd(a, b) = 1$ iff there exist $x_0, y_0 \in \mathbb{Z}$ such that $ax_0 + by_0 = 1$.

Proof 1.8 *The first part follows immediately from the existence part of the proof of Theorem 1.3. For the second part, one direction follows immediately from the first part of the corollary. Assume that there exist $x_0, y_0 \in \mathbb{Z}$ such that $ax_0 + by_0 = 1$ and set $d = gcd(a, b)$. Since $d|a$ and $d|b$ it follows that $d|(ax_0 + by_0)$ and so $d|1$ which forces d to equal 1, since $d > 0$.*

Definition 1.12 *A positive integer p is* **prime** *if p has exactly two distinct positive divisors, i.e. $p \neq 1$ and if $a > 0$ and $a|p$, then either $a = 1$ or $a = p$.*

Lemma 1.5 *An integer $p > 1$ is prime iff whenever $a, b \in \mathbb{Z}$ and $p|ab$, then either $p|a$ or $p|b$.*

Proof 1.9 *First assume that p is prime and assume that $p|ab$. If $p|a$, then we are done. Otherwise $p \nmid a$. In this case, we will show $p|b$. Since $p \nmid a$ this implies $gcd(p, a) = 1$ (convince yourself of this). By Corollary 1.1, there exist $x_0, y_0 \in \mathbb{Z}$ such that $px_0 + ay_0 = 1$. Then $b = b(px_0 + ay_0) = p(bx_0) + ab(y_0)$. Since $p|p$ and $p|ab$ we have $p|[p(bx_0) + ab(y_0)]$, i.e. $p|b$. For the converse, assume whenever $a, b \in \mathbb{Z}$ and $p|ab$, then either $p|a$ or $p|b$ and p is **not** prime. This implies $p = ab$ for some $1 < a, b < p$ (explain). Since $p|p$ we have $p|ab$ and so by assumption $p|a$ or $p|b$ which is not possible, since $1 < a, b < p$, thus a contradiction.*

Corollary 1.2 *If p is prime and $p|(a_1 a_2 \cdots a_n)$, then $p|a_i$ for some i, $1 \leq i \leq n$.*

Proof 1.10 *This follows immediately by induction.*

Next we prove the existence and uniqueness of prime factorization for integers greater than 1 called the Fundamental Theorem of Arithmetic.

Theorem 1.4 (Fundamental Theorem of Arithmetic) *Any integer $a > 1$ can be factored uniquely as a product of primes. More precisely, there exists unique primes $p_1 < p_2 < \cdots p_n$ and positive integers $e_1, e_2, \ldots, e_n$ such that $a = p_1^{e_1} p_2^{e_2} \cdots p_n^{e_n}$.*

Proof 1.11 *First, we prove existence of a prime factorization. Suppose to the contrary that there were integers greater than 1 which had no prime factorization. Set P equal to the set of all integers greater than 1 which do not have prime factorizations (which we are assuming to be non-empty). Let m be the smallest such element of P. Certainly m is not prime (else it has a trivial prime factorization), so write $m = ab$ with $1 < a, b < m$. Since m is smallest in P we know a and b are not in P and so have prime factorizations. But then multiplying these two factorizations will yield a prime factorization for $m = ab$, a contradiction.*

To prove uniqueness, again we suppose to the contrary that there are integers greater than 1 which have more than one prime factorization and set U equal to the set of all such integers. Since U is non-empty, let m be the smallest element of U. Say $m = p_1^{e_1} p_2^{e_2} \cdots p_n^{e_n}$ and $m = q_1^{f_1} q_2^{f_2} \cdots q_k^{f_k}$ where $p_1 < p_2 < \cdots p_n$ and $q_1 < q_2 < \cdots q_k$ are primes and $e_1, e_2, \ldots, e_n$ and $f_1, f_2, \ldots, f_k$ are positive integers. Equating these two factorizations we see that p_1 divides $q_1^{f_1} q_2^{f_2} \cdots q_k^{f_k}$ and by Corollary 1.2, $p_1|q_i$ for some i, $1 \leq i \leq k$. But then $p_1 = q_i$ (explain) and so by cancellation we have $p_1^{e_1-1} p_2^{e_2} \cdots p_n^{e_n} = q_1^{f_1} \cdots q_i^{f_i-1} \cdots q_k^{f_k}$. Set $r = p_1^{e_1-1} p_2^{e_2} \cdots p_n^{e_n} = q_1^{f_1} \cdots q_i^{f_i-1} \cdots q_k^{f_k}$. Since $r < m$ we know that r has a unique prime factorization, and so $p_1^{e_1-1} p_2^{e_2} \cdots p_n^{e_n}$ and $q_1^{f_1} \cdots q_i^{f_i-1} \cdots q_k^{f_k}$ must be identical factorizations. If we throw $p_1 = q_i$ back in to get the two prime factorizations of m, then $p_1^{e_1} p_2^{e_2} \cdots p_n^{e_n}$ and $q_1^{f_1} q_2^{f_2} \cdots q_k^{f_k}$ must be the same as well, which is a contradiction to $m \in P$.

EXERCISES

1 Define a relation $\sim$ on $\mathbb{Z}^{>1}$ (integers greater than 1) as follows: $n \sim m$ iff there exists a prime p such that $p|m$ and $p|n$. Decide whether or not it is an equivalence relation. If it's not, then provide a counterexample. If it is, verify

the axioms of an equivalence relation and then describe what the equivalence classes represent.

2 Prove Lemma 8.1.2.

3 Decide whether each of the following statements are true. If so, give a proof; otherwise, give a counter-example.

    a. If $a, b, c \in \mathbb{Z}$ and $a|b$, then $a|(bc)$.

    b. If $a, b \in \mathbb{Z}$ and $a|(b - 1)$, then $a|(b^2 - 1)$.

    c. If $a|c$ and $b|c$ and $gcd(a, b) = 1$, then $(ab)|c$.

4 Prove (by cases using the definition of absolute value) that for all $n \in \mathbb{Z}$ we have $n + |n| \geq 0$.

5 Prove Lemma 1.4.1.

6 For an integer a, prove if p is prime and $p|a$, then $gcd(p, a) = p$.

7 For an integer a, prove if p is prime and $p \nmid a$, then $gcd(p, a) = 1$.

8 For an integers a, b, c, prove if $a|c$, $b|c$ and $gcd(a, b) = 1$, then $(ab)|c$.

9 For an integers a, b, c, prove $gcd(a, bc) = 1$ iff $gcd(a, b) = 1$ and $gcd(a, c) = 1$.

10 For an integers a, b, c, prove if $c|ab$ and $gcd(a, c) = 1$, then $c|b$.

11 Prove that if $d = gcd(n, m)$, then we can express $n = dx$ and $m = dy$ for some integers x and y with $gcd(x, y) = 1$.

12 Prove by induction Corollary 1.2.

1.4 MODULO ARITHMETIC

Our focus in this section is the equivalence relation congruence modulo n on $\mathbb{Z}$. First, we prove some basic properties of congruence modulo n. This first property of congruence modulo n is, in fact, characteristic of a more general concept called a **congruence relation**.

Lemma 1.6 *Let $a, b, c, d \in \mathbb{Z}$. If $a \equiv_n c$ and $b \equiv_n d$, then $a + b \equiv_n c + d$ and $ab \equiv_n cd$.*

Proof 1.12 *Since $a \equiv_n c$ and $b \equiv_n d$ we have $n|(a - c)$ and $n|(b - d)$. Then $n|[(a - c) + (b - d)]$ or equivalently $n|[(a + b) - (c + d)]$ which implies $a + b \equiv_n c + d$. Furthermore, $n|[(a - c)b + (b - d)c]$ or equivalently $n|(ab - cd)$ which implies $ab \equiv_n cd$.*

Lemma 1.7 *Let $n > 1$ and a be integers. There exists $b \in \mathbb{Z}$ such that $ab \equiv_n 1$ iff $gcd(a, n) = 1$.*

Proof 1.13 *Note that $ab \equiv_n 1$ iff $n|(ab-1)$ iff $ab-1 = nk$ for some $k \in \mathbb{Z}$ iff $ab + n(-k) = 1$ for some $k \in \mathbb{Z}$ which implies $\gcd(a,n) = 1$, by Corollary 1.1.2.*

Assuming $\gcd(a,n) = 1$, by Corollary 1.1.2, there exist $x_0, y_0 \in \mathbb{Z}$ such that $ax_0 + ny_0 = 1$. Then as we can see from the previous direction, the b we are looking for such that $ab \equiv_n 1$ is x_0.

Now let's focus on the collection of equivalence classes $\mathbb{Z}/\equiv_n$. The first questions that need to be addressed is how many distinct classes are there and is there a nice way to represent them? The next result answers these questions.

Lemma 1.8 *For any positive integer n we have that $\mathbb{Z}/\equiv_n = \{[0], [1], \ldots, [n-1]\}$.*

Proof 1.14 *First note that any class is equal to one of $[0], [1], \ldots, [n-1]$, since if $m \in \mathbb{Z}$, using the Division Algorithm, we can write $m = qn + r$ for integers q and r and $0 \le r < n$. Now since $m - r = qn$ this implies $n|(m-r)$ and so $m \equiv_n r$ which implies that $[m] = [r]$ where $0 \le r \le n - 1$. Finally, note that the $[0], [1], \ldots, [n-1]$ are all distinct, since if $[r] = [s]$ for $0 \le r, s \le n - 1$, then $r \equiv_n s$ and so $n|(r-s)$ which implies $r - s = nk$ for some $k \in \mathbb{Z}$. But since $-n < r - s < n$, the only way this could be possible is if $r - s = 0$ or $r = s$.*

We would now like to take these classes of $\mathbb{Z}/\equiv_n$ and define an addition and multiplication for them. The most natural way to proceed would be to define class addition and multiplication as addition and multiplication of representatives, i.e.

$$[a] + [b] = [a+b] \qquad \text{and} \qquad [a] \cdot [b] = [a \cdot b].$$

The problem is that we have to be sure these two binary operations which are functions from $(\mathbb{Z}/\equiv_n) \times (\mathbb{Z}/\equiv_n)$ to $\mathbb{Z}/\equiv_n$ are well-defined. As it turns out there is nothing to fear as is proved below.

Lemma 1.9 *The operations of addition and multiplication in $\mathbb{Z}/\equiv_n$ defined by $[a] + [b] = [a+b]$ and $[a] \cdot [b] = [a \cdot b]$ are well-defined.*

Proof 1.15 *What we must show is that if $[a] = [c]$ and $[b] = [d]$, then $[a] + [b] = [c] + [d]$ and $[a][b] = [c][d]$. But this follows almost immediately from Lemma 1.6.*

Having the peace of mind that these two operations are well-defined, we now list some properties these operations enjoy. We will later call these properties the properties of a **commutative ring with unity**. We leave the details as a simple exercise which relies heavily on properties of the integers.

Lemma 1.10 *If $a, b, c \in \mathbb{Z}$, then*

1. $[a] + [b] = [b] + [a]$ and $[a][b] = [b][a]$.

2. $[a] + ([b] + [c]) = ([a] + [b]) + [c]$ and $[a]([b][c]) = ([a][b])[c]$.

3. $[a] + [0] = [a]$ and $[a][1] = [a]$.

4. $[a] + [-a] = [0]$.

5. $[a]([b] + [c]) = [a][b] + [a][c]$.

We point out that not every class $[a] \in \mathbb{Z}/\equiv_n$ has a class $[b]$ such that $[a][b] = [1]$. In fact, we see from Lemma 1.7 that such a class $[b]$ exists iff $gcd(a, n) = 1$ (exercise). A class $[a]$ which has such a $[b]$ is called a **unit**. Some other classes $[a]$ have the property that there is a class $[b]$ such that $[a][b] = [0]$. A class $[a] \neq [0]$ which has such a $[b]$ is called a **zero divisor**.

Example 1.15 *In $\mathbb{Z}/\equiv_{10}$ the class $[3]$ is a unit since $[3][7] = [1]$ and $[2]$ is a zero-divisor since $[2][5] = [0]$.*

In fact, the following is true:

Lemma 1.11 *Every class $[a] \in \mathbb{Z}/\equiv_n$ not equal to $[0]$ is either a unit or a zero divisor.*

Proof 1.16 *Let $[a] \neq [0]$. Either $gcd(a, n) = 1$ or $gcd(a, n) > 1$. In the former case, by Lemma 1.7, we know then that $[a]$ is a unit. In the latter case, set $d = gcd(a, n) > 1$. Since $d|a$ and $d|n$ we may write $a = dk$ and $n = dl$ for some integers $k, l \in \mathbb{Z}$ (note that $l \neq 0$ since $n \neq 0$). Notice that $[a][l] = [al] = [dkl] = [nk] = [0]$, since $nk \equiv_n 0$. Hence, we see that $[a]$ is a zero divisor.*

Note that a Corollary to this result is the fact that every non-zero class in $\mathbb{Z}/\equiv_p$ (where p is prime) is a unit.

Lemma 1.12 *[Fermat's Little Theorem] Let $a, p \in \mathbb{Z}$ with p a prime number and $p \nmid a$. In $\mathbb{Z}/\equiv_p$ we have $[a]^{p-1} = [1]$.*

Proof 1.17 *Since $[a] \neq [0]$ we know $[a]$ is a unit and so there is a class $[b]$ with $[a][b] = [1]$. Consider the following list of classes: $[1][a], [2][a], \ldots, [p-1][a]$. First note that all the classes in this list are distinct, for if $[r][a] = [s][a]$, then by multiplying on the right by $[b]$ we have $[r] = [s]$. Furthermore, these classes are all not equal to $[0]$, since if $[r][a] = [0]$, then again by multiplying on the right by $[b]$ we have $[r] = [0]$ which is not the case. Therefore, this list is simply a reordering of the classes $[1], [2], \ldots, [p-1]$ and so $[1][a][2][a] \cdots [p-1][a] = [1][2] \cdots [p-1]$. This can be rewritten as $[a]^{p-1}[2] \cdots [p-1] = [2] \cdots [p-1]$. Since $[2], \ldots, [p-1]$ are units, we can reduce this equation to $[a]^{p-1} = [1]$.*

We also have the following result:

Corollary 1.3 *Let $a, p \in \mathbb{Z}$ with p a prime number. In $\mathbb{Z}/\equiv_p$ we have $[a]^p = [p]$.*

An equivalent way to represent $\mathbb{Z}/\equiv_n$ with its two operations of addition and multiplication is to consider the following structure. Let $\mathbb{Z}_n = \{0, 1, 2, \ldots, n-1\}$ and define addition, denoted by $+_n$, and multiplication, denoted by $\cdot_n$, for these elements as follows: $m +_n k = r$ if using the Division Algorithm, $m + k = qn + r$ where $0 \leq r < n$. Similarly, $m \cdot_n k = r$ if using the Division Algorithm, $m \cdot k = qn + r$ where $0 \leq r < n$. There is a direct correspondence (which we shall call later an **isomorphism**) between $\mathbb{Z}/\equiv_n$ and $\mathbb{Z}_n$. First off a class $[a] \in \mathbb{Z}/\equiv_n$ can be equated with $a \in \mathbb{Z}_n$ by Lemma 1.8. Furthermore, for $0 \leq a, b, c \leq n - 1$, $[a] + [b] = [c]$ iff $a +_n b = c$ and $[a][b] = [c]$ iff $a \cdot_n b = c$. We prove this fact for addition (multiplication is similar). Suppose first that $[a] + [b] = [c]$. This implies that $a + b \equiv_n c$ and so $n|(a + b - c)$ which implies $a + b - c = nq$ or $a + b = nq + c$ where $0 \leq c < n$. Then by definition $a +_n b = c$. Now assume that $a +_n b = c$. By definition this implies $a + b = qn + c$ where $0 \leq c < n$. Thus, $qn = a + b - c$ and so $n|(a + b - c)$ which implies $a + b \equiv_n c$ which is equivalent to $[a] + [b] = [c]$.

As a result of this observation, all the properties we proved in this section about $\mathbb{Z}/\equiv_n$ are equally true for $\mathbb{Z}_n$ (with the square brackets removed from the statements of the properties). We will be using the structure $\mathbb{Z}_n$ from now on and dispensing with $\mathbb{Z}/\equiv_n$.

EXERCISES

1 Prove Lemma 1.10.

2 Prove that $[a] \in \mathbb{Z}/\equiv_n$ has a class $[b]$ such that $[a][b] = [1]$ iff $gcd(a, n) = 1$.

3 Consider $\mathbb{Z}/\equiv_6$

    a. Write out the addition and multiplication tables

    b. List the units and zero divisors

4 List separately the units and zero divisors of $\mathbb{Z}/\equiv_{12}$. Illustrate this explicitly for each class as we did in Example 1.15.

5 Use Fermat's Little Theorem to compute $[4^{195}]_{13}$.

6 Prove that every non-zero class in $\mathbb{Z}/\equiv_p$ (where p is prime) is a unit.

7 Prove Corollary 1.3.

CHAPTER 2

Basic Group Theory

I N THIS CHAPTER, we introduce the reader to a group structure. In Section 2.1, we present basic definitions, examples and terminology related to groups. In Section 2.2, we define a subgroup, give many classic examples and present a shortcut for verifying subgroup. In Section 2.3, we present an important class of groups called cyclic groups. Permutation groups are an essential part of group theory and are presented in Section 2.4. In Section 2.5, we create new groups and subgroups by means of a product. In Section 2.6, we introduce functions between groups called homomorphisms. As a follow up, in Section 2.7, we use isomorphisms to define what it means for groups to be essentially equal (or isomorphic). In Section 2.8, we define perhaps one of the most equivalence relations on a group whose equivalence classes are called cosets, and these cosets can sometimes form a group. In Section 2.9, we investigate exactly when cosets form a group and look at several important examples. In Section 2.10, we investigate further notmal subgroups and consider groups which have no normal subgroups. Finally, in Section 2.11, we prove some fundamental isomorphism results for groups and factor groups.

2.1 DEFINITIONS AND EXAMPLES

In this section, we introduce the reader to one of the main topics under investigation in this text, namely a group structure, which is one of several algebraic structures we shall study.

Definition 2.1 *A* **group** $(G, *)$ *consists of a set of elements G together with a binary operation $*$ satisfying the following four axioms:*

Closure *– For all $g, h \in G$ we have $g * h \in G$.*

Associativity *– For all $g, h, k \in G$ we have $g * (h * k) = (g * h) * k$.*

Identity *– There is a special element $e \in G$ having the property that for all $g \in G$ we have $g * e = g = e * g$. The element e is called the* **identity** *element of the group.*

Inverse *– For every $g \in G$ there is an $h \in G$ such that $g * h = e = h * g$. The element h is called an* **inverse** *of g.*

DOI: 10.1201/9781003335283-2

If in addition the group satisfies the **commutative** *property, i.e. for all* $g, h \in G$ *we have* $g * h = h * g$, *then we call the group an* **abelian** *group (or* **commutative** *group).*

We remark that this binary operation $*$ can be viewed as a function from $G \times G$ into G, i.e. $*(g, h) = g * h$ for $g, h \in G$. This observation will become useful in the discussion of well-definition of a binary operation.

Example 2.1 *Here are a list of some classic examples of a group. Note below that* N^* *means the collection of numbers* N *without the number* 0.

1. $(\mathbb{Z}, +)$, *the integers together with addition, forms an abelian group. The identity element is* 0 *and the inverse of* $n \in \mathbb{Z}$ *is* $-n$. *In fact,* $\mathbb{Z}$ *can be replaced by* $\mathbb{Q}$, $\mathbb{R}$ *or* $\mathbb{C}$ *to get other additive abelian groups.*

2. $(\mathbb{Z}^*, \cdot)$ *together with multiplication does not form a group, since only* ± 1 *have inverses in* $\mathbb{Z}^*$.

3. $(\mathbb{Q}^*, \cdot)$ *forms an abelian group. In fact,* $\mathbb{Q}^*$ *can be replaced by* $\mathbb{R}^*$ *or* $\mathbb{C}^*$ *to get other multiplicative abelian groups.*

4. $(\mathbb{Z}_n, +_n)$ *forms an abelian group.*

5. $(\mathbb{Z}_n^*, \cdot_n)$ *forms an (abelian) group* **iff** n *is a prime number.*

6. *For any set* A, *the collection of permutations of* A, *i.e.* $Sym(A)$, *together with the operation of composition forms a (not necessarily abelian) group.*

   An example of commutativity failing is $A = \mathbb{R}$, $f(x) = x + 1$ *and* $g(x) = x^3$. *Notice we have* $(f \circ g)(x) = x^3 + 1$ *while* $(g \circ f)(x) = x^3 + 3x^2 + 3x + 1$.

7. *The collection of real* $m \times n$ *matrices, written* $M_{mn}(\mathbb{R})$, *with matrix addition forms an abelian group.*

8. *The collection of* $n \times n$ *real invertible matrices, written* $GL_n(\mathbb{R})$ *and called the* **general linear group**, *with matrix multiplication forms a group.*

9. *The collection of* $n \times n$ *real matrices of determinant* 1, *written* $SL_n(\mathbb{R})$ *and called the* **special linear group**, *with matrix multiplication forms a group.*

10. *The collection of* $n \times n$ *real upper triangular matrices with* 1's *on the diagonal, written* $U_n(\mathbb{R})$ *and called the* **unipotent** **group**, *with matrix multiplication forms a group.*

11. *The collection of all* $n \times n$ *real diagonal matrices, written* $D_n(\mathbb{R})$, *forms an abelian group.*

12. *Consider a square and let* G *consist of all rigid motions of the square which leave the square unchanged in look. There are exactly 8 such rigid motions which can be divided into two types of equal number: rotations and reflections.*

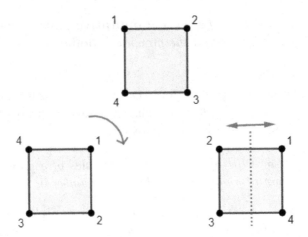

Figure 2.1 At the top is the original square. The bottom left is a 90° rotation of the square while the bottom right is a vertical reflection.

> There are the 0°, 90°, 180°, and 270° clockwise rotations, one horizontal, one vertical and two diagonal reflections. Figure 2.1 shows the square as well as a 90° rotation and a vertical reflection. This group is called the **dihedral group** and is denoted by D_4. One can easily generalize this definition to D_n (for $n \geq 3$) which consists of n rotations and n reflections of a regular n-gon.

13. Consider the set with 8 elements $\pm 1, \pm i, \pm j, \pm k$ and define an operation on this set satisfying the identities $i^2 = j^2 = k^2 = -1$, $ij = k$, $jk = i$, $ki = j$, $(-1)i = -i$, $(-1)j = -j$, $(-1)k = -k$. This group is called the **quaternions** and is denoted by Q_8.

Remark 2.1 Here are some remarks and some observations which are easily verifiable from the group axioms.

1. The groups D_4 and Q_8 are distinct (what we shall call non-isomorphic) groups.

2. There is only one identity element in a group. Indeed, if e and e' were identities, then $e = e * e' = e'$. From now on we will denote this unique identity by the notation 1 (not to be taken literally).

3. Every element of a group has only one inverse. For if h and h' were inverses of an element g in a group, then $h = h * e = h * (g * h') = (h * g) * h' = e * h' = h'$. This allows us to introduce the notation g^{-1} as representing the unique inverse of G. This notation is derived from the rational numbers, but should not be taken literally, i.e. $g^{-1} \neq 1/g$ (in fact, it makes no sense whatsoever!)

4. From now on we will drop the $*$ and simply write gh for $g * h$.

5. Oftentimes we will refer to a group $(G, *)$ simply by its set G when the operation is either understood or completely arbitrary.

$*$	1	g	h	k
1				
g				
h				
k				

Figure 2.2 A group multiplication table.

6. *Notice that even though we are dealing with an abstract group operation, we employ suggestive multiplicative notation. This is merely our convention. We could have very well represented things in an additive way, such as using 0 for the identity and $-g$ for the inverse of g. In fact, this is oftentimes used when dealing exclusively with abelian groups.*

7. *We leave it as an exercise to check that our identity and inverse axiom could be weakened to merely a **left identity** and **left inverse** and we still have a group. In other words, the identity axiom could be simply $eg = g$ and the inverse axiom could simply be $hg = e$.*

8. *A group has the **cancellation property**, i.e. if $g, h, k \in G$ and $gh = gk$ or $hg = kg$, then $h = k$.*

Definition 2.2 *The **order** of a group G, written $|G|$, is the cardinality (or size) of the set G. If G has a finite set of elements we called the group a **finite** group or a group **of finite order**. Otherwise G is called an **infinite** group.*

2.1.1 Groups of Small Order

We begin what is called a **classification** result where we seek to list all possible distinct (to be defined) *non-isomorphic* groups with a certain property. In this case, the group property is its size.

First, let us dispense with the **trivial** group $G = \{1\}$ which has only an identity element and is the only group of order 1. In presenting this material we make use of what we call a **multiplication table** reminiscent of the multiplication tables used by children to learn their *times tables*. An example of such a table is given in Figure 2.2 for a group with four elements $1, g, h$ and k.

We fill in the empty boxes by cross referencing each element of G in the first column, say h, with an element of G in the first row, say k, and fill in the cross referenced box with $h * k$ or simply hk (order is important) as we did in Figure 2.3. Note that since G satisfies the closure property, we know that hk must be one of the elements $1, g, h, k$.

*	1	g	h	k
1				
g				
h				hk
k				

Figure 2.3 Filling in a group multiplication table.

+₄	0	1	2	3
0	0	1	2	3
1	1	2	3	0
2	2	3	0	1
3	3	0	1	2

Figure 2.4 The group multiplication table for $\mathbb{Z}_4$.

*	1	g	h	k
1	1	g	h	k
g	g			
h	h			
k	k			

Figure 2.5 The identity row & column of a group multiplication table.

Example 2.2 *Consider the group $\mathbb{Z}_4$ with addition (modulo 4). Its multiplication table is given in Figure 2.4.*

By the identity axiom the second row of any group multiplcation table must be identical to the first row and the second column of the table must be identical to the first column (see Figure 2.5).

Another property of the table we can deduce using the cancellation property of G is that each of the elements of G must occur exactly once in each row and column (convince yourself of this). Hence, the rows and columns of G are permutations of the elements $1, g, h, k$ (this type of table has been studied and is called a **latin square**).

*	1	g	h
1	1	g	h
g	g	h	1
h	h	1	g

*	1	g
1	1	g
g	g	1

Figure 2.6 The group multiplication table for a group of order 2 & 3.

*	1	g	h	k
1	1	g	h	k
g	g	h	k	1
h	h	k	1	g
k	k	1	g	h

*	1	g	h	k
1	1	g	h	k
g	g	1	k	h
h	h	k	1	g
k	k	h	g	1

Figure 2.7 The group multiplication tables for a group of order 4.

If the table has symmetry across the diagonal starting at the upper lefthand corner, then G is abelian. With this in mind let us begin writing multiplication tables for small groups.

Because of the conditions we just mentioned, there is only one way to complete a multiplication table for group of order 2 or 3 (see Figure 2.6).

Note also that these two groups are abelian. Our conclusion then is that there is only one group of order 2 and only one group of order 3 and they are both abelian. When we look at groups of order 4 the situation becomes slightly more complex. One can show that there are only two possible distinct tables one can construct (see Figure 2.7).

There are, in fact, other seemingly different tables, but one can argue that they are *equivalent* to one of these two tables (later we shall say they are *isomorphic*) by simply renaming the elements of the group. The two above are clearly different tables, since for instance in the second any element multiplied by itself is the identity, which is not true for the first table. Notice that both groups are abelian. Notice also that the first table is *equivalent* (*isomorphic*) to the table for $(\mathbb{Z}_4, +_4)$ if we rename 1 as 0, g as 1, h as 2 and k as 3. The other table is the multiplication table for what we call the **Klein-4** group (named after the German mathematician Felix Klein). Our conclusion then is that there are exactly two distinct groups of order 4 and they are both abelian. We will explore higher order groups later on in the text once we have more machinery.

2.1.2 Group Exponentiation

There is a natural way to define exponentiation for a group and we do so now. It is identical to how one defines exponentiation for real numbers.

Definition 2.3 *For any $g \in G$ and n a positive integer,*

$$g^0 = 1,$$
$$g^1 = g,$$
$$g^2 = gg$$
$$\vdots$$
$$g^n = \underbrace{gg \cdots g}_{n}, \ and$$
$$g^{-n} = (g^n)^{-1}$$

One can show using induction and cases that this definition of exponentiation satisfies the same properties as the usual exponentiation for real numbers. We summarize these properties in a proposition.

Proposition 2.1 *For all $g \in G$ and integers m and n we have*

1. $g^{m+n} = g^m g^n$

2. $(g^m)^{-1} = (g^{-1})^m$

3. $(g^m)^n = g^{mn}$

A key notion in the study of groups is the notion of conjugacy.

Definition 2.4 *Given two elements $g, h \in G$ a group, we say g is **conjugate** to h, if there exists an $a \in G$ such that $h = aga^{-1}$. It's easy to check that conjugacy is an equivalence relation on G and so we may consider the equivalence classes associated with this relation, called **conjugacy** classes. For $g \in G$,*

$$[g] = \{h \in G \ : \ h = aga^{-1} \text{ for some } a \in G\} = \{aga^{-1} \ : \ a \in G\}.$$

EXERCISES

1 Verify the group axioms for the structures given in Example 2.1, namely

   a. $(\mathbb{Z}, +)$ is an abelian group.

   b. $(\mathbb{Q}, +)$ is an abelian group.

   c. $(\mathbb{R}, +)$ is an abelian group.

   d. $(\mathbb{C}, +)$ is an abelian group.

   e. $(\mathbb{Q}^*, \cdot)$ is an abelian group.

   f. $(\mathbb{R}^*, \cdot)$ is an abelian group.

g. $(\mathbb{C}^*, \cdot)$ is an abelian group.

h. $(\mathbb{Z}_n, +_n)$ forms as abelian group.

i. For any set A, the set $Sym(A)$ together with the operation of composition is a group.

j. The collection of real $m \times n$ matrices with matrix addition is an abelian group.

k. The collection of $n \times n$ real invertible matrices with matrix multiplication is a group.

l. The collection of $n \times n$ real matrices of determinant 1 with matrix multiplication is a group.

m. The collection of $n \times n$ real upper triangular matrices with 1's on the diagonal and with matrix multiplication is a group.

n. The collection of all real diagonal matrices is an abelian group.

2 As was done in Figure 2.1, draw the remaining six elements of D_4.

3 Give a counterexample to illustrate why $(\mathbb{Z}^*, \cdot)$ together with multiplication does not form a group.

4 Prove that $(\mathbb{Z}_n^*, \cdot_n)$ forms an (abelian) group iff n is a prime number.

5 Give an example of two invertible matrices which do not commute.

6 Check that the identity and inverse group axioms could be weakened to merely a **left identity** and **left inverse** and we still have a group. In other words, the identity axiom could be simply $eg = g$ and the inverse axiom could simply be $hg = e$.

7 Show that a group has the cancellation property, i.e. if $g, h, k \in G$ and $gh = gk$ or $hg = kg$, then $h = k$.

8 Explain why in a group multiplication table each element of the group must occur exactly once in every row and column of the table.

9 Write out the multiplication table for D_4 and use it to verify that D_4 is indeed a group.

10 Write out the multiplication table for Q_8 and use it to verify that Q_8 is indeed a group.

11 Using your work for the multiplication table of D_4 and Q_8 explain why they represent different groups of order 8.

12 Without using Proposition 2.1, prove that for any $g \in G$ a group we have $\left(g^{-1}\right)^{-1} = g$.

13 Prove for all $g, h \in G$ a group we have $(gh)^{-1} = h^{-1}g^{-1}$.

14 Prove for all $g_1, \ldots, g_k \in G$ a group we have

$$(g_1 \cdots g_k)^{-1} = g_k^{-1} \cdots g_1^{-1}.$$

15 Show that if G is a group and $a, b \in G$ with $o(ab) = n$, then $o(ba) = n$ as well.

16 Prove Proposition 2.1.

17 Prove that conjugacy forms an equivalence relation on a group G (see Definition 2.4).

2.2 SUBGROUPS

For every algebraic structure there is a notion of a substructure – a subset of the structure that preserves all of its axioms. For groups these substructures are called subgroups.

Definition 2.5 *Let $(G, *)$ be a group. A non-empty subset $H \subseteq G$ is a **subgroup** of G if $(H, *)$ is a group. In other words, H together with the operation of G still satisfies the four axioms of a group. The notation to signify that H is a subgroup of G is $H \leq G$.*

Example 2.3 *Here are some subgroup chains related to the examples we gave in the previous section.*

1. $(\mathbb{Z}, +) \leq (\mathbb{Q}, +) \leq (\mathbb{R}, +) \leq (\mathbb{C}, +)$.

2. $(\mathbb{Q}^*, \cdot) \leq (\mathbb{R}^*, \cdot) \leq (\mathbb{C}^*, \cdot)$.

3. $D_n(\mathbb{R}) \leq U_n(\mathbb{R}) \leq SL_n(\mathbb{R}) \leq GL_n(\mathbb{R})$.

Before we give any more examples we wish to point out some shortcuts for verifying when a non-empty subset is a subgroup of a given group.

Lemma 2.1 *Let H be a non-empty subset of a group G.*

1. $H \leq G$ iff For all $h_1, h_2 \in H$ we have $h_1 h_2^{-1} \in H$.

2. If in addition H is finite, then $H \leq G$ iff For all $h_1, h_2 \in H$ we have $h_1 h_2 \in H$ (in other words, it suffices to check the closure axiom for H).

Proof 2.1 *For the first part, assuming that $H \leq G$, take $h_1, h_2 \in H$. Since H satifies the inverse property we know that $h_2^{-1} \in H$. And since H satisfies the closure property we have that $h_1 h_2^{-1} \in H$. Now assume that for all $h_1, h_2 \in H$ we have $h_1 h_2^{-1} \in H$. We need to verify that H satisfies the four axioms of a group. To show the identity property, take any $h \in H$ (here we use the fact that H is non-empty). Then by assumption we have $hh^{-1} \in H$ (set h_1 and h_2 equal to h) – in other words*

$1 \in H$. *To verify the inverse property, take any $h \in H$. Then by assumption we have $1h^{-1} \in H$ (set $h_1 = 1$ which we now know is in H and $h_2 = h$) – in other words $h^{-1} \in H$. To verify the closure property, take and $h, k \in H$. Then by assumption we have $h(k^{-1})^{-1} \in H$ (set $h_1 = h$ and $h_2 = k^{-1}$ which we now know is in H) – in other words $hk \in H$. The associative property for H is simply "inherited" from the larger group G – in other words, when taking any three elements in H, since they are also elements in G, and G has the associative property, they must associate.*

For the second part, one direction is immediate: If $H \leq G$, then being a group it satisfies the closure property. Now assume that H satisfies the closure property. By the first part it suffices to show that H satisfies the inverse property. Take any $h \in H$ and consider the list $h, h^2, h^3, \ldots$ which is contained in H (by closure). Then some power of h must equal the identity. To see this, first note that since H is finite the list above must have repeats, i.e. there exist $1 \leq i < j$ such that $h^i = h^j$. But then $h^{j-i} = 1$. Set $m = j - i$. If $m = 1$, then $h = 1$ and certainly has an inverse (namely, itself). If $m > 1$, then $h^{m-1}h = 1$ and $h^{-1} = h^{m-1} \in H$.

Example 2.4 *Here are some additional examples of subgroups which we can now verify more readily using Lemma 2.1*

1. *To verify that $(\mathbb{Q}^*, \cdot) \leq (\mathbb{R}^*, \cdot)$ take any $\frac{a}{b}, \frac{c}{d} \in \mathbb{Q}^*$ (note that $a, b, c, d \neq 0$ and so $ad, bc \neq 0$). Then*

$$\left(\frac{a}{b}\right)\left(\frac{c}{d}\right)^{-1} = \left(\frac{a}{b}\right)\left(\frac{d}{c}\right) = \left(\frac{ad}{bc}\right) \in \mathbb{Q}^*.$$

2. *$(n\mathbb{Z}, +) \leq (\mathbb{Z}, +)$, for any $n \in \mathbb{Z}$, since if $nk, nl \in \mathbb{Z}$ we have $(nk) + (-nl) = n(k - l) \in n\mathbb{Z}$ (note that in Lemma 2.1, $h_1 h_2^{-1}$ written additively is $h_1 + (-h_2)$)*

3. *Consider the group $\mathbb{C}^*$ with multiplication and the subset*

$$H = \{z \in \mathbb{C} : z^n = 1\},$$

for some fixed positive integer n. Notice that H is finite, since it consists of all the roots of the polynomial $x^n - 1$ which is a polynomial of degree n and so has at exactly n roots in $\mathbb{C}^*$. Therefore, to show $H \leq \mathbb{C}^*$ it suffices to show that H satisfies the closure property. Take $z_1, z_2 \in H$, then $(z_1 z_2)^n = z_1^n z_2^n = 1 \cdot 1 = 1$ and so $z_1 z_2 \in H$ (here we use the fact that $\mathbb{C}^*$ is abelian).

Remark 2.2 *Some of the remarks below will be left as exercises for the reader. In what follows G always represents a group.*

1. *If $H, K \leq G$, then $H \cap K \leq G$.*

2. *If $H, K \leq G$, then it is not necessarily the case that $H \cup K \leq G$.*

3. *If $H, K \leq G$, then $H \cup K \leq G$ iff Either $H \leq K$ or $K \leq H$.*

4. *For any $g \in G$ the set $H = \{g^n \ : \ n \in \mathbb{Z}\} \leq G$, called the* **cyclic subgroup generated by** g.

5. *Any group G always has the* **trivial subgroup**, $\{1\}$, *and the* **improper subgroup**, G.

6. *For $g \in G$ define*
$$C_G(g) = \{h \in G \ : \ gh = hg\},$$
called the **centralizer of** g **in** G. *Then $C_G(g) \leq G$.*

7. *Take any subset $X \subseteq G$ and define*
$$C_G(X) = \{h \in G \ : \ gh = hg, \ for \ all \ g \in X\},$$
called the **centralizer of** X **in** G. *Then $C_G(X) \leq G$. Note, when $X = \{g\}$, $C_G(X) = C_G(g)$ and when $X = G$ we call $C_G(G)$ the* **center of** G *and use the notation $Z(G)$.*

8. *Let X be a non-empty subset of a group G. The* **subgroup generated by** X, *written $\langle X \rangle$ is the collection of all finite products of elements of X and their inverses. The set X is called the* **generating** *set of $\langle X \rangle$. Then $\langle X \rangle \leq G$*

9. *If $X = \{g_1, g_2, \ldots, g_n\}$ is a finite set, then we write $\langle g_1, g_2, \ldots, g_n \rangle$ for $\langle X \rangle$. Note that if $X = \{g\}$, then $\langle X \rangle$ is simply the cyclic subgroup generated by g.*

10. *One can show that*

   - *$\langle X \rangle$ is the "smallest" subgroup of G containing the set X, i.e. if $H \leq G$ and H contains X, then $\langle X \rangle \leq H$.*
   - *$\langle X \rangle$ is the intersection of all subgroups containing the set X.*

Sometimes it is convenient to present what is called a **Hasse Diagram** or Lattice of Subgroups of a given group.

Example 2.5 *We will illustrate the concept of a Hasse Diagram with two examples.*

1. *Consider the group $\mathbb{Z}_{12}$ with addition modulo 12. One can check that the only proper subgroups of this group are the following list of subsets: $\{0, 2, 4, 6, 8, 10\}$, $\{0, 3, 6, 9\}$, $\{0, 4, 8\}$, $\{0, 6\}$, $\{0\}$. The Lattice of Subgroups is presented as a graph with the subgroups being the vertices of the graph and there is a connection between two vertices if the set below is contained in the set above it. The Lattice of Subgroups for $\mathbb{Z}_{12}$ is given in Figure 2.8.*

2. *Consider the Quaternion group $Q_8 = \{\pm 1, \pm i, \pm j, \pm k\}$. One can check that the only proper subgroups of Q_8 are $\{1\}$, $\{\pm 1\}$, $\{\pm 1, \pm i\}$, $\{\pm 1, \pm j\}$, $\{\pm 1, \pm k\}$. The Lattice of Subgroups for Q_8 is given in Figure 2.9.*

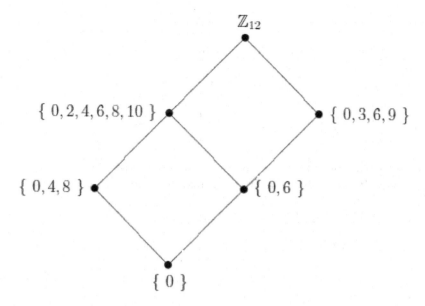

Figure 2.8 The Lattice of Subgroups for $\mathbb{Z}_{12}$.

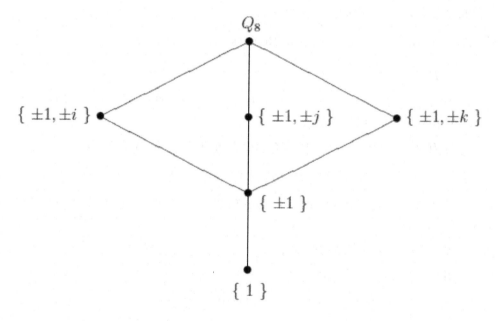

Figure 2.9 The Lattice of Subgroups for Q_8.

EXERCISES

1 Verify subgroup for each of the statements in Example 2.3 using Lemma 2.1, namely

   a. $(\mathbb{Z}, +) \leq (\mathbb{Q}, +) \leq (\mathbb{R}, +) \leq (\mathbb{C}, +)$.

   b. $(\mathbb{R}^*, \cdot) \leq (\mathbb{C}^*, \cdot)$.

   c. $D_n(\mathbb{R}) \le U_n(\mathbb{R}) \le SL_n(\mathbb{R}) \le GL_n(\mathbb{R})$.

2 Verify subgroup for each of the following subsets using Lemma 2.1:

   a. $H = \{f(x) = mx + b \ : \ m \in \mathbb{R}^*, \ b \in \mathbb{R}\} \le \mathrm{Sym}(\mathbb{R})$

   b. $H = \left\{ \begin{bmatrix} m & b \\ 0 & 1 \end{bmatrix} \ : \ m \in \mathbb{R}^*, \ b \in \mathbb{R} \right\} \le GL_n(\mathbb{R})$.

3 Decide whether or not each of the following non-empty subsets of a group G are subgroups. If so, carefully justify your answer. If not, provide a counterexample.

   a. $G = \mathrm{Sym}(\mathbb{N})$ and $H = \{\sigma \in G \ : \ \sigma(5) = 5\}$.

   b. $G = \mathbb{R} \times \mathbb{R}$ with addition coordinate-wise and $H = \{(x,y) \in G \ : \ y = x^2\}$.

4 Prove that the identity of a subgroup coincides with the identity for the entire group.

5 Prove that the inverse of an element of a subgroup coincides with the inverse of this element for the entire group.

6 Prove each of the statements in Remark 2.2, namely

   a. If $H, K \le G$, then $H \cap K \le G$.

   b. If $H, K \le G$, then it is not necessarily the case that $H \cup K \le G$ (i.e. give a counter-example).

   c. If $H, K \le G$, then $H \cup K \le G$ iff Either $H \le K$ or $K \le H$.

   d. For any $g \in G$ the set $H = \{g^n \ : \ n \in \mathbb{Z}\} \le G$.

   e. The trivial subgroup and the improper subgroup of a group G are indeed subgroups of G.

   f. If $g \in G$, then $C_G(g) \le G$.

   g. If $X \subseteq G$, then $C_G(X) \le G$.

   h. If X is a non-empty subset of G, then $\langle X \rangle \le G$.

   i. If X is a non-empty subset of G, then
- $\langle X \rangle$ is the "smallest" subgroup of G containing the set X, i.e. if $H \le G$ and H contains X, then $\langle X \rangle \le H$.
- $\langle X \rangle$ is the intersection of all subgroups containing the set X.

2.3 CYCLIC GROUPS

In this section, we study an important family of groups called cyclic groups. Later on in the text, we will classify completely cyclic groups.

Definition 2.6 *A group G is* **cyclic** *if there is an element $g \in G$ such that $G = \langle g \rangle$. In other words, every element of G can be expressed as an integer power of g. We call g the* **generator** *of G and we say G is* **generated by** *g.*

Example 2.6 *We give a finite multiplicative example and an infinite additive example of a cyclic group.*

1. *The group $\mathbb{Z}_7^*$ is cyclic, since (for instance) it is generated by 3. Indeed, if we look at successive powers of 3 we see that all the elements of the group are attained:*

$$3^1 = 3, \quad 3^2 = 2, \quad 3^3 = 6, \quad 3^4 = 4, \quad 3^5 = 5, \quad 3^6 = 1.$$

   Note that higher powers of 3 will cycle through the same values in $\mathbb{Z}_7^*$ and in the same order (hence the name cyclic group).

2. *The group $\mathbb{Z}$ with addition is cyclic, since it is generated by 1. Indeed, for any $n \in \mathbb{Z}$ we have $1^n = n$, since*

$$1^0 = 1, \qquad 1^n = \underbrace{1 + 1 + \cdots + 1}_{n} = n \quad (n > 0) \quad and$$

$$1^{-n} = \underbrace{(-1) + (-1) + \cdots + (-1)}_{n} = -n \quad (n > 0).$$

   Note that the generator of a cyclic group need not be unique. For instance, in this group -1 is also a generator of $\mathbb{Z}$.

3. *$\mathbb{Z}_n$ with addition modulo n is cyclic with generator 1 once again.*

4. *The Klein-4 group is not cyclic, since the square of any element in the group gives the identity.*

Remark 2.3 *There is a connection between cyclic and abelian as is described below.*

1. *Every cyclic group is abelian, since if $G = \langle g \rangle$, then for any $g^n, g^m \in G$ we have*

$$g^n g^m = g^{n+m} = g^{m+n} = g^m g^n.$$

2. *However, not every abelian group is cyclic. Indeed, the Klein-4 group is (the smallest) example of an abelian group which is not cyclic.*

Definition 2.7 *The* **order** *of an element g of a group, written $o(g)$, is the smallest positive power of g that attains the identity in the group (should such a power of g exist), i.e.*

$$o(g) = k, \ \text{if } g^k = 1 \text{ and for all } m, \ g^m = 1 \text{ implies } k \leq m.$$

Equivalently, if for some $0 \leq m < k$ we have $g^m = 1$, then $m = 0$. If no such integer k exists we say the element g has **infinite** *order.*

Example 2.7 *We explore the order of an element in several group settings.*

1. *Referring Example 1r, we see that $o(3) = 6$.*

2. *In $\mathbb{Z}_{12}$ with addition modulo 12, we see that $o(3) = 4$ (exercise).*

3. *In $\mathbb{Z}$, every non-zero integer has infinite order.*

4. *In $GL_n(\mathbb{Z}_p)$ with p prime and matrix multiplication modulo p being the operation, the order of the matrix $\begin{bmatrix} 1 & 1 \\ 0 & 1 \end{bmatrix}$ is p (exercise).*

Remark 2.4 *The order of a group and the order of an element although different concepts are related as follows: For any $g \in G$, $o(g) = |\langle g \rangle|$. In other words, the order of an element is equal to the order of the subgroup generated by that element. To see this we show that if $o(g) = k$, then*

$$\langle g \rangle = \{1, g, g^2, \dots, g^{k-1}\}.$$

Take any $g^n \in \langle g \rangle$. Using the Division Algorithm, divide n by k to get $n = qk + r$ where $0 \leq r \leq k - 1$. Then

$$g^n = g^{qk+r} = (g^k)^q g^r = 1^q g^r = g^r.$$

Lemma 2.2 *Let g be an element of a group G with $o(g) = k$.*

1. *If $g^n = 1$, then $k | n$.*

2. *For any integer n we have that $o(g^n) = \frac{k}{gcd(n,k)}$.*

Proof 2.2 *To prove the first part, divide n by k to get $n = qk + r$ where $0 \leq r \leq k-1$. Notice that $g^r = g^{n-qk} = g^n(g^k)^{-q} = 1$. Since $k = o(g)$ it must be that $r = 0$ and so $n = qk$ which implies that $k | n$.*

For the second part, set $d = gcd(n, k)$ so that we need to show that $o(g^n) = k/d$. Since $d = gcd(n, k)$ we can express $n = dx$ and $k = dy$ for some integers x and y with $gcd(x, y) = 1$ (see Exercise 11 in Section 1.3). First note that

$$(g^n)^{k/d} = (g^k)^{n/d} = (g^k)^x = 1^x = 1.$$

Second, suppose z is a positive integer such that $(g^n)^z = 1$. Thus, $g^{nz} = 1$ and so by the first part of the Lemma we know k divides nz. Hence, $nz = kl$ for some integer l. Then $dxz = dyl$ and by cancellation $xz = yl$ and so y divides xz. Since $gcd(x, y) = 1$ it follows that y divides z (see Exercise 10 in Section 1.3). Since y and z are both positive with $y | z$ we have $y \leq z$, i.e. $k/d \leq z$.

Example 2.8 *We apply Lemma 2.2 to the group $\mathbb{Z}_7^*$. We've seen that $o(3) = 6$. Since 3 is a generator of the group, we can use 3 to find the order of any other element in the group. For instance, let's find the order of 6. We have seen that $6 = 3^3$. Therefore,*

$$o(6) = o(3^3) = \frac{o(3)}{gcd(3, o(3))} = \frac{6}{gcd(3, 6)} = \frac{6}{3} = 2.$$

Indeed, notice that $6^2 = 1$ modulo 7.

Corollary 2.1 *If $g \in G$ generates G and G is a finite group, then g^n generates G iff $\gcd(n, o(g)) = 1$*

Proof 2.3 *This is an immediate consequence of Lemma 2.2.*

Example 2.9 *By Corollary 2.1, if we find one generator of a group, then we can easily find all of them. Consider the group $\mathbb{Z}_{11}^*$. One can check that 2 generates the group. By Corollary 2.1, 2^n generates the group iff $\gcd(n, 10) = 1$. Thus, $n = 3, 7, 9$ and so the other generators of the group are $2^3 = 8$, $2^7 = 7$ and $2^9 = 6$.*

Theorem 2.1 *Let G be a cyclic group.*

1. *Any subgroup of G is also cyclic.*

2. *If G is finite and a positive integer d divides the order of G, then there exists an element of order d and a unique subgroup $H \leq G$ of order d.*

Proof 2.4 *We're given that $G = \langle g \rangle$ for some $g \in G$. For the first statement, suppose that $H \leq G$. If H is trivial, then it is generated by 1. Otherwise, there exists positive powers of g which are in H. To see this, since H is non-trivial there exists an $m \neq 0$ such that $g^m \in H$. Since H is a group we know that $g^{-m} = (g^m)^{-1} \in H$. Certainly one of m and $-m$ is positive. Let k be the smallest positive power of g such that $g^k \in H$.*

Claim 2.1 $H = \langle g^k \rangle$.

The fact that $\langle g^k \rangle \subseteq H$ follows since $g^k \in H$ and H has the closure property. Now take any $h \in H$. Since $h \in G$ we can write $h = g^m$ for some integer m. Using the Division Algorithm, we have $m = qk + r$ where $0 \leq r < k$. Now $g^r = g^{m-qk} = g^m (g^k)^{-q} \in H$. Since k is smallest positive integer with $g^k \in H$, it must be that $r = 0$. Hence, $m = qk$ and so $g^m = (g^k)^q \in \langle g^k \rangle$ which proves the claim.

For the second statement, set $|G| = n < \infty$ and we're given that d divides n. Set $H = \langle g^{n/d} \rangle$. Then H is a subgroup of order d, since

$$o(g^{n/d}) = \frac{o(g)}{\gcd(o(g), n/d)} = \frac{n}{\gcd(n, n/d)} = \frac{n}{n/d} = d.$$

Suppose K is another subgroup of G of order d. By the first statement, K is cyclic and must be generated by some power of g. Set $K = \langle g^m \rangle$. We know $1 = (g^m)^d = g^{md}$ which implies n divides md. Thus, $md = nl$ for some integer l and so $m = (n/d)l$. Therefore, $g^m = (g^{n/d})^l \in H$ and thus $K = \langle g^m \rangle \subseteq H$. But H and K have the same order d and therefore $H = K$.

Remark 2.5 1. *Theorem 2.1.ii is not true for an arbitrary group. For instance, the Klein-4 group does not have a unique subgroup of order 2, but rather it has three such subgroups. Although we will not study these groups until the next section, for the record we state that the alternating group A_n for $n \geq 5$ which has order $n!/2$ has no group of order $n!/4$ even though $n!/4$ divides $n!/2$. We will not be able to verify this statement until much later when we show these groups are simple (to be defined).*

2. *In general, finding all the subgroups of a given group is no easy matter, however in the case of cyclic groups, Theorem 2.1 makes the task quite easy to do. Simply generate all the cyclic subgroups and isolate one of each order dividing the order of the group. Refer back to the lattice of subgroups of $\mathbb{Z}_{12}$ and we see that its distinct subgroups are generated by the elements 0, 1, 2, 3, 4 and 6 of corresponding orders 1, 12, 6, 4, 3 and 2.*

EXERCISES

1 Verify all the order statements made in Example 2.7

2 For the group $(\mathbb{Z}, +)$, verify that

   a. -1 is another generator of $\mathbb{Z}$

   b. Any other integer not equal to ± 1 is not a generator of $\mathbb{Z}$.

3 Compute the order of every element in $\mathbb{Z}_{12}$ with addition (modulo) 12.

4 Consider the group $\mathbb{Z}_7^*$ with multiplication modulo 7.

   a. Show that 3 generates the group.

   b. Use part (a) and Corollary 2.1 to find the order of 2.

   c. Use part (a) and Corollary 2.1 to find all the generators of $\mathbb{Z}_7^*$.

   d. Write out the lattice of cyclic subgroups of $\mathbb{Z}_7^*$.

5 Consider the following group with operation being matrix multiplication modulo 2:

$$G = \left\{ \begin{bmatrix} a & b \\ c & d \end{bmatrix} : a, b, c, d \in \mathbb{Z}_2 \ \& \ ad + bc \not\equiv 0 \ (mod \ 2) \right\}.$$

   a. List the elements of G.

   b. Compute $H = \left\langle \begin{bmatrix} 1 & 1 \\ 0 & 1 \end{bmatrix} \right\rangle$.

   c. Compute $\begin{bmatrix} 1 & 0 \\ 1 & 1 \end{bmatrix}^{-1}$.

6 Prove that in $GL_n(\mathbb{Z}_p)$ with p prime and matrix multiplication modulo p, the order of the matrix $\begin{bmatrix} 1 & 1 \\ 0 & 1 \end{bmatrix}$ is p (you will need a proof by induction to help verify this result).

7 Find all the generators of the cyclic group $\mathbb{Z}_{13}^*$ by finding one by hand and then applying Corollary 2.1 to find the rest.

8 Prove or disprove the following statement: For any prime p the multiplicative group $\mathbb{Z}_p^*$ is generated by 2.

9 Find all the generators of the cyclic group $\mathbb{Z}_{13}$ by finding one by hand and then applying Corollary 2.1 to find the rest.

10 Create the Lattice of Subgroups for the cyclic group $(\mathbb{Z}_{60}, +_{60})$ making use of Remark 2.5.

11 Create the Lattice of Subgroups for the cyclic group $(\mathbb{Z}_{61}^*, \cdot_{61})$ making use of Remark 2.5 (hint: first find a generator of $\mathbb{Z}_{61}^*$).

12 Prove Corollary 2.1

2.4 PERMUTATION GROUPS

We focus now in more detail on the symmetric group. This group constitutes one of the origins of group theory and is essential. Recall that for any set A the symmetric group, $Sym(A)$, is the collection of all bijections from A to A (called permutations on A) with the operation being the composition of functions. We narrow our focus to finite sets A. In fact, one can narrow further to set $A = \{1, 2, \ldots, n\}$ and the corresponding symmetric group S_n. The reason why we can do this is essentially because any permutation on $A = \{a_1, a_2, \ldots, a_n\}$ corresponds to a unique permutation in S_n. For instance, if $\sigma \in Sym(A)$ and $\sigma(a_i) = a_j$, then the corresponding permutation in S_n would send i to j. As we keep promising, this notion of two groups being essentially the same will be made more formal later in the text when we define *isomorphic* groups.

We now introduce notation for representing permutations. The first representation takes the form

$$\sigma = \begin{pmatrix} 1 & 2 & \cdots & n \\ \sigma(1) & \sigma(2) & \cdots & \sigma(n) \end{pmatrix}.$$

Example 2.10 *In S_5 if $\sigma(1) = 2$, $\sigma(2) = 4$, $\sigma(3) = 5$, $\sigma(4) = 1$ and $\sigma(5) = 3$, then*

$$\sigma = \begin{pmatrix} 1 & 2 & 3 & 4 & 5 \\ 2 & 4 & 5 & 1 & 3 \end{pmatrix}.$$

To get σ^{-1} simply flip σ over and reorder the columns according to the top row, i.e.

$$\sigma^{-1} = \begin{pmatrix} 2 & 4 & 5 & 1 & 3 \\ 1 & 2 & 3 & 4 & 5 \end{pmatrix} = \begin{pmatrix} 1 & 2 & 3 & 4 & 5 \\ 4 & 1 & 5 & 2 & 3 \end{pmatrix}.$$

Consider the permutation

$$\tau = \begin{pmatrix} 1 & 2 & 3 & 4 & 5 \\ 2 & 1 & 5 & 4 & 3 \end{pmatrix}.$$

For brevity we will write $\sigma\tau$ for $\sigma \circ \tau$. When computing the composition, in this text, the reader should note that one reads the maps from right to left just as we do with composition of functions. Hence,

$$\sigma\tau = \begin{pmatrix} 1 & 2 & 3 & 4 & 5 \\ 2 & 4 & 5 & 1 & 3 \end{pmatrix} \begin{pmatrix} 1 & 2 & 3 & 4 & 5 \\ 2 & 1 & 5 & 4 & 3 \end{pmatrix} = \begin{pmatrix} 1 & 2 & 3 & 4 & 5 \\ 4 & 2 & 3 & 1 & 5 \end{pmatrix}.$$

The second way to represent permutations is by means of k-cycles. A k-cycle in S_n is a special permutation represented as follows:

$$\sigma = (i_1 \ i_2 \ \cdots \ i_k) \quad \text{where} \quad \{i_1, \ i_2, \ \ldots, \ i_k\} \subseteq \{1, \ 2, \ \ldots, \ n\}.$$

Note that, in this text, k-cycles are read from left to right(!). Thus, σ is defined as follows:

$$\sigma(i_1) = i_2, \ \sigma(i_2) = i_3, \ \ldots, \ \sigma(i_{k-1}) = i_k, \ \sigma(i_k) = i_1.$$

If $m \notin \{i_1, \ i_2, \ \ldots, \ i_k\}$, then it is understood that $\sigma(m) = m$, i.e. any other m is fixed by σ. A 2-cycle is also called a **transposition**.

Example 2.11 *In S_7, if $\sigma = (2 \ 4 \ 3 \ 6 \ 1)$, then*

$$\sigma = \begin{pmatrix} 1 & 2 & 3 & 4 & 5 & 6 & 7 \\ 2 & 4 & 6 & 3 & 5 & 1 & 7 \end{pmatrix}.$$

To find the inverse of a k-cycle simply reverse the order of the numbers in the k-cycle. For example, $\sigma^{-1} = (1 \ 6 \ 3 \ 4 \ 2)$. Note that a transposition therefore is its own inverse, since

$$(a \ b)^{-1} = (b \ a) = (a \ b).$$

If $\tau = (2 \ 5 \ 4)(4 \ 1 \ 5 \ 2)(6 \ 5 \ 1 \ 2 \ 4)$ a composition of three cycles, then

$$\tau = \begin{pmatrix} 1 & 2 & 3 & 4 & 5 & 6 & 7 \\ 2 & 1 & 3 & 6 & 4 & 5 & 7 \end{pmatrix}.$$

Definition 2.8 *Two cycles are **disjoint** if they have no numbers in common. Formally, if $\sigma = (a_1 \ a_2 \ \cdots \ a_m)$ and $\tau = (b_1 \ b_2 \ \cdots \ b_n)$, then $A \cap B = \emptyset$ where $A = \{a_1, \ a_2, \ \ldots, \ a_m\}$ and $B = \{b_1, \ b_2, \ \ldots, \ b_n\}$.*

Lemma 2.3 *Disjoint cycles commute, i.e. if $\sigma, \tau \in S_n$ are disjoint cycles, then $\sigma\tau = \tau\sigma$.*

Proof 2.5 *Let $\sigma = (a_1 \ a_2 \ \cdots \ a_m)$ and $\tau = (b_1 \ b_2 \ \cdots \ b_n)$ be disjoint cycles with sets A and B as in the definition.*

Case 1: *If $c \notin A \cup B$, then*

$$\sigma(\tau(c)) = \sigma(c) = c = \tau(c) = \tau(\sigma(c)).$$

Case 2: *If $a_i \in A$, then $a_i, \sigma(a_i) \notin B$, and thus*

$$\sigma(\tau(a_i)) = \sigma(a_i) = \tau(\sigma(a_i)).$$

Case 3: *If $b_i \in B$, then $b_i, \tau(b_i) \notin A$, and thus*

$$\sigma(\tau(b_i)) = \tau(b_i) = \tau(\sigma(b_i)).$$

Since $\sigma\tau$ and $\tau\sigma$ agree on every input, they are therefore equal.

Example 2.12 *Consider the following permutation in S_{11}:*

$$\sigma = \begin{pmatrix} 1 & 2 & 3 & 4 & 5 & 6 & 7 & 8 & 9 & 10 & 11 \\ 3 & 1 & 4 & 7 & 8 & 5 & 2 & 6 & 11 & 10 & 9 \end{pmatrix}.$$

We shall illustrate with this example the result we wish to prove next, namely that every permutation can be written as a product of disjoint cycles. We will do this systematically so that a general algorithm is evident and can be used in the proof to follow.

Start with 1. Notice that σ sends 1 to 3. Then σ sends 3 to 4, 4 to 7, 7 to 2, and σ sends 2 back to 1. So a cycle in σ is (1 3 4 7 2). Now pick the smallest number not mentioned in the cycle we constructed, which would be 5. Now σ sends 5 to 8, which is sent to 6, and 6 is sent back to 5. Hence, a second cycle in σ is (5 8 6). The smallest number not yet mentioned in both cycles constructed is 9 which is sent to 11, which is sent back to 9. Hence, a third cycle in σ is (9 11). The last number not yet mentioned is 10 which is sent to itself. This yields a 1-cycle (10). Therefore,

$$(1\ 3\ 4\ 7\ 2)(5\ 8\ 6)(9\ 11)(10).$$

Typically, we drop any 1-cycles from the representation and simply write

$$(1\ 3\ 4\ 7\ 2)(5\ 8\ 6)(9\ 11).$$

Theorem 2.2 *Every permutation can be written as a product of disjoint cycles.*

Proof 2.6 *Let $\sigma \in S_n$. Note that for a positive integer k, σ^k means the k times composition of σ, and if $k = 0$, then σ^0 is the identity map. Consider the following infinite list:*

$$1,\ \sigma(1),\ \sigma^2(1),\ \sigma^3(1),\ \ldots.$$

Since for any k we know $\sigma^k(1) \in \{1, 2, \ldots, n\}$, then there are surely repeats in the above list. Let r be smallest such that $\sigma^r(1)$ is a repeat of an earlier value in the list.

Claim 2.2 $\sigma^r(1) = 1$.

By assumption there is a k with $0 \leq k < r$ and $\sigma^r(1) = \sigma^k(1)$. Suppose to the contrary that $k > 0$. Then $\sigma^{r-k}(1) = 1$ contradicting our assumption that $\sigma^r(1)$ is the first repeat. Hence, $k = 0$ and $\sigma^r(1) = \sigma^0(1) = 1$ which proves the claim.

Therefore, by the claim, one of the cycles in σ's representation as a product of disjoint cycles is

$$(1 \ \sigma(1) \ \sigma^2(1) \ \sigma^3(1) \ \cdots \ \sigma^{r-1}(1)).$$

Certainly, $r \leq n$. If $r = n$, then we are done with the proof with σ being represented by a single cycle. Otherwise, choose the smallest number, say i, not in the list

$$1, \ \sigma(1), \ \sigma^2(1), \ \sigma^3(1), \ \ldots, \ \sigma^{r-1}(1).$$

A similar argument to the one above shows that, for some positive integer s, another one of the cycles in σ's representation as a product of disjoint cycles is

$$(i \ \sigma(i) \ \sigma^2(i) \ \sigma^3(i) \ \cdots \ \sigma^{s-1}(i)).$$

Repeat this process of producing cycles until all of the numbers $1, 2, \ldots, n$ are used up. We leave it as an exercise to verify that we have produced disjoint cycles.

Remark 2.6 *The representation of a permutation as a product of disjoint cycles is unique up to the order of the cycles (since they commute). Although this statement might be intuitively obvious, a formal proof should really be provided, however we will skip this result.*

Definition 2.9 *The **cycle type** of a given permutation is the length and number of cycles in its unique disjoint cycle representation.*

Example 2.13 *Consider the permutation in the previous example where the disjoint cycle representation was*

$$\sigma = (1 \ 3 \ 4 \ 7 \ 2)(5 \ 8 \ 6)(9 \ 11)(10).$$

The cycle type for σ can be expressed as $(*)(**)(***)(*****)$. Typically, the cycles are arranged in increasing order of length.

Theorem 2.3 *Every permutation can be written as a product of transpositions.*

Proof 2.7 *To see this result, since we already have Theorem 2.2, it suffices to show that every cycle can be written as a product of transpositions. Therefore, consider an arbitrary k-cycle $(a_1 \ a_2 \ \cdots, \ a_k)$. Then*

$$(a_1 \ a_2 \ \cdots \ a_k) = (a_{k-1} \ a_k) \cdots (a_2 \ a_k)(a_1 \ a_k).$$

Hence, we expressed any k-cycle as a product of transpositions.

Example 2.14 *Consider in S_7 the 5-cycle $(2 \ 4 \ 1 \ 6 \ 7)$. Then*

$$(2 \ 4 \ 1 \ 6 \ 7) = (6 \ 7)(1 \ 7)(4 \ 7)(2 \ 7).$$

Now unlike the previous representation as a product of disjoint cycles, the representation of a permutation as a product of transpositions is not unique. Indeed, in the previous example one can add two transpositions of the form $(1 \ 2)$ to the end of the representation, i.e.

$$(2 \ 4 \ 1 \ 6 \ 7) = (6 \ 7)(1 \ 7)(4 \ 7)(2 \ 7)(1 \ 2)(1 \ 2).$$

For that matter we could add four such transpositions at the end, or six, or eight, ad infinitum. Although there is no unique representation as a product of transpositions, there is something that remains invariant with respect to the permutation.

Definition 2.10 *A permutation is called* **even** *if it can be represented as a product of an even number of transpositions. A permutation is* **odd** *if it can be represented as a product of an odd number of transpositions.*

Theorem 2.4 *A permutation cannot be both even and odd.*

Proof 2.8 *The majority of this result is taken up with the fact that the statement is true for the identity map. We show that the identity map is only even, i.e. if $1 = \tau_1 \tau_2 \cdots \tau_k$ where each τ_i is a transposition, then k is even. This we do now. For each number m which appears in the transpositions consider the following reasoning: Let j be largest such that m appears in τ_j. Note that $j \neq 1$ for otherwise m would not be fixed by the identity map. There are four possibilities for the product $\tau_{j-1}\tau_j$: $(m\ x)(m\ x)$, $(m\ x)(m\ y)$, $(x\ y)(m\ x)$ or $(y\ z)(m\ x)$. For the first possibility, notice that we can simply remove the two transpositions from the product. For the other three properties, we can rewrite the product so that m appears first in τ_{j-1} as follows:*

$$(m\ x)(m\ y) = (m\ y)(x\ y),$$

$$(x\ y)(m\ x) = (m\ y)(x\ y),$$

$$(y\ z)(m\ x) = (m\ x)(y\ z).$$

In summary, we either remove two transpositions from the product or we move the last occurrence of m to the left. In the first possibility repeat the process on the next largest occurrence of m (if it exists). In the other three cases repeat the process on $\tau_{j-2}\tau_{j-1}$. Note that for the number m you must eventually be in the first case, for otherwise m would appear first in τ_1 which as we have already pointed out is not possible. Therefore, ultimately we remove m and eventually all transpositions from the identity map in a two-by-two fashion until none are left. Hence, the product must be comprised of an even number of transpositions.

Now suppose that σ is any permutation and $\sigma = \tau_1 \tau_2 \cdots \tau_r$ and $\sigma = \tau_1' \tau_2' \cdots \tau_s'$ represented in two ways as a product of transpositions. Equating the two representations we have $\tau_1 \tau_2 \cdots \tau_r = \tau_1' \tau_2' \cdots \tau_s'$ and moving them all to one side by multiplying by the inverses of the transpositions we have

$$1 = \tau_r \cdots \tau_2 \tau_1 \tau_1' \tau_2' \cdots \tau_s'.$$

By the work above we know that $r + s$ must be even, but this implies that either r and s are both even or r and s are both odd.

Definition 2.11 *The* **alternating group***, written $A_n = \{\sigma \in S_n \ : \ \sigma \text{ is even}\}$. The set $O_n = \{\sigma \in S_n \ : \ \sigma \text{ is odd}\}$.*

Remark 2.7 *We wish to point out several observation regarding* A_n.

1. *As suggested by the name,* A_n *is a subgroup of* S_n. *To see this, since* A_n *is finite, it is enough to show that* A_n *satisfies the closure property. But this is apparent, since if* $\sigma = \tau_1 \tau_2 \cdots \tau_r$ *is a product of an even number of transpositions and so is* $\tau = \tau_1' \tau_2' \cdots \tau_s'$, *then*

$$\sigma\tau = \tau_1 \tau_2 \cdots \tau_r \tau_1' \tau_2' \cdots \tau_s',$$

with $r + s$ *also even.*

2. *By Theorem 2.4, we know that* S_n *is a disjoint union of* A_n *and* O_n.

3. *Exactly half of the permutations in* S_n *(n* $\geq$ *2) are even. To see this it is enough to show that* $|A_n| = |O_n|$ *which we do by defining a bijection between the two sets. Define a function* $f : A_n \to O_n$ *by* $f(\sigma) = \sigma(1\ 2)$. *Certainly* f *maps into* O_n *since it adds one transposition to the end of any even permutation, thus making it odd. The function is one-to-one, since if* $f(\sigma_1) = f(\sigma_2)$, *then* $\sigma_1(1\ 2) = \sigma_2(1\ 2)$ *and by cancellation in the group* S_n *we have* $\sigma_1 = \sigma_2$. *Finally,* f *maps onto* O_n, *since for any* $\tau \in O_n$ *notice that* $f(\tau(1\ 2)) = \tau(1\ 2)(1\ 2) = \tau$.

We leave as an exercise the following result:

Lemma 2.4 *In the permutation group* S_n,

1. *the order of a k-cycle is k.*

2. *if* σ *equals a product of disjoint cycles, say* $\sigma = \sigma_1 \sigma_2 \cdots \sigma_m$, *then the order of* σ *is the least common multiple of the orders of* $\sigma_1, \sigma_2, \ldots, \sigma_m$.

Example 2.15 *We shall illustrate Lemma 2.4 with several examples.*

1. *Consider the permutation*

$$\sigma = \begin{pmatrix} 1 & 2 & 3 & 4 & 5 & 6 & 7 & 8 & 9 \\ 2 & 5 & 4 & 3 & 8 & 7 & 9 & 1 & 6 \end{pmatrix} = (1\ 2\ 5\ 8)(3\ 4)(6\ 7\ 9).$$

Therefore, $o(\sigma) = lcm(4, 2, 3) = 12$.

2. *Let* $\sigma = (1\ 3\ 7)(2\ 7\ 3)(1\ 4) \in S_7$. *Now* σ *is not represented as a product of disjoint cycles, so we cannot yet use the second part of Lemma 2.4. We first will need to rewrite* σ *as a product of disjoint cycles. One can compute*

$$\sigma = \begin{pmatrix} 1 & 2 & 3 & 4 & 5 & 6 & 7 \\ 4 & 1 & 2 & 3 & 5 & 6 & 7 \end{pmatrix} = (1\ 4\ 3\ 2).$$

Therefore, σ *is a 4-cycle and as such its order is 4.*

3. *Using cycle types, we are now in a position to compute the orders of all the elements in S_n (for a given n) and the number of elements of each order. Let's do this for S_5:*

Cycle Type	Order	Number
$(*)(*)(*)(*)(*)$	1	1
$(*)(*)(*)(**)$	2	$\frac{5!}{3! \cdot 2} = 10$
$(*)(**)(**)$	2	$\frac{5!}{2! \cdot 2 \cdot 2} = 15$
$(*)(*)(* * *)$	3	$\frac{5!}{2! \cdot 3} = 20$
$(**)(* * *)$	6	$\frac{5!}{2 \cdot 3} = 20$
$(*)(* * **)$	4	$\frac{5!}{4} = 30$
$(* * * * *)$	5	$\frac{5!}{5} = 24$

*Some explanation is required to understand how the cycle types decided and were counted. To determine the cycle types, simply consider all different ways (up to commutativity) of expressing 5 as a sum of positive integers (these are called the **partitions** of the number 5). There are two things to consider. First, there are k different ways to represent the same k-cycle. To see this, consider the example 3-cycle*

$$(1\ 2\ 3) = (3\ 1\ 2) = (2\ 3\ 1).$$

Second, if the cycle type contains m disjoint k-cycles, then there are m! ways to order them all of which yield the same result (since they are disjoint). Now let's count the cycle type $(*)(**)(**)$. There are 5! ways to fill in the asterisks. Each 2-cycle can be represented in two ways, thus we divide by 2 twice. Furthermore, the two 2-cycles can be ordered in 2! different ways, thus we also divide by 2!.

Example 2.16 *We now make the connection between the dihedral group and the symmetric group.*

1. *For n = 3, the dihedral group consists of three rotations and three reflections of an equilateral triangle. Label the three vertices with the numbers 1, 2 and 3 (see Figure 2.10).*

The rotations 0°, 120° and 240° will be denoted respectively by ρ_0, ρ_1 and ρ_2. The reflections will be denoted by μ_1, μ_2 and μ_3 where for $i = 1, 2, 3$ the reflection μ_i fixes vertex i and swaps the other two vertices. So we can consider the elements of D_3 as permutations of the numbers $1, 2, 3$. Then the elements of D_3 are

$$\rho_0 = \begin{pmatrix} 1 & 2 & 3 \\ 1 & 2 & 3 \end{pmatrix} \qquad \rho_1 = \begin{pmatrix} 1 & 2 & 3 \\ 2 & 3 & 1 \end{pmatrix} \qquad \rho_2 = \begin{pmatrix} 1 & 2 & 3 \\ 3 & 1 & 2 \end{pmatrix}$$

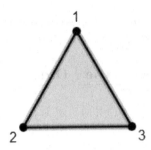

Figure 2.10 Symmetries of the triangle.

$$\mu_1 = \begin{pmatrix} 1 & 2 & 3 \\ 1 & 3 & 2 \end{pmatrix} \qquad \mu_2 = \begin{pmatrix} 1 & 2 & 3 \\ 3 & 2 & 1 \end{pmatrix} \qquad \mu_3 = \begin{pmatrix} 1 & 2 & 3 \\ 2 & 1 & 3 \end{pmatrix}.$$

Note first that $|D_3| = |S_3|$ *and so for* $n = 3$ *we have* $D_3 = S_3$. *Let's decide which elements are even and which are odd. We know* ρ_0 *is even and*

$$\rho_1 = (1\ 2\ 3) = (2\ 3)(1\ 3) \qquad\qquad \rho_2 = (1\ 3\ 2) = (3\ 2)(1\ 2)$$

$$\mu_1 = (2\ 3) \qquad \mu_2 = (1\ 3) \qquad \mu_3 = (1\ 2).$$

Thus, we see that the rotations are all even and the reflections are all odd and so $A_3 = \{\rho_0,\ \rho_1,\ \rho_2\}$.

2. *For the group* D_4 *we label the vertices of a square with the numbers 1 through 4. Thus, the elements of* D_4 *are*

$$\rho_0 = \begin{pmatrix} 1 & 2 & 3 & 4 \\ 1 & 2 & 3 & 4 \end{pmatrix} \qquad\qquad \rho_1 = \begin{pmatrix} 1 & 2 & 3 & 4 \\ 2 & 3 & 4 & 1 \end{pmatrix}$$

$$\rho_2 = \begin{pmatrix} 1 & 2 & 3 & 4 \\ 3 & 4 & 1 & 2 \end{pmatrix} \qquad\qquad \rho_3 = \begin{pmatrix} 1 & 2 & 3 & 4 \\ 4 & 1 & 2 & 3 \end{pmatrix}$$

$$\mu_1 = \begin{pmatrix} 1 & 2 & 3 & 4 \\ 1 & 4 & 3 & 2 \end{pmatrix} \qquad\qquad \mu_2 = \begin{pmatrix} 1 & 2 & 3 & 4 \\ 3 & 4 & 1 & 2 \end{pmatrix}$$

$$\mu_3 = \begin{pmatrix} 1 & 2 & 3 & 4 \\ 3 & 2 & 1 & 4 \end{pmatrix} \qquad\qquad \mu_4 = \begin{pmatrix} 1 & 2 & 3 & 4 \\ 4 & 3 & 2 & 1 \end{pmatrix}.$$

Note that for $n = 4$ *the subscripts of the* μ_i *do not have the nice relationship to the vertices as they did for* $n = 3$. *Furthermore, the reader can check that half of the rotations are even and half are odd. The same holds true for the reflections. Since* $|D_4| = 8$ *and* $|S_4| = 4! = 24$, D_4 *is a proper subgroup of* S_4 *and since* $|A_4| = 12$ *and* D_4 *has only four even permutations, these four form a proper subgroup of* A_4.

3. *For any $n \geq 3$ there are several general statements we can make. First of all since $|D_n| = 2n$ and $|S_n| = n!$ it is only for $n = 3$ that the two groups coincide. Otherwise D_n is a proper subgroup of S_n and the even permutations in D_n form a proper subgroup of A_n. Secondly, It is always the case that the rotations are generated by ρ_1.*

EXERCISES

1 Explain why the size of S_n is $n!$.

2 Consider the following permutations in S_6:

$$\sigma = \begin{pmatrix} 1 & 2 & 3 & 4 & 5 & 6 \\ 6 & 2 & 4 & 5 & 1 & 3 \end{pmatrix} \text{ and } \tau = \begin{pmatrix} 1 & 2 & 3 & 4 & 5 & 6 \\ 2 & 1 & 4 & 3 & 6 & 5 \end{pmatrix}.$$

   a. Compute $\sigma\tau$, $\tau\sigma$, σ^{-1} and τ^{-1}.

   b. Express each of σ and τ as a product of disjoint cycles.

   c. Express each of σ and τ as a product of transpositions.

   d. Decide if each of σ and τ is even or odd.

3 Consider the permutation group S_{10} and the following two elements:

$$\sigma = \begin{pmatrix} 1 & 2 & 3 & 4 & 5 & 6 & 7 & 8 & 9 & 10 \\ 4 & 2 & 5 & 8 & 7 & 10 & 3 & 9 & 1 & 6 \end{pmatrix} \qquad \tau = (1\ 2\ 4)(2\ 5\ 6)(2\ 6\ 8)$$

   a. Write σ as a product of disjoint cycles.

   b. Use part a to compute the order of σ.

   c. Decide whether σ is even or odd.

   d. Compute $\sigma\tau$ and write your answer in the same form as σ was given.

4 Verify that the cycles produced in Theorem 2.2 are disjoint.

5 Prove Lemma 2.4.

6 Apply Lemma 2.4 to find the order of σ in Example 2.12.

7 In S_{11}, apply Lemma 2.4 to find the order of $(1\ 3\ 2\ 6\ 8\ 11)(5\ 7\ 9\ 10)$

8 Decide which of the elements in D_4 are even and which are odd.

9 For $\sigma, \tau \in S_n$, prove the following statements:

   a. $\tau(a_1\ a_2\ \cdots\ a_k)\tau^{-1} = (\tau(a_1)\ \tau(a_2)\ \cdots\ \tau(a_k))$.

   b. σ and $\tau\sigma\tau^{-1}$ have the same cycle type.

   c. Cycle types correspond exactly to conjugacy classes for the group S_n. Hence, the table we constructed earlier for S_5 is, in fact, the table of conjugacy classes for S_5.

2.5 PRODUCTS OF GROUPS

In this section, in two different settings, we will construct new groups from old ones in each case by means of a product (to be defined). In the first case we will consider two arbitrary groups and in the second case we will consider two subgroups of a given group.

Definition 2.12 *Let $(G, *)$ and $(G', *')$ be two groups. The (**external**) **product** of G and G', has as its objects the elements of the Cartesian product of G and G',*

$$G \times G' = \{(g, g') \ : \ g \in G, \ g' \in G'\},$$

and has as its operation the coordinate-wise product,

$$(g_1, g_1')(g_2, g_2') = (g_1 * g_2, g_1' *' g_2').$$

One can easily generalize this construction to any finite number of groups. For brevity we replace $G_1 \times G_2 \times \cdots \times G_n$ by

$$\prod_{i=1}^{n} G_i = \{(g_1, g_2, \ldots, g_n) \ : \ g_1 \in G_1, \ g_2 \in G_2, \ \ldots, g_n \in G_n\}.$$

Example 2.17 *Consider the external product $\mathbb{Z}_4 \times \mathbb{Z}_7^* \times Q_8$. For illustration we compute the product $(2, 3, k)(3, 4, j) = (2 +_4 3, 3 \cdot_7 4, kj) = (1, 5, -i)$. The identity in this group is $(0, 1, 1)$.*

Proposition 2.2 *The order of $(g_1, g_2, \ldots, g_n) \in \prod_{i=1}^{n} G_i$ is the least common multiple of the orders of the individual $g_i \in G_i$ for $i = 1, 2, \ldots, n$.*

Proof 2.9 *We shall prove it for two groups and the result easily generalizes by induction on n. Let $(g, g') \in G \times G'$ and set $m = o(g)$, $n = o(g')$ and $l = lcm(m, n)$. Thus, $l = ma$ and $l = nb$ for some $a, b \in \mathbb{Z}$. First of all,*

$$(g, g')^l = (g^l, (g')^l) = ((g^m)^a, ((g')^n)^b) = (1^a, (1')^b) = (1, 1').$$

Now suppose that for some positive integer we have $(g, g')^r = (1, 1')$. Then $(g^r, (g')^r) = (1, 1')$ and so $g^r = 1$ and $(g')^r = 1'$. By Lemma 2.2.i, it follows that m and n both divide r. In other words, r is a common multiple of m and n and so by definition the least common multiple $l \leq r$.

Example 2.18 *We illustrate Proposition 2.2 with several examples.*

1. *The order of $(2, 4, i) \in \mathbb{Z}_4 \times \mathbb{Z}_7^* \times Q_8 = lcm(2, 3, 4) = 12$.*

2. *Consider the group $\mathbb{Z}_4 \times \mathbb{Z}_6$. Since both groups are cyclic, $\mathbb{Z}_4$ has elements of orders 1, 2 and 4 and $\mathbb{Z}_6$ has elements of orders 1, 2, 3 and 6. Since in $G \times G'$ we have $o(g, g') = lcm(o(g), o(g'))$ this group has elements of orders 1, 2, 3, 4, 6 and 12. Note, therefore, that $\mathbb{Z}_4 \times \mathbb{Z}_6$ is not cycle since it has 24 elements.*

Now we present the second construction promised in this section.

Definition 2.13 *Let $(G, *)$ be a group and X, Y two subsets of G. The product set, written*

$$XY = \{x * y \ : \ x \in X, \ y \in Y\}.$$

If $X = \{x\}$ we write xY for $\{x\}Y$.

Proposition 2.3 *For any two finite subgroups H, K of a group G, we have*

$$|HK| = \frac{|H||K|}{|H \cap K|}.$$

Proof 2.10 *Define the following relation on $H \times K$: $(h, k) \sim (h', k')$ iff $hk = h'k'$. One can easily verify that this is an equivalence relation.*

Claim 2.3 *The class of (h, k), i.e. $[(h, k)] = \{(hz^{-1}, zk) \ : \ z \in H \cap K\}$.*

First note that $(hz^{-1}, zk) \in [(h, k)]$, since $hz^{-1}zk = hk$ and so $(hz^{-1}, zk) \sim (h, k)$. On the other hand if $(h'k') \sim (h, k)$, then $hk = h'k'$ and so $(h')^{-1}h = k'k^{-1}$. Set $z = (h')^{-1}h = k'k^{-1} \in H \cap K$ so that $(h', k') = (hz^{-1}, zk)$.

The Claim shows that the equivalence classes all have the same size as $H \cap K$. Therefore, the size of the quotient set $(H \times K)/ \sim$ is the size of $H \times K$ divided by $H \cap K$. Now consider the map

$$f : (H \times K)/ \sim \ \to HK \quad by \quad f([(h, k)]) = hk.$$

This map is well-defined and one-to-one, since

$$f([(h, k)]) = f([(h', k')]) \Leftrightarrow hk = h'k' \Leftrightarrow (h, k) \sim (h', k') \Leftrightarrow [(h, k)] = [(h', k')].$$

Now this map certainly maps onto HK and so

$$|HK| = |(H \times K)/ \sim| = \frac{|H \times K|}{|H \cap K|} = \frac{|H||K|}{|H \cap K|}.$$

Example 2.19 *Now it is not always the case that for $H, K \leq G$ that $HK \leq G$. For instance, take $G = S_3$, $H = \langle \mu_1 \rangle$ and $K = \langle \mu_2 \rangle$. Then*

$$HK = \{\rho_0, \mu_1, \mu_2, \mu_1\mu_2\} = \{\rho_0, \mu_1, \mu_2, \rho_1\}.$$

Then HK is not a subgroup of G, since for instance $\mu_2\mu_1 = \rho_2 \notin HK$. Indeed, the fact that $HK \neq KH$ is the reason why it is not a subgroup, as we shall see in the next result.

Theorem 2.5 *For subgroups H, K of a group G, the following are equivalent:*

1. $HK \leq G$.

2. $HK = KH$.

3. $KH \leq G$

Proof 2.11 *It's enough to show that the first two statements are equivalent, since the second statement is symmetric. First assume that $HK \leq G$. Take $kh \in KH$, then $kh = ((kh)^{-1})^{-1} = (h^{-1}k^{-1})^{-1} \in HK$, since $HK \leq G$ and $h^{-1}k^{-1} \in HK$. Now take an $hk \in HK$. Note that $k^{-1}h^{-1} = (hk)^{-1} \in HK$, since $HK \leq G$. So we can write $k^{-1}h^{-1} = h'k' \in HK$. Then*

$$hk = ((hk)^{-1})^{-1} = (k^{-1}h^{-1})^{-1} = (h'k')^{-1} = (k')^{-1}(h')^{-1} \in KH.$$

Now assume that $HK = KH$. Take $hk, h'k' \in HK$. Note that $h'(k'k^{-1}) \in HK = KH$, so we can write $h'k'k^{-1} = k''h''$ Then

$$(h'k')(hk)^{-1} = h'k'k^{-1}h^{-1} = k''h''h^{-1} \in KH = HK.$$

Remark 2.8 *A few remarks are in order for the product HK.*

1. *HK is called the **internal product** of H and K, since the construction occurs within the group G, whereas the external product of two groups yields a new group which lies outside both of the two groups in its construction.*

2. *An immediate Corollary to Theorem 2.5 is that in an abelian group the product HK is always a subgroup of G. In this case the product is often represented as a sum, i.e. $H + K$, and is called the **internal sum**.*

Example 2.20 *We give two examples of an internal product/sum.*

1. *Consider the multiplicative group $G = \mathbb{Z}_{19}^*$ with subgroups $H = \langle 4 \rangle = \{4, 16, 7, 9, 17, 11, 6, 5, 1\}$ and $K = \langle 8 \rangle = \{8, 7, 18, 11, 12, 1\}$. Note that $H \cap K = \{7, 11, 1\}$. Since G is abelian we know that HK is a subgroup of G and the size of HK, by Proposition 2.3, is $\frac{9 \cdot 6}{3} = 18$. Thus, we know, in fact, that $HK = G$.*

2. *Consider the additive group $G = \mathbb{Z}$ with subgroups $H = m\mathbb{Z}$ and $K = n\mathbb{Z}$. We show that the internal sum $H + K$ is the subgroup $d\mathbb{Z}$ where $d = \gcd(m, n)$. So, for instance, $4\mathbb{Z} + 6\mathbb{Z} = 2\mathbb{Z}$. To see this take $x \in H + K$ and write $x = ma + nb$. Since d divides both a and b, then d divides $ma + nb = x$. Hence, $x = dy \in d\mathbb{Z}$. Now take $x \in d\mathbb{Z}$ and write $x = dy$. We know there exists integers x_0 and y_0 such that $d = mx_0 + ny_0$ and so*

$$x = dy = m(yx_0) + n(yy_0) \in H + K.$$

Definition 2.14 *Let G be an abelian group.*

1. *When $H, K \leq G$ and $H \cap K = 0$ (by this we mean $H \cap K = \{0\}$), we write $H \oplus K$ instead of $H + K$ to signify this fact and it is called an internal **direct sum** of H and K.*

2. More generally, if $H_1, H_2, \ldots, H_n \leq G$ and for each $i = 1, 2, \ldots, n$ we have $H_i \cap (H_1 + \cdots + H_{i-1} + H_{i+1} + \cdots + H_n) = 0$ we write $H \oplus H_2 \oplus \cdots \oplus H_n$ instead of $H_1 + H_2 + \cdots + H_n$ to signify this fact and it is called an internal **direct** sum of $H_1, H_2, \ldots, H_n$.

Lemma 2.5 *The following are equivalent for G an abelian group with subgroups $H_1, H_2, \ldots, H_n$:*

1. $G = H_1 \oplus H_2 \oplus \cdots \oplus H_n$

2. *For every $g \in G$ there exist unique $h_1 \in H_1$, $h_2 \in H_2$, $\ldots$, $h_n \in H_n$ such that $g = h_1 + h_2 + \cdots + h_n$.*

Proof 2.12 *We show that the first statement implies the second. Let $g \in G$. The existence of $h_1 \in H_1$, $h_2 \in H_2$, $\ldots$, $h_n \in H_n$ such that $g = h_1 + h_2 + \cdots + h_n$ follows immediately, since we know $G = H_1 + H_2 + \cdots + H_n$. To see uniqueness, suppose in addition $g = k_1 + k_2 + \cdots + k_n$ for some $k_1 \in H_1$, $k_2 \in H_2$, $\ldots$, $k_n \in H_n$. Then for each $i = 1, 2, \ldots, n$ we have*

$$k_i - h_i = (h_1 - k_1) + \cdots (h_{i-1} - k_{i-1}) + (h_{i+1} - k_{i+1}) + \cdots + (h_n - k_n),$$

which is in $H_i \cap (H_1 + \cdots + H_{i-1} + H_{i+1} + \cdots + H_n) = 0$.

Therefore for each $i = 1, 2, \ldots, n$ we have $h_i - k_i = 0$ or $h_i = k_i$ which proves uniqueness.

Now we show that the second statement implies the first. The fact that $G = H_1 + H_2 \cdots + H_n$ follows immediately. Thus, it remains to be proved that for each $i = 1, 2, \ldots, n$ we have $H_i \cap (H_1 + \cdots + H_{i-1} + H_{i+1} + \cdots + H_n) = 0$. Suppose $g \in H_i \cap (H_1 + \cdots + H_{i-1} + H_{i+1} + \cdots + H_n$. Now g can be represented as an element in $H_1 + H_2 + \cdots + H_n$ in two different ways.

$$g = 0 + \cdots + 0 + g + 0 + \cdots + 0, \text{ where } g \text{ is in the ith position, and}$$

$$g = h_1 + \cdots + h_{i-1} + 0 + h_{i+1} + \cdots + h_n,$$

for some $h_1 \in H_1$, $\cdots$, $h_{i-1} \in H_{i-1}$, $h_{i+1} \in H_{i+1}$, $\ldots$, $h_n \in H_n$. By unique representation we can equate the corresponding entries in each of the sums, so in particular we get $g = 0$.

EXERCISES

1 Verify that the external product of any two groups is again a group.

2 Using induction and the proof in the case of two groups, prove Proposition 2.2 in general.

3 Consider the group $D_4 \times Q_8$.

   a. Compute the orders of all the elements in D_4 and Q_8.

b. Compute the order of (ρ_1, i).

c. What are the possible orders of elements in $D_4 \times Q_8$?

4 Verify that the relation defined in the proof of Proposition 2.3 is an equivalence relation.

5 Consider the multiplicative group $G = \mathbb{Z}_{17}^*$ with subgroups $H = \langle 4 \rangle$ and $K = \langle 6 \rangle$.

a. Compute the elements of H, K and $H \cap K$.

b. Compute the size of HK and decide if $HK = G$.

6 Let G be an abelian group and $H, K \leq G$. Prove that if $gcd(|H|, |K|) = 1$, then $H + K = H \oplus K$.

7 Let G be a group and $H, K \leq G$. Suppose every element in G can be expressed as a product hk, where $h \in H$ and $k \in K$.

Prove that if $H \cap K = \{1\}$, then every in G can be expressed **uniquely** as a product hk, i.e. for all $g \in G$ there exists a unique $h \in H$ and unique $k \in K$ such that $g = hk$.

2.6 HOMOMORPHISMS

Every algebraic structure has its associated functions which in some sense *respect* the algebraic operation(s) of the structure, and groups are no exception. In this section, we explore these functions and provide ample examples to illustrate how they work.

Definition 2.15 *Let $(G, *)$ and $(G', *')$ be two groups. A function $\phi : G \to G'$ is a* **group homomorphism** *if for all $g, h \in G$, we have*

$$\phi(g * h) = \phi(g) *' \phi(h).$$

One says that the function **respects** *the group operations. In more words, it says that the image of the product of two elements is the same as the product of the two images.*

Example 2.21 *We give several examples of a group homomorphism with their verifications.*

1. *Consider the groups $(\mathbb{R}^{>0}, \cdot)$ and $(\mathbb{R}, +)$ (note that $\mathbb{R}^{>0}$ signifies positive real numbers). The map $\phi : \mathbb{R}^{>0} \to \mathbb{R}$ by $\phi(a) = \ln a$ is a group homomorphism, since*

$$\phi(ab) = \ln(ab) = \ln a + \ln b = \phi(a) + \phi(b).$$

2. *Consider the groups $GL_n(\mathbb{R})$ with matrix multiplication and $(\mathbb{R}^*, \cdot)$. The map $\phi : GL_n(\mathbb{R}) \to \mathbb{R}^*$ by $\phi(A) = |A|$ is a homomorphism, since*

$$\phi(AB) = |AB| = |A||B| = \phi(A)\phi(B).$$

We point out that ϕ maps into $\mathbb{R}^*$, since our inputs are invertible matrices and as such have non-zero determinant.

3. *Consider the additive groups $\mathbb{Z}$ and $\mathbb{Z}/\equiv_n$. The map $\phi : \mathbb{Z} \to \mathbb{Z}/\equiv_n$ by $\phi(k) = [k]_n$ is a homomorphism, since*

$$\phi(k + m) = [k + m]_n = [k]_n + [m]_n = \phi(k) + \phi(m).$$

4. *Consider the multiplicative groups $\mathbb{C}^*$ and $\mathbb{R}^{>0}$. The map $\phi : \mathbb{C}^* \to \mathbb{R}^{>0}$ by $\phi(z) = |z|$ is a homomorphism (see Exercise 1 at the end of this section), since*

$$\phi(z_1 z_2) = |z_1 z_2| = |z_1||z_2| = \phi(z_1)\phi(z_2).$$

5. *Consider an infinite cyclic group $G = \langle g \rangle$ (i.e. the order of g is infinite) and the additive group $\mathbb{Z}$. The map $\phi : G \to \mathbb{Z}$ by $\phi(g^n) = n$ is a homomorphism, since*

$$\phi(g^m g^n) = \phi(g^{m+n}) = m + n = \phi(g^m) + \phi(g^n).$$

6. *For any abelian group G and integer n, the map $\phi : G \to G$ by $\phi(g) = g^n$ is a homomorphism (left as an exercise).*

7. *For any group G, the identity map $1_G : G \to G$ by $1_G(g) = g$ for all $g \in G$ is a homomorphism.*

8. *For any two groups G and G' with identities 1 and $1'$ respectively, the **trivial** map $\phi : G \to G'$ by $\phi(g) = 1'$ for all $g \in G$ is a a homomorphism.*

Lemma 2.6 *Let G, G' and G'' be groups and $\phi : G \to G'$, $\psi : G' \to G''$ be group homomorphisms. Then*

1. *$\phi(1) = 1'$.*

2. *For all $g \in G$ we have $\phi(g^{-1}) = \phi(g)^{-1}$.*

3. *The composition $\psi \circ \phi : G \to G''$ is a group homomorphism.*

4. *If $H \leq G$, then $\phi(H) \leq G'$, i.e. the image of a subgroup is a subgroup.*

5. *If $H' \leq G''$, then $\phi^{-1}(H') \leq G$, i.e. the inverse image of a subgroup is a subgroup.*

6. *If G is cyclic, then so is $\phi(G)$.*

7. *If G is abelian, then so is $\phi(G)$.*

Proof 2.13 *We shall prove the first two statements and leave the rest as exercises. To see the first statement is true, notice that*

$$\phi(1) = \phi(1 \cdot 1) = \phi(1)\phi(1),$$

and so multiplying both sides by $\phi(1)^{-1}$ yields $1' = \phi(1)$.

For the second statement, notice that

$$\phi(g^{-1})\phi(g) = \phi(g^{-1}g) = \phi(1) = 1',$$

and likewise $\phi(g)\phi(g^{-1}) = 1'$. Thus, $\phi(g^{-1})$ is the inverse of $\phi(g)$ which is exactly what we wish to show.

We now introduce some terminology regarding homomorphisms which shall be used in the remainder of the text.

Definition 2.16 *A group homomorphism $\phi : G \to G'$ is called*

1. *a **monomorphism** if ϕ is one-to-one.*

2. *an **epimorphism** if ϕ maps onto G'.*

3. *an **isomorphism** if ϕ is both one-to-one and maps onto G'.*

Definition 2.17 *A homomorphism from a group into itself is called an **endomorphism** and an isomorphism in this case is called an **automorphism**. A monomorphism is sometimes referred to as an **embedding** and we say such a map **embeds** a copy of G in G'.*

Example 2.22 *An important class of automorphisms are called **inner automorphisms** and are defined as follows: Let $g \in G$ a group and define the map $i_g : G \to G$ by $i_g(a) = gag^{-1}$ which is an automorphism, since*

$$i_g(ab) = g(ab)g^{-1} = gag^{-1}gbg^{-1} = i_g(a)i_g(b).$$

We now define an important substructure associated with homomorphism.

Definition 2.18 *Let $\phi : G \to G'$ be a group homomorphism. The **kernel** of ϕ, written*

$$ker\phi = \{g \in G \ : \ \phi(g) = 1'\}.$$

Figure 2.11 represents a generic picture of a homomorphism ϕ from a group $(G, 1)$ to another group $(G', 1')$ and the associated structures just introduced.

Example 2.23 *Consider the homomorphisms already presented in Example 2.21. Bear in mind it is important to keep track of what the identity element is in the codomain in order to compute the kernel of a homomorphism.*

1. *Recall $\phi : \mathbb{R}^{>0} \to \mathbb{R}$ by $\phi(a) = \ln a$. A positive real number*

$$a \in ker\phi \quad iff \quad \phi(a) = 0 \quad iff \quad \ln a = 0 \quad iff \quad a = 1.$$

Therefore, $ker\phi = \{1\}$.

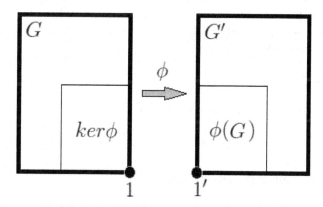

Figure 2.11 Visual representation of a homomorphism and its associated structures.

2. *Recall $\phi : GL_n(\mathbb{R}) \to \mathbb{R}^*$ by $\phi(A) = |A|$. An invertible matrix*

$$A \in ker\phi \quad iff \quad \phi(A) = 1 \quad iff \quad |A| = 1.$$

Therefore, $ker\phi = SL_n(\mathbb{R})$.

3. *Recall $\phi : \mathbb{Z} \to \mathbb{Z}/ \equiv_n$ by $\phi(k) = [k]_n$. An integer*

$$k \in ker\phi \quad iff \quad \phi(k) = [0]_n \quad iff \quad [k]_n = [0]_n \quad iff \quad k \equiv_n 0 \quad iff \quad n|k.$$

Therefore, $ker\phi = n\mathbb{Z}$.

4. *Recall $\phi : \mathbb{C}^* \to \mathbb{R}^{>0}$ by $\phi(z) = |z|$. A complex number*

$$z \in ker\phi \quad iff \quad \phi(z) = 1 \quad iff \quad |z| = 1.$$

Therefore, geometrically, the kernel of ϕ is the unit circle in the complex plane.

Lemma 2.7 *Let $\phi : G \to G'$ be a homomorphism. Then*

1. *$1 \in ker\phi$.*

2. *$ker\phi \leq G$.*

3. *ϕ is a monomorphism iff $ker\phi = \{1\}$ (written more simply $ker\phi = 1$)*
   *In which case we say the kernel is **trivial**.*

Proof 2.14 *The first statement follows immediately from Lemma 2.6.1 and the second statement is left as an exercise. To prove the third statement first assume that ϕ is one-to-one and take any $k \in ker\phi$. Thus, $\phi(k) = 1'$. By the first part, we know that $\phi(1) = 1'$ as well. Therefore, $k = 1$ and the kernel is trivial. Now assume that the kernel is trivial and suppose $\phi(g) = \phi(h)$. Then $\phi(g)\phi(h)^{-1} = 1'$ and, using properties of a homomorphism, $\phi(gh^{-1}) = 1'$. Since the kernel is trivial it must be that $gh^{-1} = 1$ and so $g = h$.*

Example 2.24 *Consider again the homomorphisms already presented in Example 2.21. The homomorphism $\phi : \mathbb{R}^{>0} \to \mathbb{R}$ by $\phi(a) = \ln a$ is one-to-one for we found that the kernel of ϕ was trivial. The remaining three homomorphisms have non-trivial kernels and therefore are not one-to-one.*

Definition 2.19 *Let G be a group.*

1. *The **automorphism group** of G, written*

$$Aut(G) = \{\phi : G \to G \; : \; \phi \text{ is an automorphism}\}.$$

2. *The **inner automorphism group** of G, written*

$$Inn(G) = \{i_g : G \to G \; : \; g \in G\}.$$

Proposition 2.4 *Let G be a group. Then*

1. *$Inn(G) \leq Aut(G) \leq Sym(G)$ (where the operation is composition)*

2. *The map $\iota : G \to Inn(G)$ by $\iota(g) = i_g$ is an epimorphism.*

Proof 2.15 *To prove $Aut(G) \leq Sym(A)$, by earlier work we know that if $\phi, \psi \in Aut(G)$, then so is $\psi^{-1} \in Aut(G)$ and thus so is $\phi \circ \psi^{-1} \in Aut(G)$. To show $Inn(G) \leq Aut(G)$ requires a bit more work. We show for any $g, h \in G$ that $i_g^{-1} = i_{g^{-1}}$ and $i_g \circ i_h = i_{gh}$. First, for all $a \in G$,*

$$i_g(i_{g^{-1}})(a) = i_g[(g^{-1})a(g^{-1})^{-1}] = i_g(g^{-1}ag) = g(g^{-1}ag)g^{-1} = a.$$

Similarly, $i_{g^{-1}}(i_g(a)) = a$ and so $i_{g^{-1}}$ is the inverse of i_g as required. Second, for all $a \in G$,

$$i_g(i_h(a)) = i_g(hah^{-1}) = g(hah^{-1})g^{-1} = (gh)a(gh)^{-1} = i_{gh}(a).$$

Hence, the result has been proved, since for all $g, h \in G$ we have $i_g i_h^{-1} = i_{gh^{-1}} \in Inn(G)$.

The map ι certainly maps onto $Inn(G)$ by its very definition. To show it is a homomorphism, notice that $\iota(gh) = \iota(g)\iota(h)$ since we have shown that $i_{gh} = i_g i_h$.

We will now prove Cayley's Theorem which states that any group G can be embedded in a symmetric group, namely $Sym(G)$.

For each $g \in G$ consider the map $\lambda_g : G \to G$ by $\lambda_g(a) = ga$. It is easy to check (see exercise) that each λ_g is a bijection and hence an element of $Sym(G)$. The Greek letter λ stands for **left** multiplication by an element of G.

Example 2.25 *Consider the multiplication table derived earlier for $\mathbb{Z}_4$ (see Figure 2.7 of Section 2.1). Each row of the body of the table corresponds to the outputs of the functions we just defined, namely $\lambda_1, \lambda_g, \lambda_h$ and λ_k. For instance,*

$$\lambda_h = \begin{pmatrix} 1 & g & h & k \\ h & k & 1 & g \end{pmatrix} = (1 \; h)(g \; k).$$

Proposition 2.5 (Cayley's Theorem) *A group G can be embedded in $Sym(G)$ and thus every group is a subgroup of a permutation group.*

Proof 2.16 *Consider the map $\lambda : G \to Sym(G)$ by $\lambda(g) = \lambda_g$. To show λ is a homomorphism we need to show for all $g, h \in G$ that $\lambda(gh) = \lambda(g)\lambda(h)$ or equivalently $\lambda_{gh} = \lambda_g \lambda_h$. But this is true, since for all $a \in G$,*

$$\lambda_{gh}(a) = (gh)a = g(ha) = \lambda_g(ha) = \lambda_g(\lambda_h(a)).$$

Finally, λ is a monomorphism, since its kernel is trivial. Indeed, if $g \in ker\lambda$, then $\lambda(g) = 1_G$, the identity map, so for all $a \in G$ we have $\lambda_g(a) = a$. In particular, $\lambda_g(1) = 1$ or equivalently $g = 1$.

EXERCISES

1 For any real complex number $z = a + bi \in \mathbb{C}$ we define the **magnitude** of z, written $|z| = \sqrt{a^2 + b^2}$. As a vector $\langle a, b \rangle$ in the complex plane, $|z|$ is simply the length of that vector. Prove that for any $z_1, z_2 \in \mathbb{C}$ we have $|z_1 z_2| = |z_1||z_2|$.

2 Let $\phi \in Aut(G)$. For each of the following subgroups of $H \leq G$, show that $\phi(H) \subseteq H$:

   a. $H = \{g \in G : ga = ag, \forall a \in G\}$, the **center** of G.

   b. H is the only subgroup of order n for some fixed value n.

3 Prove that for any abelian group G and integer n, the map $\phi : G \to G$ by $\phi(g) = g^n$ is a homomorphism.

4 Prove that for any group G, the identity map $1_G : G \to G$ by $1_G(g) = g$ for all $g \in G$ is a homomorphism.

5 Prove that for any two groups G and G' with identities 1 and $1'$ respectively, the **trivial** map $\phi : G \to G'$ by $\phi(g) = 1'$ for all $g \in G$ is a homomorphism.

6 Prove the remaining parts of Lemma 2.6.

7 Prove Lemma 2.7.2.

8 For each $g \in G$ consider the map $\lambda_g : G \to G$ by $\lambda_g(a) = ga$. Prove that λ_g is a bijection.

9 As we did in Example 2.25 express all the λ_g maps as permutations with their corresponding cyclic representation for the Klein-4 group.

10 Consider the element $i \in Q_8$.

   a. Via Cayley's Theorem identify $i \in Q_8$ with a permutation, then express this permutation as a product of disjoint cycles and determine its order.

   b. Using your work in part a, express the permutation as a product of transpositions and decide whether it is an even or odd permutation.

2.7 ISOMORPHIC GROUPS

We now look at isomorphisms more closely and the notion of isomorphic groups which is a way of identifying two groups as essentially equal. To begin, we note that certain properties of the domain of an isomorphism are carried over and preserved in the codomain.

Lemma 2.8 *Let $\phi : G \to G'$ be a group isomorphism.*

1. *If G is abelian, then so is G'.*

2. *If G is cyclic, then so is G'.*

3. *If $g \in G$ has order k, then so does $\phi(g)$.*

Proof 2.17 *We have seen already in Lemma 2.6 that the image of a homomorphism is abelian (or cyclic) if the domain is abelian (or cyclic) and since in our case ϕ maps onto G' the first two statements hold. To prove the third statement, first note that*

$$\phi(g)^k = \phi(g^k) = \phi(1) = 1'.$$

Second, if for some positive integer n we have $\phi(g)^n = 1'$, then $\phi(g^n) = 1'$. Thus, $g^n \in \ker\phi$ which is assumed to be trivial. Therefore, $g^n = 1$. Since $o(g) = k$ it must be that $k \leq n$.

Definition 2.20 *Two groups G and G' are said to be **isomorphic**, written $G \cong G'$, if there exists an isomorphism $\phi : G \to G'$.*

Example 2.26 *We have seen some of the content of these examples earlier in the text.*

1. *Recall the map $\phi : \mathbb{R}^{>0} \to \mathbb{R}$ by $\phi(a) = \ln a$ which we have already shown is a monomorphism. Furthermore, ϕ maps onto $\mathbb{R}$, since for any real number y the positive real number e^y is mapped by ϕ onto y, i.e. $\phi(e^y) = \ln e^y = y$. Therefore, $\mathbb{R}^{>0} \cong \mathbb{R}$.*

2. *Consider the following set of matrices:*

$$G = \left\{ \begin{bmatrix} a & b \\ -b & a \end{bmatrix} : a, b \in \mathbb{R}, \text{ but not both equalling zero} \right\}.$$

We leave it as an exercise to show that G together with matrix multiplication forms a group. We show that this group is isomorphic to the non-zero complex numbers under multiplication. To do this we have to come up with a map from $\mathbb{C}^*$ to G which is an isomorphism. This can be quite an creative process, but in this case the map is easy to find. Define $\phi : \mathbb{C}^* \to G$ by

$$\phi(a + bi) = \begin{bmatrix} a & b \\ -b & a \end{bmatrix}.$$

Now we need to check three things: That ϕ is indeed a homomorphism, that $\ker\phi$ is trivial and that ϕ maps onto G. This we leave as an exercise.

3. *A messy but true fact is that if A is a set of size n, then $Sym(A) \cong S_n$.*

Remark 2.9 *Isomorphic groups are basically equal in the sense that they share all the same group properties and have the same multiplication table – the names of the elements in each group might be different and the operations might be different, but the way in which the operation relates to the group elements are identical for both groups. More formally, isomorphism is an equivalence relation on the collection of groups and each equivalence class of this relation consists of the subcollection of groups which have been identified as isomorphic.*

Therefore, an easy way to determine when two groups are **not** isomorphic is to exhibit a particular group property they **do not** share. We summarize some of these properties in a lemma the proof of which follows immediately.

Lemma 2.9 *Let G and G' be two groups.*

1. *If one group is abelian and the other is not, then $G \not\cong G'$.*

2. *If one group is cyclic and the other is not, then $G \not\cong G'$.*

3. *If $|G| \neq |G'|$, then $G \not\cong G'$.*

4. *If one group has an element of order k and the other does not, then $G \not\cong G'$.*

5. *If G has n elements of order k and G' does not, then $G \not\cong G'$.*

Example 2.27 *Consider some of the groups we have encountered up to this point in the text.*

1. *The groups $\mathbb{Z}_4$ and the Klien-4 group are not isomorphic, since $\mathbb{Z}_4$ is cyclic, but the Klien-4 group is not.*

2. *The groups $\mathbb{Z}_6$ and S_3 are not isomorphic, since $\mathbb{Z}_6$ is abelian, but S_3 is not.*

3. *$\mathbb{Z}_6 \not\cong \mathbb{Z}_5$, since $|\mathbb{Z}_6| = 6 \neq 5 = |\mathbb{Z}_5|$.*

4. *The Dihedral Group, D_4, is not isomorphic to the Quaternions, Q_8, since D_4 has elements $\rho_0, \rho_1, \rho_2, \rho_3, \mu_1, \mu_2, \mu_3, \mu_4$ of orders $1, 4, 2, 4, 2, 2, 2, 2$ (respectively) and Q_8 has elements $1, -1, i, -i, j, -j, k, -k$ of orders $1, 2, 4, 4, 4, 4, 4, 4$ (respectively). So for instance, D_4 has five elements of order 2 while Q_8 has only one.*

One important aspect of abstract algebra is the notion of **classification**. We consider groups sharing a certain property and try to discover the complete list of specific groups which have that property – more precisely we list the distinct equivalence classes under the relation *isomorphism*. Therefore if $\mathcal{P}$ is some property, then the classification of groups with property $\mathcal{P}$ will consist of all the distinct non-isomorphic groups which have property $\mathcal{P}$ (ones says *distinct up to isomorphism*). We will do this now for cyclic groups.

Theorem 2.6 *Consider a cyclic group G.*

1. *If G is a finite cyclic group of order n, then $G \cong \mathbb{Z}_n$.*

2. *If G is an infinite cyclic group, then $G \cong \mathbb{Z}$.*

Proof 2.18 *For the first statement we shall show that any two cyclic groups of order n are isomorphic and thus by transitivity of the equivalence relation isomorphism the result follows. Let $G = \langle g \rangle$ and $G' = \langle g' \rangle$ be two cyclic groups of order n. Define the map $\phi : G \to G'$ by $\phi(g^k) = (g')^k$. Since elements of G can be represented in multiple ways as a power of g we need to check that the map is well-defined. In other words, if $g^k = g^m$, then $\phi(g^k) = \phi(g^m)$. To see this, since $g^k = g^m$, then $g^{m-k} = 1$ and so n divides $m - k$. Write $m - k = nr$ for some integer r. Note that*

$$(g')^{m-k} = (g')^{nr} = ((g')^n)^r = (1')^r = 1'.$$

Therefore, $(g')^k = (g')^m$ and so $\phi(g^k) = \phi(g^m)$. Now we show that ϕ is a homomorphism.

$$\phi(g^k g^m) = \phi(g^{k+m}) = (g')^{k+m} = (g')^k (g')^m = \phi(g^k)\phi(g^m).$$

The kernel of this map is trivial, since if $\phi(g^m) = 1'$, then $(g')^m = 1$ and so n divides m. Write $m = nk$. Then $g^m = g^{nk} = (g^n)^k = 1^k = 1$. Finally, ϕ maps onto G', since for any $(g')^m \in G'$ the element g^m maps onto $(g')^m$. Hence, ϕ is an isomorphism and thus $G \cong G'$.

Now suppose that $G = \langle g \rangle$ is an infinite cyclic group. We leave as an exercise the fact that the map $\phi : G \to \mathbb{Z}$ by $\phi(g^n) = n$ is a homomorphism. Since g has infinite order, the kernel of ϕ is trivial and the fact that ϕ maps onto $\mathbb{Z}$ is immediate from the definition of ϕ.

Theorem 2.6 says there is only one cyclic group – up to isomorphism – of a given cardinality. Therefore, all cyclic groups have been classified.

Example 2.28 *Let's apply this classification result to some concrete examples.*

1. *Consider the following three cyclic groups of order four: $\mathbb{Z}_4$, $\mathbb{Z}_5^*$ and the subgroup of D_4 consisting of all the rotations of the square (generated by the $90°$ rotation). By the theorem just proved we know that all three are isomorphic.*

2. *The cyclic groups $\mathbb{Z}/\equiv_n$ and $\mathbb{Z}_n$ are isomorphic, since they have the same order.*

3. *Consider the infinite cyclic subgroup of $\mathbb{R}^*$ defined by $H = \{\pi^n : n \in \mathbb{Z}\}$. By the second part of the theorem just proved we know that $H \cong \mathbb{Z}$.*

A fact about finite abelian groups which is useful to note and which we shall prove later (its proof requires theory we do not yet have) is the following:

Theorem 2.7 (Classification of Finite Abelian Groups I) *If G is a non-trivial finite abelian group, then G is isomorphic to a direct sum of non-trivial cyclic subgroups of prime power order. Furthermore, this direct sum representation is unique up to isomorphism and order.*

Example 2.29 *Let's apply Theorem 2.7 to a couple of examples.*

1. *The abelian groups of order 8 are $\mathbb{Z}_8$, $\mathbb{Z}_4 \oplus \mathbb{Z}_2$ and $\mathbb{Z}_2 \oplus \mathbb{Z}_2 \oplus \mathbb{Z}_2$, since these are the only groups (up to isomorphism) of order 8 which are direct sums of cyclic groups which have orders a power of 2. Therefore, there are five groups (up to isomorphism) of order 8 – the three just mentioned as well as D_4 and Q_8 (one really should check there are no other non-abelian groups of order 8, but this is non-trivial). Thus, we have classified groups of order eight (give or take a few verifications).*

2. *Consider groups of order $72 = 2^3 \cdot 3^2$. Considering that subgroups of such a group can only have orders $2, 2^2, 2^3, 3, 3^2$ the only abelian groups (up to isomorphism) of order 72 are*

$$\mathbb{Z}_8 \oplus \mathbb{Z}_9, \quad \mathbb{Z}_8 \oplus \mathbb{Z}_3 \oplus \mathbb{Z}_3, \quad \mathbb{Z}_4 \oplus \mathbb{Z}_2 \oplus \mathbb{Z}_9, \quad \mathbb{Z}_4 \oplus \mathbb{Z}_2 \oplus \mathbb{Z}_3 \oplus \mathbb{Z}_3.$$

$$\mathbb{Z}_2 \oplus \mathbb{Z}_2 \oplus \mathbb{Z}_2 \oplus \mathbb{Z}_9, \quad \mathbb{Z}_2 \oplus \mathbb{Z}_2 \oplus \mathbb{Z}_2 \oplus \mathbb{Z}_3 \oplus \mathbb{Z}_3.$$

Lemma 2.10 *The following are equivalent for G an abelian group with subgroups $H_1, H_2, \ldots, H_n$ and $G = H_1 + H_2 + \cdots + H_n$:*

1. *$G = H_1 \oplus H_2 \oplus \cdots \oplus H_n$*

2. *For every $g \in G$ there exist unique $h_1 \in H_1$, $h_2 \in H_2$, $\ldots$, $h_n \in H_n$ such that $g = h_1 + h_2 + \cdots + h_n$.*

3. *$G \cong H_1 \times H_2 \times \cdots \times H_n$ via the map $(h_1 + h_2 + \cdots + h_n) \mapsto (h_1, h_2, \ldots, h_n)$.*

Proof 2.19 *In Lemma 2.5, we proved the first two statements are equivalent. Now we show that the second statement is equivalent to the third statement. Assuming the second statement, it is straightforward to show that the map given in the third statement is an isomorphism. In fact, the uniqueness in statement two insures that the map is well-defined.*

Assuming now that the map in the third statement is an isomorphism we show the second statement holds. Again existence of a representation follows, since $G = H_1 + H_2 + \cdots + H_n$. Finally, we show unique representation. Suppose there is a $g \in G$ and $h_1 \in H_1$, $h_2 \in H_2$, $\ldots$, $h_n \in H_n$ such that $g = h_1 + h_2 + \cdots + h_n$ and $k_1 \in H_1$, $k_2 \in H_2$, $\ldots$, $k_n \in H_n$ such that $g = k_1 + k_2 + \cdots + k_n$. Since the map is well-defined, it must be that $h_1 + h_2 + \cdots + h_n$ and $k_1 + k_2 + \cdots + k_n$ are sent to the same place, i.e. $(h_1, h_2, \ldots, h_n) = (k_1, k_2, \ldots, k_n)$ and so $h_1 = k_1$, $h_2 = k_2$, $\ldots$, $h_n = k_n$.

EXERCISES

1 Consider the following set of matrices:

$$G = \left\{ \begin{bmatrix} a & b \\ -b & a \end{bmatrix} : a, b \in \mathbb{R}, \text{ but not both equalling zero} \right\}.$$

a. Show that G together with matrix multiplication forms a group.

b. Define $\phi : \mathbb{C}^* \to G$ by

$$\phi(a + bi) = \begin{bmatrix} a & b \\ -b & a \end{bmatrix}.$$

Verify that ϕ is an isomorphism and thus conclude that $G \cong \mathbb{C}^*$.

2 Prove that each of the following pairs of groups are isomorphic:

a. $(\mathbb{Z}_{16}, +_{16})$ and $(\mathbb{Z}_{17}^*, \cdot_{17})$

b. $\left\{ \begin{bmatrix} 1 & n \\ 0 & 1 \end{bmatrix} : n \in \mathbb{Z} \right\}$ with matrix multiplication and $(\mathbb{Z}, +)$.

c. Q_8 and the following matrix group G: Let

$$A = \begin{bmatrix} 0 & 1 \\ -1 & 0 \end{bmatrix} \qquad \text{and} \qquad B = \begin{bmatrix} 0 & i \\ i & 0 \end{bmatrix}.$$

Set $G = \langle A, B \rangle$, the subgroup of $M_{22}(\mathbb{C})$ (under matrix multiplication) generated by A and B.

d. The two subgroups listed in Exercise 2 for Section 2.2.

3 Prove that if A is a set of size n, then $Sym(A) \cong S_n$

4 Prove that the notion of isomorphic groups forms an equivalence relation on the collection of groups.

5 Suppose that $G = \langle g \rangle$ is an infinite cyclic group and define the map $\phi : G \to \mathbb{Z}$ by $\phi(g^n) = n$. Prove that ϕ is a homomorphism.

2.8 COSETS OF A GROUP

We now step up the level of complexity a bit by defining a very important equivalence relation on a group. This results ultimately in a new group whose elements consist of the equivalence classes of this special equivalence relation. This is one of the reasons why equivalence classes were gone over so carefully in the first chapter – to make the understanding of this chapter easier.

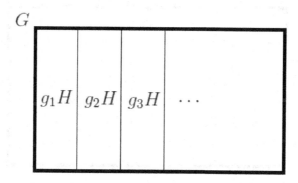

Figure 2.12 The cosets of an arbitrary group G.

Consider a subgroup H of a group G. This equivalence relation will be dependent on our choice of H and is defined as follows: For $g_1, g_2 \in G$,

$$g_1 \sim g_2 \qquad \text{iff} \qquad g_1^{-1}g_2 \in H.$$

The fact that this relation is an equivalence relation is strongly contingent on the fact that H is a subgroup of G. Indeed, to show $\sim$ is reflexive, that $g \sim g$ relies on the fact that $1 \in H$. To show symmetry, we need the fact that H is closed under inverse. To show transitivity, we need the fact that H has the closure property. We leave it to the reader to verify these three properties.

Now let's look at the equivalence classes formed by this equivalence relation. For $g \in G$,

$$[g] = \{g' \in G \ : \ g \sim g'\} = \{g' \in G \ : \ g^{-1}g' \in H\}$$
$$= \{g' \in G \ : \ g^{-1}g' = h \ \text{ for some } \ h \in H\}$$
$$= \{g' \in G \ : \ g' = gh \ \text{ for some } \ h \in H\}$$
$$= \{gh \ : \ \text{for all } \ h \in H\}.$$

So the class of g consists of g times every element in H. We shall employ the product notation gH for this class and it shall be called a **left coset of G with respect to** or **modulo** H. In a similar manner one can define right cosets.

Since cosets are equivalence classes we know by our work in Chapter 1 that they partition G. In other words, G is a union of left cosets and any two cosets are either disjoint or coincide (Figure 2.12).

Example 2.30 *We construct the cosets of a group in some specific settings.*

1. *Consider the multiplicative group $G = \mathbb{Z}_7^*$ and the subgroup $H = \{1,6\}$. The subgroup H which is the left coset $1H$ is one left coset. In order to form a new coset choose an element of G not in $1H$, say 2 (otherwise we will get H again – this is a property of equivalence classes which we saw in Lemma 1.1 of Section 1.1). Then $2H = \{2,5\}$ is a second left coset. Now choose an element in G not in the first two cosets already formed, say 3. Then $3H = \{3,4\}$. At this point we have exhausted all the elements in G and so we know these three left cosets are the complete list of unique cosets of G modulo H.*

μ_1	μ_2	μ_3	μ_4
ρ_0	ρ_1	ρ_2	ρ_3

Figure 2.13 The cosets of D_4 modulo $\langle \mu_1 \rangle$.

2. *Consider the additive group* $G = \mathbb{Z}$ *and the subgroup* $3\mathbb{Z} = \{0, \pm 3, \pm 6, \dots\}$. *We will use additive notation for these cosets. The first coset is the subgroup* $3\mathbb{Z} = 0 + 3\mathbb{Z}$. *Choose* $1 \notin 3\mathbb{Z}$ *and compute the second coset* $1 + 3\mathbb{Z} = \{\dots, -5, -2, 1, 4, \dots\}$. *Choose* 2 *which is not in the first two cosets already formed and compute* $2 + 3\mathbb{Z} = \{\dots, -4, -1, 2, 5, \dots\}$. *At this point we have exhausted* $\mathbb{Z}$ *and so these three cosets are the complete list of cosets modulo* $3\mathbb{Z}$. *Notice that we obtained the same three cosets when we computed the equivalence classes of* $\mathbb{Z}$ *with the equivalence relation congruence modulo 3. Indeed,*

$$3\mathbb{Z} = [0]_3 \qquad 1 + 3\mathbb{Z} = [1]_3 \qquad 2 + 3\mathbb{Z} = [2]_3.$$

It is precisely because of this connection to congruence modulo n that the cosets are referred to as cosets **modulo** H.

3. *Consider the group* D_4 *and the subgroup* $\langle \mu_1 \rangle = \{\rho_0, \mu_1\}$. *Once again the subgroup itself,* $\langle \mu_1 \rangle = \rho_0 \langle \mu_1 \rangle$ *is one coset. Since* μ_2 *is not in the subgroup we form the second coset*

$$\mu_2 \langle \mu_1 \rangle = \{\mu_2 \rho_0, \mu_2 \mu_1\} = \{\mu_2, (\,1\,2\,)(\,3\,4\,)(\,2\,4\,)\}$$

$$= \{\mu_2, (\,1\,2\,3\,4\,)\} = \{\mu_2, \rho_1\}.$$

Now choose an element in D_4 *not in the first two cosets already formed, say* μ_3. *Then the third coset is*

$$\mu_3 \langle \mu_1 \rangle = \{\mu_3 \rho_0, \mu_3 \mu_1\} = \{\mu_3, \rho_2\}.$$

Finally, choose an element in D_4 *not in the first three cosets already formed, say* μ_4. *Then the third coset is*

$$\mu_4 \langle \mu_1 \rangle = \{\mu_4 \rho_0, \mu_4 \mu_1\} = \{\mu_4, \rho_3\}.$$

At this point we have exhausted all the elements of D_4 *and have the complete list of cosets of* D_4 *modulo* $\langle \mu_1 \rangle$ *(see Figure 2.13).*

Here are some useful properties of the left cosets of G modulo H which are easy to verify:

1. The subgroup H is always one of the cosets.

2. For any $g_1, g_2 \in G$ we have $g_1 H = g_2 H$ iff $g_2 \in g_1 H$. In other words, if you choose an element of a coset and form its coset you will get the same one. Any other choice will get you a new one.

3. For any $g \in G$, $g \in H$ iff $gH = H$.

Definition 2.21 *Let $H \leq G$. We denote the collection of cosets of G modulo H by the notation G/H and the number of distinct cosets of G modulo H by $[G : H]$ which is called the* **index of H in G**.

Example 2.31 *For the following examples the reader should refer to Example 2.30.*

1. $\mathbb{Z}_7^*/H = \{H, 2H, 3H\}$ and $[\mathbb{Z}_7^* : H] = 3$.

2. $\mathbb{Z}/3\mathbb{Z} = \{3\mathbb{Z}, 1 + 3\mathbb{Z}, 2 + 3\mathbb{Z}\}$ and $[\mathbb{Z} : 3\mathbb{Z}] = 3$.

3. $D_4/\langle \mu_1 \rangle = \{\langle \mu_1 \rangle, \mu_2 \langle \mu_1 \rangle, \mu_3 \langle \mu_1 \rangle, \mu_4 \langle \mu_1 \rangle\}$ and $[D_4 : \langle \mu_1 \rangle] = 4$.

You may have noticed in our examples that cosets of a given group always had the same size. This observation hints at a theorem which is attributed to the Italian mathematician Joseph-Louis Lagrange and is a fundamental result in group theory.

Theorem 2.8 (Lagrange's Theorem) *For any finite group G with subgroup H we have $|G| = [G : H]|H|$.*

Proof 2.20 *First note that any coset of G modulo H has the same size as H (and thus all cosets have the same size). To see this take any $g \in G$ and consider the map $f : H \to gH$ by $f(h) = gh$. It is easily seen that this map is a bijection between the two sets and hence they have the same size. Now since G is finite and is the union of its cosets, it must be the case that the number of cosets $[G : H]$ is finite. Set $n = [G : H]$ and write G as a disjoint union of cosets, i.e. $G = g_1 H \sqcup g_2 H \sqcup \cdots \sqcup g_n H$. Then*

$$|G| = |g_1 H| + |g_2 H| + \cdots + |g_n H| = \underbrace{|H| + |H| + \cdots + |H|}_{n} = n|H| = [G : H]|H|.$$

Example 2.32 *We give some simple illustrations of Lagrange's Theorem (refer to Example 2.30).*

1. $|\mathbb{Z}_7^*| = [\mathbb{Z}_7^* : H]|H|$, since $6 = 3 \cdot 2$.

2. $|D_4| = [D_4 : \langle \mu_1 \rangle]|\langle \mu_1 \rangle|$, since $8 = 4 \cdot 2$.

One can already see the importance of this theorem by the results listed below which immediately follow from it.

Corollary 2.2 *If G be a finite group, then*

1. *The order of any subgroup $H \leq G$ divides the order of the group G.*

2. *The order of any element $g \in G$ divides the order of G.*

3. *For any element $g \in G$ we have $g^{|G|} = 1$.*

4. *For any prime p and any integer $n \neq 0$ we have $n^{p-1} \equiv 1 \pmod{p}$.*

5. *For any chain of subgroups $K \leq H \leq G$ we have $[G : K] = [G : H][H : K]$.*

6. *If the order of G is prime, then G is cyclic.*

Proof 2.21 *The first statement is an immediate consequence of Lagrange's Theorem which says that $|G|$ is an integer multiple of $|H|$. The second statement then follows from the first, since $o(g)$ equals the order of the subgroup it generates, i.e. $o(g) = |\langle g \rangle|$. The third statement follows from the second, since we can express $|G| = o(g)k$ for some $k \in \mathbb{Z}$ and so*

$$g^{|G|} = g^{o(g)k} = (g^{o(g)})^k = 1^k = 1.$$

To see the fourth statement consider the multiplicative group $\mathbb{Z}_p^*$ and the result follows from the third statement. For the fifth statement we apply Lagrange's Theorem on the following three subgroup relations: $H \leq G$, $K \leq H$ and $K \leq G$ to get

$$|G| = [G : H]|H| \qquad |H| = [H : K]|K| \qquad |G| = [G : K]|K|.$$

Putting these three equations together we have

$$[G : K]|K| = |G| = [G : H]|H| = [G : H][H : K]|K|.$$

Cancelling $|K|$ from both sides of the above equation yields the result. For the sixth statement, since the order of the group is a prime, p say, we can take a $g \neq 1$ in the group. Now since $o(g) \neq 1$ and we know $o(g)$ divides the order of the group, it follows that $o(g)$ equals p and so the cyclic subgroup $\langle g \rangle$ has the same size as the entire group, thus making the group cyclic.

Remark 2.10 *The last statement in Corollary 2.2 leads to an important classification result. It says that there is only one finite group (up to isomorphism) of any given prime order and that group must be cyclic. Hence, we have completely classified groups of prime order.*

EXERCISES

1 Verify that the relation defined at the beginning of this section and which produced cosets as equivalence classes is indeed an equivalence relation.

2 Verify the following properties of the left cosets of G modulo H:

 a. The subgroup H is always one of the cosets.

b. For any $g_1, g_2 \in G$ we have $g_1H = g_2H$ iff $g_2 \in g_1H$.

c. For any $g \in G$, $g \in H$ iff $gH = H$.

3 For each group G and subgroup H, compute the cosets of G modulo H.

a. G is the Klein-4 group and $H = \{1, g\}$ (refer to its multiplication table given in Figure 2.7).

b. $G = S_3$ and $H = \langle \mu_1 \rangle$.

c. $G = D_4$ and $H = \langle \mu_2 \rangle$.

d. $G = Q_8$ and $H = \{\pm 1\}$.

4 For $H \leq G$ and $g \in G$, verify that $f : H \to gH$ by $f(h) = gh$ is a bijection.

2.9 FACTOR GROUPS AND NORMAL SUBGROUPS

We return now to the cosets we constructed in Section 2.8. Our goal is to take this cosets and make them the elements of a new group. We will only be able to do this if we add a condition on the subgroup under which we are making these cosets (to be presented shortly). There are many good reasons to want to construct this group of cosets. One reason among many is that since the group of cosets will have smaller size than the original group we can thus utilize this group when we are proving facts by induction on the size of a group. Another reason is that the group of cosets can sometimes give us insight into other groups to which it might be isomorphic.

Recall that if $H \leq G$, then $G/H = \{gH \ : \ g \in G\}$, the set of cosets. The most natural way to define an operation on cosets would be to multiply the representatives using the original group operation, i.e. $(g_1H)(g_2H) = (g_1g_2)H$. This seems to work quite effectively, since it easily satisfies the four axioms of a group. Indeed, this follows since the original group satisfies the four axioms:

1. The product of two cosets is another coset.

2. The identity coset is $1H$ or simply H, since $(gH)(1H) = (g1)H = gH$.

3. $(gH)^{-1} = g^{-1}H$, since $(gH)(g^{-1}H) = (gg^{-1})H = 1H$.

4. Associativity follows, since the representatives associate:

$$[(g_1H)(g_2H)](g_3H) = (g_1g_2)H(g_3H) = [(g_1g_2)g_3]H$$

$$= [g_1(g_2g_3)]H = (g_1g_2g_3)H = (g_1H)[(g_2H)(g_3H)].$$

The reader may be therefore thinking what is the trouble with always defining this operation on the cosets? The potential problem lies in the fact that, as equivalence classes, cosets have multiple representations and since our operation relies on the representation, we want to be sure that it is not dependent on it. More formally, this operation can be viewed as a potential function from $G/H \times G/H$ to G/H and we need to be sure that this function is well-defined. In other words, if $g_1 H = g_1' H$ and $g_2 H = g_2' H$, then we want to guarantee that $(g_1 H)(g_2 H) = (g_1' H)(g_2' H)$.

Let's first explore an example where things go terribly wrong in order to motivate the need for a necessary and sufficient condition for the coset operation to be well-defined.

Example 2.33 *Consider the group $G = S_3$ and the subgroup $H = \langle \mu_1 \rangle = \{1, \mu_1\}$. One can check that $G/H = \{H, \mu_2 H, \mu_3 H\}$ where $\mu_2 H = \{\mu_2, \rho_2\} = \rho_2 H$ and $\mu_3 H = \{\mu_3, \rho_1\} = \rho_1 H$. Notice that*

$$(\mu_2 H)(\mu_3 H) = (\mu_2 \mu_3)H = \rho_1 H \quad while \quad (\rho_2 H)(\rho_1 H) = (\rho_2 \rho_1)H = 1H = H.$$

So we see in this case that the resulting coset produced by the product of two cosets is dependent on the way in which we represent them. Hence, these cosets cannot form a group using the definition of multiplication of cosets which was just defined.

We now explore this coset multiplication more carefully in order to extract a necessary and sufficient condition on H for it to be well-defined. Now since $g_1 H = g_1' H$ and $g_2 H = g_2' H$, this implies $g_1' \in g_1 H$ and $g_2' \in g_2 H$ and so $g_1' = g_1 h_1$ and $g_2' = g_2 h_2$ for some $h_1, h_2 \in H$. In order to have $(g_1 H)(g_2 H) = (g_1' H)(g_2' H)$ or equivalently $(g_1 g_2)H = (g_1' g_2')H$, we need $g_1' g_2' \in (g_1 g_2)H$ or equivalently $g_1' g_2' = g_1 g_2 h$ for some $h \in H$. By substitution, we need $g_1 h_1 g_2 h_2 = g_1 g_2 h$ or $h_1 g_2 h_2 = g_2 h$ or $g_2^{-1} h_1 g_2 = h h_2^{-1} \in H$. So it would appear that the condition we need for the coset operation to be well-defined is the following property which we introduce formally.

Definition 2.22 *A subgroup H of a group G is **normal**, written $H \lhd G$, if for all $g \in G$ and $h \in H$, we have $g^{-1} h g \in H$.*

Example 2.34 *Consider the group*

$$G = \left\{ \begin{bmatrix} a & b \\ 0 & a^{-1} \end{bmatrix} : a, b \in \mathbb{Z}_5, \ a \neq 0 \right\}$$

and the subgroup $H = \left\{ \begin{bmatrix} 1 & c \\ 0 & 1 \end{bmatrix} : c \in \mathbb{Z}_5 \right\}.$

We demonstrate that $H \lhd G$. This follows, since

$$\begin{bmatrix} a & b \\ 0 & a^{-1} \end{bmatrix}^{-1} \begin{bmatrix} 1 & c \\ 0 & 1 \end{bmatrix} \begin{bmatrix} a & b \\ 0 & a^{-1} \end{bmatrix} = \begin{bmatrix} a^{-1} & -b \\ 0 & a \end{bmatrix} \begin{bmatrix} 1 & c \\ 0 & 1 \end{bmatrix} \begin{bmatrix} a & b \\ 0 & a^{-1} \end{bmatrix}$$

$$= \begin{bmatrix} a^{-1} & a^{-1}c - b \\ 0 & a \end{bmatrix} \begin{bmatrix} a & b \\ 0 & a^{-1} \end{bmatrix} = \begin{bmatrix} 1 & (a^{-1})^2 c \\ 0 & 1 \end{bmatrix} \in H.$$

We now prove the necessary and sufficient condition for coset multiplication to be well-defined. A majority of the proof comes from the exploration we just did.

	H	$2H$	$3H$
H	H	$2H$	$3H$
$2H$	$2H$	$3H$	H
$3H$	$3H$	H	$2H$

Figure 2.14 The quotient group multiplication table for Examples 2.35.1.

Theorem 2.9 *Let H be a subgroup of G. The set of cosets under the operation $(g_1H)(g_2H) = (g_1g_2)H$ forms a group iff $H \lhd G$.*

Proof 2.22 *First assume that $H \lhd G$. As we have seen above it is enough to show that the coset operation is well-defined for the cosets to form a group. Therefore, suppose that $g_1H = g_1'H$ and $g_2H = g_2'H$. Then $g_1' \in g_1H$ and $g_2' \in g_2H$ and so $g_1' = g_1h_1$ and $g_2' = g_2h_2$ for some $h_1, h_2 \in H$. Since $H \lhd G$ we know that $g_2^{-1}h_1g_2 \in H$ and so $g_2^{-1}h_1g_2 = h$ for some $h \in H$. Thus, $h_1g_2 = g_2h$ and so $g_1h_1g_2h_2 = g_1g_2hh_2$ or equivalently $g_1'g_2' = g_1g_2hh_2 \in g_1g_2H$. Therefore, $g_1'g_2'H = g_1g_2H$ or equivalently $(g_1'H)(g_2'H) = (g_1H)(g_2H)$.*

Now assume that the cosets form a group under the coset operation. Take $g \in G$ and $h \in H$. Then

$$g^{-1}hg = g^{-1}hg1 \in g^{-1}HgH = (g^{-1}g)H = 1H = H.$$

Therefore, $H \lhd G$.

Remark 2.11 *We point out that in an abelian group the problem of well-definition of the coset operation is not an issue, since in an abelian group every subgroup is normal. Indeed, if $H \leq G$ abelian, $g \in G$ and $h \in H$, we have $g^{-1}hg = hg^{-1}g = h \in H$.*

Definition 2.23 *Suppose the $H \lhd G$. The set of cosets under the coset operation forms a group called the **factor group** or **quotient group** of G with respect to H.*

Example 2.35 *We present several examples of factor groups.*

1. *Let $G = \mathbb{Z}_7^*$ with multiplication modulo 7 and $H = \{1, 6\}$. Since G is abelian we know that $H \lhd G$ and so we may consider the factor group G/H. The cosets for this factor group are H, $2H = \{2, 5\}$ and $3H = \{3, 4\}$. The multiplication group for this factor group is presented in Figure 2.14. Since there is only one group (up to isomorphism) of order three, it follows that $G/H \cong \mathbb{Z}_3$.*

2. *Consider the group*

$$G = \left\{ \begin{bmatrix} a & b \\ 0 & a^{-1} \end{bmatrix} \; : \; a, b \in \mathbb{Z}_5, \; a \neq 0 \right\}$$

	H	xH	yH	zH
H	H	xH	yH	zH
xH	xH	zH	H	yH
yH	yH	H	zH	xH
zH	zH	yH	xH	H

Figure 2.15 The quotient group multiplication table for Examples 2.35.2.

$$\text{and the subgroup } H = \left\{ \begin{bmatrix} 1 & c \\ 0 & 1 \end{bmatrix} \ : \ c \in \mathbb{Z}_5 \right\}.$$

Now G is not abelian, so one needs to first check that H is normal in G, however in Example 2.34 we proved this very thing. We now construct the elements of the factor group G/H. Note that $|G/H| = [G : H] = |G|/|H| = (4)(5)/5 = 4$, so there will be four cosets in this factor group. There is, of course the identity coset H. The other cosets are

$$\begin{bmatrix} 2 & 0 \\ 0 & 2^{-1} \end{bmatrix} H = \begin{bmatrix} 2 & 0 \\ 0 & 3 \end{bmatrix} H = \left\{ \begin{bmatrix} 2 & b \\ 0 & 3 \end{bmatrix} : b \in \mathbb{Z}_5 \right\},$$

$$\begin{bmatrix} 3 & 0 \\ 0 & 3^{-1} \end{bmatrix} H = \begin{bmatrix} 3 & 0 \\ 0 & 2 \end{bmatrix} H = \left\{ \begin{bmatrix} 3 & b \\ 0 & 2 \end{bmatrix} : b \in \mathbb{Z}_5 \right\} \text{ and}$$

$$\begin{bmatrix} 4 & 0 \\ 0 & 4^{-1} \end{bmatrix} H = \begin{bmatrix} 4 & 0 \\ 0 & 4 \end{bmatrix} H = \left\{ \begin{bmatrix} 4 & b \\ 0 & 4 \end{bmatrix} : b \in \mathbb{Z}_5 \right\}.$$

For brevity, set

$$x = \begin{bmatrix} 2 & 0 \\ 0 & 3 \end{bmatrix}, \quad y = \begin{bmatrix} 3 & 0 \\ 0 & 2 \end{bmatrix} \text{ and } z = \begin{bmatrix} 4 & 0 \\ 0 & 4 \end{bmatrix}.$$

We leave it to the reader to verify the multiplication table for this group presented in Figure 2.15. From our knowledge of the multiplcation tables for groups of order four it is evident from the table that $G/H \cong \mathbb{Z}_4$.

3. *Let $G = \mathbb{Z}$ with addition and consider the subgroup $H = 3\mathbb{Z}$. Now $3\mathbb{Z} \triangleleft \mathbb{Z}$, since G is abelian. We have already computed*

$$\mathbb{Z}/3\mathbb{Z} = \{3\mathbb{Z}, 1 + 3\mathbb{Z}, 2 + 3\mathbb{Z}\},$$

and the coset operation is addition of representatives used in the underlying group $\mathbb{Z}$. The addition table is given in Figure 2.16. Once again, we have a group

	$3\mathbb{Z}$	$1 + 3\mathbb{Z}$	$2 + 3\mathbb{Z}$
$3\mathbb{Z}$	$3\mathbb{Z}$	$1 + 3\mathbb{Z}$	$2 + 3\mathbb{Z}$
$1 + 3\mathbb{Z}$	$1 + 3\mathbb{Z}$	$2 + 3\mathbb{Z}$	$3\mathbb{Z}$
$2 + 3\mathbb{Z}$	$2 + 3\mathbb{Z}$	$3\mathbb{Z}$	$1 + 3\mathbb{Z}$

Figure 2.16 The quotient group multiplication table for Examples 2.35.3.

of order three and so we know $\mathbb{Z}/3\mathbb{Z} \cong \mathbb{Z}_3$, the only group (up to isomorphism) of order three. But, in fact, in this case the identification of the groups is much more apparent, since coset addition of representatives is exactly addition modulo three. More generally, we have the identification of three representations of the same group, namely, for any positive integer n,

$$\mathbb{Z}_n \cong \mathbb{Z}/\equiv_n \cong \mathbb{Z}/n\mathbb{Z}.$$

The next result is important in group theory and has analogous results when studying other algebraic structures. It makes a strong connection between groups and factor groups.

Theorem 2.10 (Fundamental Theorem of Group Homomorphisms) *Let $\phi : G \to G'$ be a homomorphism and set $K = \ker\phi$. Then $G/K \cong \phi(G)$. Furthermore, if ϕ is an epimorphism, then $G/K \cong G'$.*

Proof 2.23 *The proof is straightforward and direct. We simply produce the map which makes the two groups isomorphic. Define $\Psi : G/K \to \phi(G)$ by $\Psi(gK) = \phi(g)$. First of all, the following shows that Ψ is both well-defined and one-to-one:*

$$gK = hK \iff gh^{-1} \in K \iff \phi(gh^{-1}) = 1' \iff \phi(g)\phi(h)^{-1} = 1' \iff \phi(g) = \phi(h).$$

The map Ψ is a homomorphism, since

$$\Psi(gKhK) = \Psi(ghK) = \phi(gh) = \phi(g)\phi(h) = \Psi(gK)\Psi(hK).$$

The fact that Ψ maps onto $\phi(G)$ is immediate from its definition, and certainly if ϕ is an epimorphism, then $\phi(G) = G'$ and so $G/K \cong G'$.

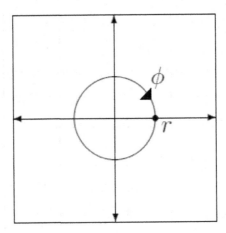

Figure 2.17 The map ϕ for Example 2.36.3.

Example 2.36 *We illustrate Theorem 2.10 with several examples.*

1. *Recall the epimorphism $\phi : GL_n(\mathbb{R}) \to \mathbb{R}^*$ by $\phi(A) = |A|$ for which we computed $\ker \phi = SL_n(\mathbb{R})$. Since ϕ maps onto $\mathbb{R}^*$ we know $\phi(GL_n(\mathbb{R})) = \mathbb{R}^*$. Therefore, by the Fundamental Theorem of Group Homomorphisms (FTH), we have $GL_n(\mathbb{R})/SL_n(\mathbb{R}) \cong \mathbb{R}^*$. This isomorphism turns out to be a useful result for counting the size of $SL_n(F)$ (over a different finite set of scalars F) and this we do later on in the text.*

2. *Recall the epimorphism $\phi : \mathbb{Z} \to \mathbb{Z}/\equiv_n$ by $\phi(k) = [k]_n$ for which we computed $\ker \phi = n\mathbb{Z}$. By FTH, we have once again that $\mathbb{Z}/n\mathbb{Z} \cong \mathbb{Z}/\equiv_n$.*

3. *Recall the epimorphism $\phi : \mathbb{C}^* \to \mathbb{R}^{>0}$ by $\phi(z) = |z|$ for which we computed $\ker \phi = S^1$, the unit circle in the complex plane. By FTH, we have $\mathbb{C}^*/S^1 \cong \mathbb{R}^{>0}$. In this case the identification of the two groups as being isomorphic is intuitive in the sense that the factor group has cosets which consist of complex numbers of the same length and visually what is happening is that the punctured (without the origin) complex plane is being collapsed onto the positive real axis circle-by-circle, i.e. for all real numbers $r > 0$, all the points on a circle of radius r in the complex plane are sent to r on the positive real axis (see Figure 2.17).*

4. *Recall the epimorphism $\iota : G \to Inn(G)$ by $\iota(g) = i_g$ where $i_g(a) = gag^{-1}$ for all $a \in G$. Let's find the kernel of this map.*

$$g \in \ker \iota \iff \iota(g) = 1_G \iff \text{For all } a \in G, \ \iota(g)(a) = a$$

$$\iff \text{For all } a \in G, \ i_g(a) = a \iff \text{For all } a \in G, \ gag^{-1} = a$$

$$\iff \text{For all } a \in G, \ ga = ag \iff g \in Z(G).$$

Hence, $\ker \iota = Z(G)$ and by FTH we have $G/Z(G) \cong Inn(G)$.

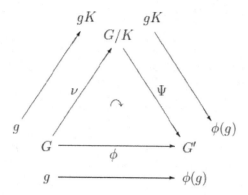

Figure 2.18 A commutative diagram illustrating that $\Psi \circ \nu = \phi$.

5. *Consider the identity map 1_G and FTH to see that $G/\{1\} \cong G$ and consider the trivial map (the map that sends every input to the identity) and FTH to see that G/G is isomorphic to the trivial group.*

Remark 2.12 *We make some additional remarks about the Fundamental Theorem of Homomorphisms (FTH).*

1. *When G is finite, $|G| = |ker\phi||\phi(G)|$.*

2. *For those who have taken a linear algebra course, there is a similar result for vector space dimension and a linear transformation T with domain a vector space V.*

$$\dim(V) = \dim(kerT) + \dim(T(V)).$$

But notice that the linear algebra result is the logarithmic equivalent of the group theory result.

We now present another map which connects factor groups and groups and which shows up in the study of any algebraic structure.

Definition 2.24 *Let $N \triangleleft G$ a group. The map $\nu : G \to G/N$ by $\nu(g) = gN$ is called the* **canonical** *(or* **natural***) map.*

Recall we have already defined this map in a more general context when we introduced equivalence classes – it was a map which set every element in the set to its equivalence class. The map just defined is simply a specific example in which the set is a group and the equivalence classes are cosets. It is a simple matter to verify that the canonical map is an epimorphism and we leave it as an exercise. This map plays an underlying role in two important facts. The first is in FTH, for we can relate the maps $\Psi : G/K \to \phi(G)$ and $\phi : G \to G'$ in the proof as follows: $\Psi \circ \nu = \phi$, and we have the picturesque commutative diagram in Figure 2.18.

The second fact involving the canonical map is Theorem 2.11.

Theorem 2.11 *A subgroup H of a group G is normal iff H is the kernel of some homomorphism with domain G.*

Proof 2.24 *Assuming $H \lhd G$, consider the canonical map $\nu : G \to G/H$ by $\nu(g) = gH$. Then the kernel of this map is H, since*

$$g \in \ker \nu \iff \nu(g) = H \iff gH = H \iff g \in H.$$

If H is the kernel of some homomorphism, it is certainly normal, since the kernel of a homomorphism is always a normal subgroup (left as an exercise).

2.9.1   Semidirect Products

One way to generalize the direct product is the semidirect product which has internal and external types just as the direct product does. This topic is slightly off the beaten path, and may be skipped, however it is included for completeness.

Definition 2.25 *Let $H \leq G$ and $N \lhd G$ with $N \cap H = 1$. The **internal semidirect product** of N and H, written $N \rtimes H$, represents the subgroup $NH \leq G$.*

Remark 2.13 *A couple of remarks are in order at this point.*

1. *Since $N \lhd G$, we have $hN = Nh$ for all $h \in H$, and so $NH = HN$, from which it follows that $N \rtimes H \leq G$.*

2. *We can describe explicitly the product of two elements in $N \rtimes H$. If $n_1 h_1, n_2 h_2 \in N \rtimes H$, then*
$$(n_1 h_1)(n_2 h_2) = [n_1(h_1 n_2 h_1^{-1})](h_1 h_2).$$

   Notice that multiplication in the semidirect product is very close to the multiplication in the direct product except that n_2 is conjugated by h_1.

Definition 2.26 *A group G **splits** if there exists $H \leq G$ and $N \lhd G$ such that $G = N \rtimes H$.*

Example 2.37 *These examples will explore the idea of a group splitting.*

1. *Let G be the group of all invertible upper triangular matrices with matrix multiplication. Let U be the subgroup of unipotent matrices and T be the subgroup of invertible diagonal matrices. Note that*

$$G = \left\{ \begin{bmatrix} a & b \\ 0 & a^{-1} \end{bmatrix} : a, b \in \mathbb{R}, a \neq 0 \right\},$$

$$U = \left\{ \begin{bmatrix} 1 & b \\ 0 & 1 \end{bmatrix} : b \in \mathbb{R} \right\}, \qquad T = \left\{ \begin{bmatrix} a & 0 \\ 0 & a^{-1} \end{bmatrix} : a \in \mathbb{R}^* \right\}.$$

   We leave it to the reader as an exercise to verify that $T \leq G$, $U \lhd G$, $G = UT$ and $U \cap T = 1$ and so G splits as $G = U \rtimes T$.

2. D_n, reflections and rotations on a regular n-gon, splits as $D_n = \langle \rho \rangle \rtimes \langle \mu \rangle$, where $\rho = (1\ 2\ \cdots\ n)$ and μ is any reflection. Now $\langle \rho \rangle \lhd D_n$ since it has index 2 in D_n (see Exercise 14). Since $\langle \rho \rangle$ consists of all the rotations, its intersection with $\langle \mu \rangle$ is trivial. To show $D_n = \langle \rho \rangle \langle \mu \rangle$, it is necessary to show that $D_n = \langle \rho, \mu \rangle$ and use that fact that $\mu \rho \mu = \rho^{-1}$. We leave it as an exercise for the reader to work out the details.

3. It's easy to check that $S_n = A_n \rtimes \langle \mu \rangle$.

4. The cyclic group $\mathbb{Z}_{p^n}$ does not split.

5. The quaternions do not split.

Remark 2.14 The semidirect product $N \rtimes H$ induces a homomorphism $\phi : H \to Aut(N)$ defined by $[\phi(h)](n) = hnh^{-1}$ (left as an exercise). This observation allows us to generalize the semidirect product and produce the external semidirect product.

Definition 2.27 Let G and G' be groups and suppose we have a homomorphism $\phi : G' \to Aut(G)$. The **external semidirect product** of G and G' via ϕ, written $G \rtimes_\phi G'$ has as its underlying set the cartesian product $G \times G'$ with multiplication defined as follows: For $g_1, g_2 \in G$ and $g'_1, g'_2 \in G'$,

$$(g_1, g'_1)(g_2, g'_2) = (g_1[\phi(g'_1)](g_2), g'_1 g'_2).$$

Remark 2.15 The reader should verify the following remarks.

1. It's easy to check that $G \rtimes_\phi G'$ is a group. The identity is $(1, 1')$ and $(g, g')^{-1} = ([\phi(g')]^{-1}(g^{-1}), (g')^{-1})$.

2. If ϕ is the trivial homomorphism, then $G \rtimes_\phi G'$ is just the usual external direct product, namely $G \times G'$.

Example 2.38 Set $G = \langle x \rangle$ be a cyclic group of order 3 and $G' = \langle y \rangle$ be a cyclic group of order 2.

1. Consider the homomorphism $\phi : G' \to Aut(G)$ defined by $[\phi(y)](x^k) = x^{-k}$ (which is an automorphism on G, since G is abelian). The semidirect product $G \rtimes_\phi G'$ in this case is not abelian. Indeed,

$$(x, y)(x^2, y) = (x[\phi(y)](x^2), y^2) = (x^2, 1) \qquad while$$

$$(x^2, y)(x, y) = (x^2[\phi(y)](x), y^2) = (x, 1).$$

Since $|G \rtimes_\phi G'| = |G||G'| = 6$, it must be that $G \rtimes_\phi G' \cong S_3$. Indeed, to get S_3 directly take N to be the subgroup generated by ρ, the 120° rotation of the equilateral triangle and H be the subgroup generated by any reflection μ. Then $\phi : H \to Aut(N)$ maps each element of H to the corresponding inner automorphism, i.e. $[\phi(\mu)](\rho) = \mu \rho \mu = \rho^{-1}$.

2. Now let $\phi : G' \to Aut(G)$ be the trivial homomorphism. As stated in the remarks above this semi-direct product $G \rtimes_\phi G'$ is just the direct product $G \times G' \cong \mathbb{Z}_3 \times \mathbb{Z}_2 \cong \mathbb{Z}_6$ the other possible group of order 6.

The above construction can be generalized to $G = \mathbb{Z}_n = G'$ for some positive integer n and one can show the two groups you get are D_n and $\mathbb{Z}_{2n}$.

Remark 2.16 *Just as in the case of abelian groups every internal direct sum of subgroups is isomorphic to the external direct product of the same subgroups, so too is it true in the case of semidirect products. Indeed, any external semidirect product of groups can be viewed as an internal direct product of isomorphic copies of these groups.*

We now prove the details of this remark. Suppose that $G = N \rtimes_\phi H$. Consider the embeddings (i.e. monomorphisms) $\alpha : N \to G$ and $\beta : H \to G$ defined by $\alpha(n) = (n, 1_H)$ and $\beta(h) = (1_N, h)$. Set $N' = \alpha(N)$ and $H' = \beta(H)$. First, $G = N' \rtimes H'$, an internal semi-direct product of N' and H'. The only thing that really needs to be checked is the normality of N' in G, but this is clear since

$$(n, h)^{-1}(n', 1_H)(n, h) = ([\phi(h)]^{-1}(n^{-1}), h^{-1})(n'[\phi(1_H)](n), 1_H h)$$

$$= ([\phi(h)]^{-1}(n^{-1}), h^{-1})(n'n, h) = ([\phi(h)]^{-1}(n^{-1})[\phi(h^{-1})](n'n), 1_H) \in N'.$$

Notice also that conjugating an $(n, 1_H)$ in N' by and element $(1_N, h)$ in H' yields

$$(1_N, h)(n, 1_H)(1_N, h)^{-1} = (1_N[\phi(h)](n), h1_H)([\phi(h)]^{-1}(1_N^{-1}), h^{-1})$$

$$= ([\phi(h)](n), h)(1_N, h^{-1}) = ([\phi(h)](n)[\phi(h)](1_N), 1_H) = ([\phi(h)](n), 1_H).$$

Hence, conjugating N' by H' in the internal semidirect product induces the original automorphism which defined the external semidirect product.

EXERCISES

1 Consider the group D_4 and the subgroup $H = \{\iota, \rho\}$, where ρ is a 180° rotation of the square. Verify that $H \lhd D_4$.

2 Verify the multiplication table given in Example 2.35.2

3 For $G = Q_8$ and $H = \{\pm 1, \pm i\}$ what is the value of $[G : H]$?

4 Consider the subgroup $H = \{\pm 1\}$.

   a. Compute the elements of Q_8/H and the value of $[Q_8 : H]$.

   b. Write out the multiplication table for Q_8/H and decide to what group Q_8/H is isomorphic? (justify)

5 Let $G = \mathbb{Z}_{13}^*$ and $H = \langle 3 \rangle$.

a. Why do we know that $H \lhd G$?

b. Compute the elements of G/H.

c. Make the multiplication table for G/H.

d. What group is G/H isomorphic to?

6 Consider the group $\mathbb{Z}_3^* \times \mathbb{Z}_3^*$ (the operation is multiplication).

a. Construct the lattice of subgroups.

b. What well-known group must $\mathbb{Z}_3^* \times \mathbb{Z}_3^*$ be isomorphic to? (explain why)

c. Consider the following group:

$$G = \left\{ \begin{bmatrix} a & b \\ 0 & c \end{bmatrix} : a, c \in \mathbb{Z}_3^* \ \& \ b \in \mathbb{Z}_3 \right\}.$$

Define the map $\phi : G \longrightarrow \mathbb{Z}_3^* \times \mathbb{Z}_3^*$ by

$$\phi \begin{bmatrix} a & b \\ 0 & c \end{bmatrix} = (a, c).$$

Check that ϕ is a homomorphism and then apply FTH in this setting.

d. Let $K = ker\phi$ in part c. List the elements of G/K and write its lattice of subgroups.

7 Consider the map $\phi : \mathbb{Z} \times \mathbb{Z} \to \mathbb{Z}$ by $\phi(m, n) = m - 3n$.

a. Verify that ϕ is an epimorphism.

b. Compute $ker \phi$.

c. Express $ker \phi$ as a cyclic subgroup of $\mathbb{Z} \times \mathbb{Z}$.

d. Apply the Fundamental Theorem of Homomorphisms to obtain an isomorphism between two groups.

8 Consider the group A_4.

a. List all cycle types, the number of each type and their orders.

b. Let $H = \{1, (1\ 2)(3\ 4), (1\ 3)(2\ 4), (1\ 4)(2\ 3)\}$. What well-known group is H isomorphic to? (explain why)

c. Form the elements of A_4/H and create the multiplication table (you may assume that $H \lhd A_4$).

d. What well-known group is A_4/H isomorphic to? (explain why)

9 Consider the following group:

$$G = \left\{ \begin{pmatrix} 1 & 2 & 3 & 4 \\ 1 & 2 & 3 & 4 \end{pmatrix}, \begin{pmatrix} 1 & 2 & 3 & 4 \\ 2 & 3 & 4 & 1 \end{pmatrix}, \begin{pmatrix} 1 & 2 & 3 & 4 \\ 3 & 4 & 1 & 2 \end{pmatrix}, \begin{pmatrix} 1 & 2 & 3 & 4 \\ 4 & 1 & 2 & 3 \end{pmatrix}, \right.$$

$$\left. \begin{pmatrix} 1 & 2 & 3 & 4 \\ 1 & 4 & 3 & 2 \end{pmatrix}, \begin{pmatrix} 1 & 2 & 3 & 4 \\ 2 & 1 & 4 & 3 \end{pmatrix}, \begin{pmatrix} 1 & 2 & 3 & 4 \\ 3 & 2 & 1 & 4 \end{pmatrix}, \begin{pmatrix} 1 & 2 & 3 & 4 \\ 4 & 3 & 2 & 1 \end{pmatrix} \right\}.$$

a. Compute the cyclic decomposition for each element of G.

b. Compute $H = \left\langle \begin{pmatrix} 1 & 2 & 3 & 4 \\ 3 & 4 & 1 & 2 \end{pmatrix} \right\rangle$

c. Compute G/H.

10 Verify that the canonical map is a group epimorphism.

11 Prove that for any homomorphism $\phi : G \to G'$ that the kernel of ϕ is normal in G.

12 Prove the following three statements are equivalent for $H \leq G$.

a. $H \triangleleft G$

b. For all $g \in G$ we have $g^{-1}Hg = H$.

c. For all $g \in G$ we have $gH = Hg$.

13 Prove that if $H \leq G$ and $[G : H] = 2$, then $H \triangleleft G$ using the following steps:

a. If $g_1, g_2 \in G$, then $g_1 g_2 \in H$ (consider the cosets $g_1^{-1}H$ and $g_2 H$).

b. Define the map $\phi : G \to \{\pm 1\}$ by

$$\phi(g) = \begin{cases} 1, & \text{if } g \in H \\ -1, & \text{if } g \notin H \end{cases}$$

Prove by cases that ϕ is a homomorphism.

c. Consider the kernel of ϕ to conclude that $H \triangleleft G$.

14 Prove that if $H \leq G$ and $[G : H] = 2$, then $H \triangleleft G$ using the following steps:

a. Define an equivalence relation on G which produces **right** cosets. You must verify that it is an equivalence relation.

b. Prove in two cases that for all $g \in G$ that $gH = Hg$.

c. Conclude that $H \triangleleft G$.

15 In Definition 2.4 we defined a particular equivalence relation on a group G. Prove that if H is a union of equivalence classes of this relation (i.e. a union of conjugacy classes), then $H \triangleleft G$.

16 Referring to Example 2.37.1, verify that $T \leq G$, $U \triangleleft G$, $G = UT$ and $U \cap T = 1$ and so G splits as $G = U \rtimes T$.

17 Work out the details of Example 2.37.2.

18 Verify the following statements:

    a. $S_n = A_n \rtimes \langle \mu \rangle$.

    b. The cyclic group $\mathbb{Z}_{p^n}$ does not split.

    c. The quaternions do not split.

19 Verify that the map definied in REmark 2.14 is indeed a homomorphism.

20 Verify the statements in Remark 2.14.

2.10 NORMAL AND SIMPLE GROUPS

We first summarize the many equivalent ways to represent the normal property for a subgroup, most of which was already proved in Section 2.9.

Lemma 2.11 *The following are equivalent for a subgroup H of a group G:*

    1. $H \triangleleft G$.

    2. For all $g \in G$ we have $g^{-1}Hg \subseteq H$.

    3. For all $g \in G$ we have $g^{-1}Hg = H$.

    4. For all $g \in G$ we have $Hg \subseteq gH$.

    5. For all $g \in G$ we have $Hg = gH$.

Example 2.39 *We now give some examples of normal subgroups. All of these examples have been previously verified as being subgroups, so we need only check the normality property.*

    1. In any group G the trivial group and the entire group are always normal subgroups. Note that $G/\{1\} = \{\{g\} : g \in G\} \cong G$ and $G/G = \{G\}$ is isomorphic to the trivial group.

    2. Any subgroup of an abelian group is normal.

    3. $SL_n(\mathbb{R}) \triangleleft GL_n(\mathbb{R})$, since for $A \in GL_n(\mathbb{R})$ and $B \in SL_n(\mathbb{R})$ we have $|A^{-1}BA| = |A|^{-1}|B||A| = |A|^{-1}|A||B| = |B| = 1$ and so $A^{-1}BA \in SL_n(\mathbb{R})$.

    4. $A_n \triangleleft S_n$, since if $\sigma \in S_n$ and $\tau \in A_n$, then $\sigma^{-1}\tau\sigma$ will always be a product of an even number of transpositions and so in A_n. Now for $n \geq 2$, $[S_n : A_n] = 2$ so there will be two cosets in the factor group S_n/A_n, namely the even permutations A_n and the odd permutations $(1\ 2)A_n$. Since there is only one group up to isomorphism of order two we also know that $S_n/A_n \cong \mathbb{Z}_2$.

5. *For any group homomorphism $\phi : G \to G'$ it is always the case that $\ker\phi \lhd G$, since for all $g \in G$ and $k \in \ker\phi$ we have*

$$\phi(g^{-1}kg) = \phi(g^{-1})\phi(k)\phi(g) = \phi(g)^{-1}1'\phi(g) = 1',$$

and so $g^{-1}kg \in \ker\phi$.

6. *$Z(G) \lhd G$, since for all $g \in G$ and $z \in Z(G)$ we have $g^{-1}zg = zg^{-1}g = z \in Z(G)$.*

Next we present a wonderfully surprising and enlightening result related to the center of a group.

Lemma 2.12 *If $G/Z(G)$ is cyclic, then G is abelian.*

Proof 2.25 *Set $Z = Z(G)$ and we are assuming that $G/Z = \langle gZ \rangle$ for some $g \in G$. Take any $a, b \in G$. Since $aZ, bZ \in G/Z$ this implies that $aZ = (gZ)^m = g^m Z$ and $bZ = (gZ)^n = g^n Z$ for some integers m and n. Therefore, $a \in g^m Z$ and $b \in g^n Z$ and so $a = g^m z_1$ and $b = g^n z_2$ for some $z_1, z_2 \in Z$. But then*

$$ab = g^m z_1 g^n z_2 = g^m g^n z_1 z_2 = g^{m+n} z_2 z_1 = g^{n+m} z_2 z_1 = g^n g^m z_2 z_1 = g^n z_2 g^m z_1 = ba.$$

We now introduce a family of groups that can be considered the building blocks of groups in a similar way to which prime numbers are the building blocks for natural numbers.

Definition 2.28 *A non-trivial group G is called **simple** if it has no non-trivial proper normal subgroups.*

Hence, you can in addition view normal subgroups as group factors (and hence the name *factor* group for the collection of corresponding cosets). To summarize, a simple group is a group whose only normal subgroups are the trivial subgroup and the improper subgroup, just as a prime number has factors only 1 and itself.

Example 2.40 *For any prime p the group $\mathbb{Z}_p$ is simple. To see this, suppose $H \lhd \mathbb{Z}_p$. Then in particular $H \leq \mathbb{Z}_p$ and so $|H|$ divides $|\mathbb{Z}_p| = p$. Since p is prime, this implies $|H| = 1$ or p, i.e. $H = 1$ or $\mathbb{Z}_p$. In Theorem 2.12, we show that these are, in fact, the only abelian simple groups.*

Theorem 2.12 *If G is an abelian simple, then $G \cong \mathbb{Z}_p$ for some prime p.*

Proof 2.26 *Suppose that G is abelian simple. First note that as such G has no non-trivial proper subgroups, since in an abelian group every subgroup is normal. First, we show G must be cyclic. Indeed, take any $g \neq 1$ in G and consider the subgroup $\langle g \rangle$. Since $g \neq 1$ it must be the case that $\langle g \rangle = G$, and thus G is cyclic. Second we note that G must be finite, for if G were infinite, then $\langle g^2 \rangle$ would be a non-trivial proper subgroup of G. Hence, $|G| = n < \infty$, and so $G \cong \mathbb{Z}_n$ (see Theorem 2.6.1). Finally, n must be prime, for otherwise n would have a proper non-trivial positive divisor $d \neq 1, n$ which in a cyclic group implies there exists a non-trivial proper subgroup of G of order d (see Theorem 2.1.ii).*

Simple groups come in several varieties. There are first of all the abelian simple groups which we just classified. The non-abelian simple groups for the most part are clustered into families. We shall explore some of these families later in the text, namely A_n is simple for $n \neq 1, 2, 4$ and $SL_n(K)/Z(SL_n(K))$ where K is a field (most of the time). The latter simple groups are one family of simple groups arising from a larger family of simple groups of **Lie type**. Then there are the non-abelian simple groups which do not fall into any family of simple groups. These miscellaneous simple groups are called the **sporadic** groups. The largest of these groups is called the **monster** group (the second largest is called the **little monster**) and has order

$$2^{46} \cdot 3^{20} \cdot 5^9 \cdot 7^6 \cdot 11^2 \cdot 13^3 \cdot 17 \cdot 19 \cdot 23 \cdot 29 \cdot 31 \cdot 41 \cdot 47 \cdot 59 \cdot 71 \approx 8.08 \times 10^{53}!!!$$

EXERCISES

1 Prove that in any group G the trivial group and the entire group are always normal subgroups.

2 Let G be a finite abelian group such that p divides $|G|$, but p^2 does not divide $|G|$, for some prime p. Set $N = \{g \in G : g^p = 1\}$

   a. Prove that $N \triangleleft G$.

   b. Prove that G/N has no elements of order p.

3 Prove that if $H, K \triangleleft G$ and $K \leq H$, then the following is a well-defined homomorphism:
$$\phi : G/K \to G/H \text{ by } \phi(gK) = gH.$$

4 Show that if $|G| = p^3$ where p is a prime and $|Z(G)| \geq p^2$, then G is abelian.

2.11   THE GROUP ISOMORPHISM THEOREMS

In this section, we explore some of the fundamental structural properties of quotient groups. We have already seen such a result when we presented the Fundamental Theorem of Homomorphisms. This result is, in fact, a portion of the First Isomorphism Theorem which we shall see in this section. There are three Isomorphism Theorems which we present herein. First, we need a preliminary result.

Lemma 2.13 *Let N be a normal subgroup of a group G.*

   1. If H is a subgroup of G such that $N \leq H \leq G$, then H/N is a subgroup of G/N. If in addition H is normal in G, then so is H/N normal in G/N.

   2. If $\overline{H}$ is a subgroup of G/N, then there is a subgroup H of G with $N \leq H \leq G$ and $\overline{H} = H/N$. If in addition $\overline{H}$ is normal in G/N, then so is H normal in G.

Proof 2.27 *It is easy to see that $H/G \leq G/N$, since for $h_1N, h_2N \in H/N$ we have $(h_1N)(h_2N)^{-1} = (h_1h_2^{-1})N \in H/N$ because $H \leq G$. Assuming $H \triangleleft G$ we get*

$H/N \lhd G/N$ since for $hN \in H/N$ and $gN \in G/N$ we have $(gN)^{-1}(hN)(gN) = (g^{-1}hg)N \in H/N$, because $H \lhd G$.

For the second statement, let $\overline{H} \leq G/N$. We need to construct H which will satisfy the conclusions of this result. To this end, set $H = \{g \in G : gN \in \overline{H}\}$. First note that $H \leq G$, since $H = \nu^{-1}(\overline{H})$ (recall ν is the canonical map which sends g to gN) and we know the inverse image of a subgroup is again a subgroup. Second, H contains N since $N = \ker \nu$ and $N \in \overline{H}$. Finally, we show that $\overline{H} = H/N$. If $gN \in \overline{H}$, then $\nu(g) \in \overline{H}$ and so $g \in \nu^{-1}(\overline{H}) = H$. Hence, $gN \in H/N$. If $gN \in H/N$, then $g \in H$ which implies $\nu(g) \in \overline{H}$ and so $gN \in \overline{H}$.

The following illustrates how Lemma 2.13 can be applied to prove nice results. But first we prove another lemma. Note that the proof of this lemma will illustrate how, in the case of finite groups, we can use factor groups and induction to achieve a result. This is a classic type of argument in finite group theory.

Lemma 2.14 *If G is a finite abelian group whose order is divisible by a prime p, then G must have a element of order p.*

Proof 2.28 *We use induction to prove this lemma, i.e. assume that all groups of smaller order than G have the desired property of this lemma (note that the lemma holds vacuously for the trivial group). If G is also cyclic, then the result follows from an earlier result on cyclic groups(Theorem 2.1.1). So assume that G is not cyclic. Take any $1 \neq g \in G$ and set $H = \langle g \rangle$. Since G is assumed not cyclic, we know that H is a proper subgroup of G. Note that $|G| = |G/H||H|$, so since the prime p divides $|G|$, it must divide $|G/H|$ or $|H|$. If p divides $|H|$, then by induction H, and therefore G, has an element of order p and we are done. However, if p divides $|G/H|$ we have a bit more work to do. Again, by induction, G/H has an element, say aH, of order p. This means $(aH)^p = H$ and $aH \neq H$ or equivalently $a^p \in H$ and $a \notin H$. Consider the map $\phi : H \to H$ by $\phi(h) = h^p$ which is a homomorphism, since G is abelian. Now if the kernel of ϕ were nontrivial, then there would an $1 \neq h \in H$ with $\phi(h) = 1$. Then this h would be an element of order p and we would be done. So assume the kernel of ϕ is trivial. This implies that ϕ is one-to-one and so, since H is finite, ϕ also maps onto H. Therefore, in particular, there is an $h \in H$ which ϕ maps onto a^p, i.e. $h^p = a^p$. But then ah^{-1} is an element of G of order p, since $(ah^{-1})^p = a^p(h^p)^{-1} = 1$ and $ah^{-1} \neq 1$ (since otherwise $a = h$, but $h \in H$ while $a \notin H$).*

Corollary 2.3 *If G is a finite abelian group and d divides the order of G, then G has a subgroup of order d.*

Proof 2.29 *Lemma 2.14 does most of the work for us. Once again we prove this result by induction on the order of the group. If G is trivial, then the result follows immediately ($d = 1$ only). Assume that G is non-trivial and let p be a prime divisor of d. By Lemma 2.14, there exists $g \in G$ of order p. Set $K = \langle g \rangle$, a subgroup of order p. Consider the factor group G/K which has smaller order than G. Since d/p divides the order of G/K, by induction, there exists a subgroup $\overline{H} \leq G/K$ of order d/p. By Lemma 2.13, there exists a subgroup H of G with $K \leq H \leq G$ and $\overline{H} = H/K$. But then H is the subgroup we seek, since $|H| = |\overline{H}||K| = (d/p)p = d$.*

Theorem 2.13 (First Isomorphism Theorem) *The following statements are true:*

1. **Fundamental Theorem of Homomorphisms:** *If $\phi : G_1 \to G_2$ is a group homomorphism, then $G_1/ker(\phi) \cong \phi(G_1)$. If in addition ϕ is an epimorphism, then $G_1/ker(\phi) \cong G_2$.*

2. **Correspondence Theorem:** *For any normal subgroup N of a group G there is an inclusion preserving one-to-one correspondence between (normal) subgroups of G containing N and (normal) subgroups of G/N.*

3. *If $\phi : G_1 \to G_2$ is a group homomorphism, there is an inclusion preserving one-to-one correspondence between (normal) subgroups of G containing $ker(\phi)$ and (normal) subgroups of $G/ker(\phi)$.*

Proof 2.30 *The first statement is the Fundamental Theorem of Homomorphisms which we have already proved (see Theorem 2.10).*

To show the second statement, set $\mathcal{H}$ to be the subgroups of G containing N and $\overline{\mathcal{H}}$ to be the subgroups of G/N. Define the map $f : \mathcal{H} \to \overline{\mathcal{H}}$ by $f(H) = H/N$. By Lemma 2.13, we know that f maps both into and onto $\overline{\mathcal{H}}$ when H is simply a subgroup or when H is a normal subgroup. We now show that f is inclusion preserving. First, if we take $H_1, H_2 \in \mathcal{H}$ with $H_1 \subseteq H_2$, then certainly $H_1/N \subseteq H_2/N$ and so $f(H_1) \subseteq f(H_2)$. Second, if we have $f(H_1) \subseteq f(H_2)$, then $H_1/N \subseteq H_2/N$. So for $h_1 \in H_1$ we have $h_1 N \in H_2/N$ and so there is an $h_2 \in H_2$ such that $h_1 N = h_2 N$. Hence, $h_1 h_2^{-1} \in N \subseteq H_2$ which implies that $h_1 \in h_2 H_2 \subseteq H_2$. To show f is one-to-one, notice that if $f(H_1) = f(H_2)$, then certainly $f(H_1) \subseteq f(H_2)$ and $f(H_2) \subseteq f(H_1)$. Therefore, by the work above, we get $H_1 \subseteq H_2$ and $H_2 \subseteq H_1$ and so $H_1 = H_2$.

The third statement follows from the second with $N = ker(\phi)$.

Example 2.41 *In the following examples we illustrates aspects of Theorem 2.13.*

1. *Consider the group $\mathbb{Z}$ with normal subgroup $8\mathbb{Z}$. Figure 2.19 illustrates the correspondence between the subgroups of $\mathbb{Z}$ containing $8\mathbb{Z}$ and the subgroups of $\mathbb{Z}/8\mathbb{Z}$.*

2. *Consider the quaternions $Q_8 = \{\pm 1, \pm i, \pm j, \pm k\}$ and the normal subgroup $N = \{\pm 1\}$. Figure 2.20 illustrates the correspondence between the subgroups of Q_8 containing N and the subgroups of Q_8/N.*

Theorem 2.14 (Second Isomorphism Theorem) *If G is a group with $H \leq G$ and $N \triangleleft G$, then*

1. *$HN \leq G$.*

2. *$H \cap N \triangleleft H$.*

3. *$HN/N \cong H/H \cap N$.*

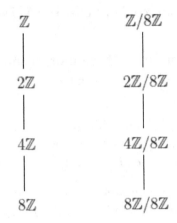

Figure 2.19 First example illustrating the Correspondence Theorem.

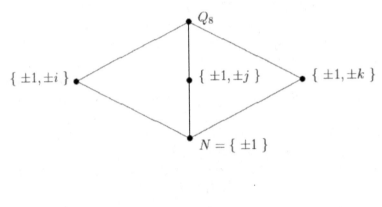

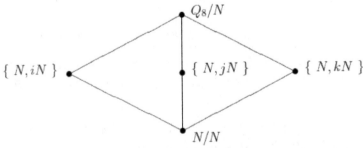

Figure 2.20 Second example illustrating the Correspondence Theorem.

Proof 2.31 *To prove the first statement, by Theorem 2.5, it is enough to show that*
$HN = NH$. *Now this is evident since* $N \lhd G$ *which implies, in particular, that for*
any $h \in H$ *we have* $hN = Nh$.

The second statement is simply a verification of normality, since we know that
$H \cap N \le G$. *Therefore, take an* $h \in H$ *and an* $x \in H \cap N$. *Since* $h, x \in H$ *we have*
$h^{-1}xh \in H$ *and since* $x \in N \lhd G$ *we also have* $h^{-1}xh \in N$ *and so* $h^{-1}xh \in H \cap N$.

For the third statement, first note that since $N \triangleleft HN$ *and* $H \cap N \triangleleft H$ *the quotient groups* HN/N *and* $H/H \cap N$ *make sense. Define the map* $\phi : H \rightarrow HN/N$ *by* $\phi(h) = hN$. *The map* ϕ *is certainly a homomorphism, since*

$$\phi(h_1 h_2) = (h_1 h_2)N = (h_1 N)(h_2 N) = \phi(h_1)\phi(h_2).$$

In addition, ϕ *is an epimorphism, since for any* $(hn)N \in HN/N$ *we have* $(hn)N = hN$ *and so* $\phi(h) = hN = (hn)N$. *Finally, we show the kernel of* ϕ *is* $H \cap N$. *Indeed,* $h \in H$ *is in the kernel of* ϕ *iff* $\phi(h) = N$ *iff* $hN = N$ *iff* $h \in N$ *and so* $h \in H \cap N$. *Now we invoke the Fundamental Theorem of Homomorphisms to get* $H/H \cap N = H/\ker(\phi) \cong HN/N$.

In this final Isomorphism Theorem we see that quotient groups act very much like fractions.

Theorem 2.15 (Third Isomorphism Theorem) *If* G *is a group with* $H, K \triangleleft G$ *and* $K \subseteq H$, *then*

1. $H/K \triangleleft G/K$

2. $(G/K)/(H/K) \cong G/H$.

Proof 2.32 *Consider the map* $\phi : G/K \rightarrow G/H$ *by* $\phi(gK) = gH$. *We first need to verify that* ϕ *is a well-defined mapping. To see this take* $g_1 K = g_2 K$ *which implies* $g_1 g_2^{-1} \in K \subseteq H$ *and so* $g_1 H = g_2 H$, *i.e.* $\phi(g_1 K) = \phi(g_2 K)$. *Certainly,* ϕ *is an epimorphism. Next we show that the kernel of* ϕ *is* H/K. *The coset* gK *is in the kernel of* ϕ *iff* $\phi(gK) = H$ *iff* $gH = H$ *iff* $g \in H$ *iff* $gK \in H/K$. *This gives us the first statement, since* $H/K = \ker(\phi) \triangleleft G/K$ *and the second statement follows from the Fundamental Theorem of Homomorphisms.*

We will later illustrate the importance of these Isomorphism Theorems when we prove some results about solvable and nilpotent groups.

EXERCISES

---

1 As was done in Example 2.41, illustrate the Correspondence Theorem for each of the following groups:

   a. The group $G = \mathbb{Z}_{18}$ and the normal subgroup $H = \{0, 6, 12\}$.

   b. The group $G = \mathbb{Z}_{13}^*$ and the normal subgroup $H = \{1, 3, 9\}$.

   c. The group $G = D_4$ and the normal subgroup $H = \{\iota, \rho\}$, where ρ is a 180° rotation of the square.

2 Let G be a finite group.

   a. Show that if $H \leq K \leq G$, then $[G : K] \leq [G : H]$.

b. Let $H \leq G$ and define $H^G = \{H^g \ : \ g \in G\}$. Let $f : H^G \longrightarrow G/N_G(H)$ by $f(H^g) = gN_G(H)$ (recall that $N_G(H) = \{g \in G \ : \ H^g = H\}$). Show that f is a well-defined one-to-one map (and so, it maps onto $G/N_G(H)$ as well).

c. Use parts a. and b. to show that if $[G : H] < \infty$, then $|H^G| < \infty$.

d. Let $H, K \leq G$ and define $f : H/(H \cap K) \longrightarrow G/K$ by $f(h(H \cap K)) = hK$. Show that f is a well-defined one-to-one map (and so, it maps onto G/K onto as well).

e. Let $H, K \leq G$ and $[G : H], [G : K] < \infty$. Use part d. to show that $[G : H \cap K] < \infty$.

f. We have seen that $N = \bigcap_{g \in G} H^g$ is normal in G. Use parts c. and e. to show that if $H \leq G$ and $[G : H] < \infty$, then $[G : N] < \infty$.

CHAPTER 3

Simple Groups

I N THIS CHAPTER, we shall look at two families of non-abelian groups: The alternating group and the projective linear group. We will prove that they are both families of simple groups.

3.1   THE ALTERNATING GROUP

The first family of simple groups, which is the topic of discussion in this lesson, is A_n with $n = 3$ or $n \geq 5$. First, we prove a useful lemma.

Lemma 3.1 *If X is the collection of all 3-cycles in S_n for $n \geq 3$, then $A_n = \langle X \rangle$.*

Proof 3.1 *First note that any 3-cycle $(k \; l \; m) = (k \; m)(k \; l) \in A_n$ and so $X \subseteq A_n$. Since $\langle X \rangle$ is the smallest subgroup containing X, we have that $\langle X \rangle \subseteq A_n$ (see Exercise 6i. in Section 2.2).*

To show the reverse inclusion is suffices to point out that $(k \; m)(k \; l) = (k \; l \; m)$ and $(k \; m)(n \; l) = (k \; l \; n)(k \; m \; n)$, since this implies any product of an even number of transpositions can be pairwise rewritten as 3-cycles and thus in total as an element of $\langle X \rangle$.

Theorem 3.1 *If $n \neq 1, 2, 4$, then A_n is simple.*

Proof 3.2 *First, we dispense with the small alternating groups: $A_1 = 1$ which is not simple, $A_2 = 1$ which is not simple, and $A_3 \simeq \mathbb{Z}_3$ which is simple. Lastly, A_4 is not simple, since one can check that the following is a normal subgroup of A_4:*

$$N = \{1, (1 \; 2)(3 \; 4), (1 \; 3)(2 \; 4), (1 \; 4)(2 \; 3)\}.$$

Now assume that $n \geq 5$ and we show that A_n is simple. Suppose, to the contrary, that we had a proper normal subgroup $N \lhd A_n$.

Claim 3.1 *N has no 3-cycles.*
Suppose, to the contrary, that there was a $(k \; l \; m) \in N$. Now take any $(k' \; l' \; m') \in A_n$ (see Lemma 3.1) and let $\sigma \in S_n$ be such that

$$\sigma(k') = k, \quad \sigma(l') = l \quad and \quad \sigma(m') = m.$$

DOI: 10.1201/9781003335283-3

Then $(k' \ l' \ m') = \sigma^{-1}(k \ l \ m)\sigma$. Therefore, if $\sigma \in A_n$, we would then have $(k' \ l' \ m') \in N$, since $N \lhd A_n$. If $\sigma \notin A_n$, then choose $\tau = (r \ s) \in S_n$ with $r, s \notin \{k, l, m\}$ (here is where we need the fact that $n \geq 5$). Then, since disjoint cycles commute, we have

$$(k' \ l' \ m') = \sigma^{-1}(k \ l \ m)\sigma = \sigma^{-1}\tau(k \ l \ m)\tau\sigma = (\tau\sigma)^{-1}(k \ l \ m)(\tau\sigma) \in N.$$

So, in either case we have this arbitrary 3-cycle $(k' \ l' \ m') \in N$. Hence, N contains all the 3-cycles, which implies by Lemma 3.1 that $N = \langle X \rangle = A_n$, a contradiction.

Claim 3.2 *Any $\sigma \in N$ has a disjoint cyclic decomposition made up entirely of cycles of length ≤ 3.*

Suppose, to the contrary, that this claim were false and take any $\sigma \in N$. Then the disjoint cycle decomposition of σ would include an s-cycle with $s > 3$, say $\tau = (k \ l \ m \ r \cdots)$. Set $\rho = (m \ l \ k) \in A_n$ and

$$\tau' = \rho^{-1}\tau\rho = (k \ l \ m)(k \ l \ m \ r \cdots)(m \ l \ k) = (l \ m \ k \ r \cdots),$$

which differs from τ only in the first 3 places. Note that $\rho^{-1}\sigma\rho \in N \lhd A_n$ and has a disjoint cycle decomposition which differs from σ in that τ is replaced by τ'. Therefore,

$$(\rho^{-1}\sigma\rho)\sigma^{-1} = \tau'\tau^{-1} = (l \ m \ k \ r \cdots)(\cdots r \ m \ l \ k) = (l \ r \ k).$$

But then $(l \ r \ k) \in N$ which contradicts the first claim we proved.

Claim 3.3 *Any $\sigma \in N$ has a disjoint cyclic decomposition made up entirely of transpositions.*

Suppose, to the contrary, that this claim were false and take $\sigma \in N$ having 3-cycles in its disjoint cycle decomposition. Then using the second claim, the disjoint cycle decomposition of σ would include at least two 3-cycles, say $(k \ l \ m)(k' \ l' \ m')$. Indeed, if σ contained a only single 3-cycle $(k \ l \ m)$, then $\sigma^2 \in N$ would equal a single 3-cycle $(k \ l \ m)(k \ l \ m) = (k \ m \ l)$, contradicting the first claim. Set $\sigma' = (k' \ l' \ m)^{-1}\sigma(k' \ l' \ m) \in N \lhd A_n$. Notice that

$$\sigma' = (m \ l' \ k')(k \ l \ m)(k' \ l' \ m')(k' \ l' \ m) \cdots = (m' \ m \ k')(k \ l \ l') \cdots,$$

so that

$$\sigma'\sigma^{-1} = (m' \ m \ k')(k \ l \ l')(m \ l \ k)(m' \ l' \ k') = (k \ k' \ m \ l' \ m') \in N,$$

but this contradicts the second claim.

Claim 3.4 *Any $\sigma \in N$ has a disjoint cyclic decomposition made up entirely of transpositions which uses all the elements of $\{1, 2, \ldots, n\}$.*

Suppose, to the contrary, that this claim were false. Then there is a $\sigma \in N$ and an $x \in \{1, 2, \ldots, n\}$ which does not show up in the disjoint cyclic decomposition of σ into transpositions. Say $\sigma = (k \ l)(m \ r) \cdots$ and set $\sigma' = (l \ x \ k)^{-1}\sigma(l \ x \ k) \in N \lhd A_n$. Note that

$$\sigma' = (k \ x \ l)\sigma(l \ x \ k) = (k \ x \ l)(k \ l)(l \ x \ k)(m \ r) \cdots = (x \ k)(m \ r) \cdots,$$

which differs from σ only in the first transposition. But then

$$\sigma'\sigma = (x\ k)(k\ l) = (k\ l\ x) \in N,$$

but this contradicts the first claim.

Having established these four claims, we can now achieve a final contradiction to the assumption that A_n is not simple. Take any $\sigma \in N \triangleleft A_n$. Since σ is even and the fourth claim, we know $\sigma = (k\ l)(m\ n)(r\ s)\cdots$ and $n = 6,8,10,\ldots$. Set $\sigma' = [(m\ r)(k\ m)]^{-1}\sigma[(m\ r)(k\ m)] \in N \triangleleft A_n$. However,

$$\sigma' = (k\ m)(m\ r)(k\ l)(m\ n)(r\ s)\cdots(m\ r)(k\ m) = (k\ s)(l\ m)(r\ n)\cdots,$$

which differs from σ only in these first three transpositions so that

$$\sigma'\sigma = (k\ s)(l\ m)(r\ n)(k\ l)(m\ n)(r\ s) = (k\ m\ r)(l\ s\ n),$$

but this contradicts the third claim, since $\sigma'\sigma \in N$.

To prove the next result, which is a consequence of the theorem just proved, we first need to point out the $Z(S_n) = 1$ for $n \geq 3$. Indeed, take any $\sigma \neq 1$ in S_n. As such, there is an i with $\sigma(i) \neq i$. Since $n \geq 3$, we can construct a $\tau \in S_n$ with $\tau(i) = i$, $\tau(\sigma(i)) = j$ where $j \neq i, \sigma(i)$. Notice then that

$$\sigma\tau(i) = \sigma(i) \quad \text{while} \quad \tau\sigma(i) = j \neq \sigma(i).$$

Thus, $\sigma\tau \neq \tau\sigma$ and so $\sigma \notin Z(S_n)$.

Corollary 3.1 *If $n \neq 4$, then the only proper non-trivial normal subgroup of S_n is A_n.*

Proof 3.3 *As in the proof of Theorem 3.1, one sees that the statement is true for $n = 1, 2, 3$ and false for $n = 4$. Therefore, we assume $n \geq 5$. First note that certainly $A_n \triangleleft S_n$, since $[S_n : A_n] = 2$ (see Exercise 14 in Section 2.9). Now take any $N \triangleleft S_n$ with $N \neq 1$. Then $N \cap A_n \triangleleft A_n$ and so by Theorem 3.1, either $N \cap A_n = A_n$ or $N \cap A_n = 1$ which in turn implies either $A_n \leq N$ or $N \cap A_n = 1$. In the former case, since $|S_n| = 2|A_n|$ either $|N| = |A_n|$ or $|N| = |S_n|$ and so either $N = A_n$ or $N = S_n$.*

It suffices to show the latter case when $N \cap A_n = 1$ is not possible. Since both $A_n \triangleleft S_n$ and $N \triangleleft S_n$ we know that $A_n N \leq S_n$. And since $|A_n N| > |A_n|$ and $|S_n| = 2|A_n|$, it must be that $|A_n N| = |S_n|$ and so $S_n = A_n N$. Furthermore, since $N \cap A_n = 1$, we have $2|A_n| = |S_n| = |A_n N| = |A_n||N|$, by Proposition 2.3, and so $|N| = 2$. Let $1 \neq \tau \in N$. Since $N \triangleleft S_n$ this means for all $\sigma \in S_n$ we have $\sigma^{-1}\tau\sigma = \tau$. But then $\tau\sigma = \tau\sigma$ and so $\tau \in Z(S_n) = 1$, yielding a contradiction.

Remark 3.1 *Now that we have shown A_n to be simple for $n \neq 1, 2, 4$ we can exhibit a family of groups which illustrate the fact that if d divides the order of a finite group, there does not necessarily exist a subgroup of order d. The family of groups are A_n for $n \geq 5$. Each of these groups has order divisible by $n!/4$, yet has no subgroup of order $n!/4$. Indeed, if it did, then the index of that subgroup in A_n would be 2 thus making it a normal subgroup of A_n, contradicting the fact that A_n is simple.*

EXERCISES

1 Verify that the following is a normal subgroup of A_4:

$$N = \{1, (1\ 2)(3\ 4), (1\ 3)(2\ 4), (1\ 4)(2\ 3)\}.$$

3.2 THE PROJECTIVE LINEAR GROUPS

In this section, we shall look at second family of simple groups. This family is called Projective Linear Groups which are factor groups of certain matrix groups. The reader is forewarned that some of the topics in this section do not necessarily flow from earlier sections. These topics include fields, finite fields which will be discussed later in the text; topics in linear algebra such as bases and vector space linear transformations; as well as knowledge of the commutator subgroup presented later on in the text.

Definition 3.1 *Let K be a field either infinite (this includes the rational, real and complex numbers), or finite (in which case $|K| = q$ where $q = p^k$ with p a prime number and k a natural number).*

1. *The* **n-by-n general linear group over a field** K, *written $GL_n(K)$ is the collection of all invertible $n \times n$ matrices with entries from K.*

2. *The* **n-by-n special linear group over a field** K, *written $SL_n(K)$ is the collection of all invertible $n \times n$ matrices with entries from K with determinant 1.*

3. *The* **n-by-n upper triangular matrices over a field** K, *written $B_n(K)$, are the collection of matrices $A = [\alpha_{ij}]$ with entries from K with $a_{ij} = 0$ if $i > j$.*

4. *The* **n-by-n unipotent matrices over a field** K, *written $U_n(K)$, are the collection of upper triangular matrices entries from K which have 1's on the diagonal.*

5. *The* **general linear group of a vector space V over a field** K, *written $GL_V(K)$, is the collection of all vector space automorphisms of V.*

6. *If $|K| = q$ where $q = p^k$ with p prime, then for the above structures we employ the notation $GL_n(q)$, $SL_n(q)$, $B_n(q)$, $U_n(q)$ and $GL_V(q)$.*

For our purposes, the vector space V is assumed to be finite dimensional, say of dimension n. The linear algebraic fact that there is a one-to-one correspondence between vector space automorphisms and their matrix representation with respect to a fixed basis leads to the fact that $GL_n(K) \cong GL_V(K)$. Note also that $GL_n(K)/SL_n(K) \simeq K^*$ via the homomorphism which takes an element of $GL_n(K)$ to its determinant (and the use of the Fundamental Theorem of Homomorphisms). We now give some facts in the case when $|K| = q$.

Theorem 3.2 *Suppose K is a field with q elements where $q = p^k$ and p prime, then*

1. $|GL_n(q)| = (q^n - 1)(q^n - q) \cdots (q^n - q^{n-1})$.

2. $|SL_n(q)| = (q^n - 1)(q^n - q) \cdots (q^n - q^{n-1})/(q-1)$.

3. $|U_n(q)| = q^{n(n-1)/2}$.

4. $q^{n(n-1)/2}$ is the largest power of p which divides the order of $GL_n(q)$.

Proof 3.4 *For the first statement, we will, in fact, compute the size of $GL_V(q)$ and appeal to the fact that $GL_n(q) \cong GL_V(q)$. First fix a basis for V, say $v_1, v_2, \ldots, v_n$. Recall that a linear transformation is completely determined by where it sends the basis $v_1, v_2, \ldots, v_n$ and to be an isomorphism it should send the basis $v_1, v_2, \ldots, v_n$ to another basis, say $w_1, w_2, \ldots, w_n$, for V. Hence, computing the size of $GL_n(q)$ is reduced to counting the number of distinct bases for V. For our first vector w_1 in any basis for V, we have $q^n - 1$ choices (since we cannot choose the zero vector). Now the span of w_1 contains q vectors. So to choose our second linearly independent vector w_2, we must choose outside of the span of w_1, hence leaving us with $q^n - q$ choices. The span of w_1, w_2 contains q^2 vectors, so we have $q^n - q^2$ choices for w_3, and so on. This argument can be made more formal using induction. Therefore, the number of bases for V, and hence the size of $GL_n(q)$, is $(q^n - 1)(q^n - q) \cdots (q^n - q^{n-1})$.*

For the second statement, notice that

$$q - 1 = |K^*| = |GL_n(q)/SL_n(q)| = |GL_n(q)|/|SL_n(q)|,$$

and so $|SL_n(q)| = |GL_n(q)|/(q-1) = (q^n - 1)(q^n - q) \cdots (q^n - q^{n-1})/(q-1)$.

The third statement is a simple counting argument. The number of entries above the diagonal in an $n \times n$ matrix is $1 + 2 + \cdots + (n-1) = n(n-1)/2$ and the number of choices for values in each of these entries is q.

For the fourth statement, since $U_n(q) \leq GL_n(q)$ we have $q^{n(n-1)/2} = |U_n(q)|$ divides $|GL_n(q)|$. An elegant way to show this is the largest such power of p requires Sylow Theory (which we haven't covered yet), however another way to see part iv is to notice that

$$|GL_n(q)| = (q^n - 1)(q^n - q) \cdots (q^n - q^{n-1})$$
$$= q q^2 \cdots q^{n-1}(q^n - 1)(q^{n-1} - 1) \cdots (q - 1)$$
$$q^{1+2+\cdots+(n-1)}(q^n - 1)(q^{n-1} - 1) \cdots (q - 1)$$
$$= q^{n(n-1)/2}(q^n - 1)(q^{n-1} - 1) \cdots (q - 1),$$

and so we see that $q^{n(n-1)/2}$ divides $|GL_n(q)|$. Furthermore, the remaining factors of $|GL_n(q)|$, namely $(q^n - 1), (q^{n-1} - 1), \ldots, (q - 1)$, are all not divisible by the prime p and so p cannot divide the product.

The next result states that the center of invertible matrices consists of scalar matrices. This center is an ingredient in the family of simple groups we wish to define.

Theorem 3.3 *Each scalar matrix aI with $a \neq 0$ commutes with all the elements of $GL_n(K)$ and any element of $GL_n(K)$ which commutes with all the elements of $GL_n(K)$ must be a scalar matrix aI with $a \neq 0$, i.e.*

$$Z(GL_n(K)) = \{aI_n \; : \; a \in K^*\}.$$

Proof 3.5 *For any matrix A and scalar matrix aI, we have $A(aI) = a(AI) = aA = (aI)A$, so then*

$$\{aI_n \; : \; a \in K^*\} \subseteq Z(GL_n(K)).$$

Now suppose B is a matrix in $GL_n(K)$ which commutes with every element of $GL_n(K)$. Choose any $v_1 \in K^n$ and extend to $v_1, v_2, \ldots, v_n$ a basis for K^n. For $i = 2, 3, \ldots, n$, take $A_i \in GL_n(K)$ to be the matrices having the property that $A_i v_j = v_j$ for $j \neq i$ and $A_i v_i = v_1 + v_i$ (note that A_i maps a basis to a basis and therefore is invertible). Consider the subspace U_i of K^n consisting of the vectors which are fixed by A_i. It's easy to check that $U_i = \mathrm{span}(v_1, \ldots, v_{i-1}, v_{i+1}, \ldots, v_n)$. Note that for $i = 2, 3, \ldots, n$ and every $u_i \in U_i$ we have that $A_i(Bu_i) = (A_i B)u_i = BA_i u_i = Bu_i$ which puts $Bu_i \in U_i$ for all i. If we set $U = \bigcap_{i=2}^n U_i$, this in turn means that for $u \in U$ we have $Bu \in U$. Now $U = \mathrm{span}(v_1)$ so that $Bv_1 = a_1 v_1$ for some $a_1 \in K$. Since v_1 is chosen arbitrarily in K^n we get that for any $v \in K^n$ we have $Bv = av$ for some $a \in K$. In particular, for each i we have $Be_i = a_i e_i$ for some $a_i \in K$ (where $e_1, e_2, \ldots, e_n$ is the standard basis). This makes B a diagonal matrix. But notice that for any $i \neq j$, we have on the one hand that $B(v_i + v_j) = Bv_i + Bv_j = a_i v_i + a_j v_j$ while on the other hand $B(v_i + v_j) = a(v_i + v_j) = av_i + av_j$, for some $a \in K$. By unique representation for a basis we get $a_i = a$ and $a_j = a$ and so $a_i = a_j$ for $i \neq j$. Therefore, $B = aI$ for some $a \in K$ and certainly $a \neq 0$, since $B \in GL_n(K)$.

Definition 3.2 *An element of $SL_n(K)$ is called a **transvection** if it is not the identity matrix yet it fixes pointwise some subspace of K^n of dimension $n - 1$ (called a **hyperplane**).*

Example 3.1 *Any matrix of the form $I_n + aE_{ij}$ where $a \in K^*$ and $i \neq j$ is a transvection, where E_{ij} is a matrix filled with zeros except that the ij-th entry is 1.*

Remark 3.2 *Some consequences of Theorem 3.3 are the following:*

1. *It follows from the proof of Theorem 3.3 that $Z(GL_n(K)) \simeq K^*$ and if $|K| = q < \infty$, then $|Z(GL_n(K))| = q - 1$.*

2. *$Z(SL_n(K)) = \{aI_n \; : \; a^n = 1\}$, since we require that $\det(aI) = 1$. Therefore, if $|K| = q < \infty$, then $|Z(SL_n(K))| = \gcd(n, q - 1)$. Note, this follows from a fact we will see later in the text that K^* is a cyclic group of order $q - 1$.*

3. *Each A_i as defined in the proof of Theorem 3.3 is a transvection of the form $I_n + E_{1i}$ where $i = 2, 3, \ldots, n$.*

4. *From the proof of Theorem 3.3, we see that a matrix in $GL_n(K)$ will commute with all the elements of $GL_n(K)$ iff it commutes with the collection of transvections of the form $I_n + E_{1i}$ where $i = 2, 3, \ldots, n$.*

Definition 3.3 *For any field K,*

1. *The* **n-by-n projective general linear group over a field** K, *written*
   $PGL_n(K) = GL_n(K)/Z(GL_n(K))$.

2. *The* **n-by-n projective special linear group over a field** K, *written*
   $PSL_n(K) = SL_n(K)/Z(SL_n(K))$.

It follows then from our work in this section that if K is a finite field of order q, then

$$|PGL_n(q)| = \frac{|GL_n(q)|}{|Z(GL_n(q))|} = (q^n - 1)(q^n - q)\cdots(q^n - q^{n-1})/(q - 1) = |SL_n(q)|.$$

$$|PSL_n(q)| = \frac{|SL_n(q)|}{|Z(SL_n(q))|} = (q^n - 1)(q^n - q)\cdots(q^n - q^{n-1})/(q - 1)\gcd(n, q - 1).$$

Having introduced the family of groups $PGL_n(K)$ and $PSL_n(K)$ we look in detail at what makes many of these groups simple. First, we look more carefully at transvections. As we have seen already, it is the transvections that are the key to all our results about this family of groups.

Lemma 3.2 *Every transvection is conjugate to the transvection $I_n + E_{12}$. In other words, if $A \in SL_n(K)$ is a transvection, then there exists a $P \in GL_n(K)$ such that $I_n + E_{12} = P^{-1}AP$. Furthermore, if $n \geq 3$, then every transvection is conjugate to $I_n + E_{12}$ by an element of $SL_n(K)$.*

Proof 3.6 *Let A be a transvection fixing pointwise some subspace U of dimension $n - 1$. Using the linear algebraic fact regarding similarity of matrix representations, if we can find a basis for K^n such that the matrix representation of A with respect to that basis is $I_n + E_{12}$, then we will have proved the lemma. Take any basis $u_1, u_2, \ldots, u_{n-1}$ for U and any $v \in K^n - U$. Now, we can express $Av = a_1u_1 + \cdots + a_{n-1}u_{n-1} + av$ where $a_1, \ldots, a_{n-1}, a \in K$. If $a_1 = \cdots = a_{n-1} = 0$, then the matrix representation of A with respect to the basis $u_1, u_2, \ldots, u_{n-1}, v$ is $\mathrm{diag}(1, \ldots, 1, a)$ and since $A \in SL_n(K)$, this implies that $a = |\mathrm{diag}(1, \ldots, 1, a)| = 1$. Thus, the matrix representation is the identity matrix I_n which contradicts the definition of transvection. Therefore, it must be the case that some $a_i \neq 0$. For this i with $a_i \neq 0$, replace u_i in our basis for U by the vector $a_1u_1 + \cdots + a_{n-1}u_{n-1} \in U$ which will be linearly independent from $u_1, \ldots, u_{i-1}, u_{i+1}, \ldots, u_{n-1}$ (left as an exercise). Now re-label the basis for U so that $a_1u_1 + \cdots + a_{n-1}u_{n-1}$ is u_1. Hence, $Av = u_1 + av$ and then the matrix representation of A with respect to the basis $u_1, v, u_2, \ldots, u_{n-1}$ is $\mathrm{diag}(1, a, 1\ldots, 1) + E_{1,2}$. As before, $a = 1$ and so the matrix representation of A with respect to the basis $u_1, v, u_2, \ldots, u_{n-1}$ is $I_n + E_{12}$ and we're done.*

Let A be any transvection with $n \geq 3$. By the work above, there is a $P \in GL_n(K)$ such that $P^{-1}AP = I_n + E_{12}$. Set $D = \mathrm{diag}(1, 1, \ldots, 1, |P|)$. One can check that for $n \geq 3$ we have $D^{-1}(I_n + E_{12})D = I_n + E_{12}$ so that $(DP^{-1})A(PD^{-1}) = I_n + E_{12}$ and thus A is conjugate to $I_n + E_{12}$ via PD^{-1}. Now notice that $|PD^{-1}| = |P||P|^{-1} = 1$ and so $PD^{-1} \in SL_n(K)$.

Corollary 3.2 *All transvections are conjugate to each other in $GL_n(K)$. Furthermore, if $n \geq 3$, then all transvections are conjugate to each other in $SL_n(K)$.*

Proof 3.7 *By Lemma 3.2, since every transvection is conjugate to $I_n + E_{12}$, they are therefore conjugate to each other and if $n \geq 3$ they are conjugate via an element of $SL_n(K)$.*

Theorem 3.4 *Every element of $SL_n(K)$ can be expressed as a product of transvections of the form $I_n + aE_{ij}$ where $a \in K^*$ and $i \neq j$ i.e., $SL_n(K)$ is generated by these transvections. Every element of $GL_n(K)$ can be expressed as a product of transvections of the form $I_n + aE_{ij}$ where $a \in K^*$ and $i \neq j$ and a diagonal matrix of the form $\mathrm{diag}(1, 1, \ldots, b)$ for some $b \in K^*$ i.e., $GL_n(K)$ is generated by these transvections and diagonal matrices.*

Proof 3.8 *Take any $A = [a_{ij}] \in SL_n(K)$. Since A is invertible, we can row reduce it to the identity matrix. We need to show explicitly how we row reduce A to I_n to prove the theorem.*

We may assume that $a_{12} \neq 0$ (otherwise add some row i with $a_{1i} \neq 0$ to row 2). Now $a_{12}^{-1}(1 - a_{11})$ times row 2 added to row 1 changes a_{11} into 1. Using this pivot 1 and row operations of the form $aR_1 + R_i$ with $i > 1$ and $a \in K^*$ we can put 0's below the pivot 1. Now the $(1, 1)$th minor of the resulting matrix is in $SL_{n-1}(K)$ so we may repeat the process just mentioned on this minor. Continuing in this way using the same type of elementary row operations we can effectively row reduce A to I_n.

Now notice that we did this using only elementary row operations of the form $aR_j + R_i$ with $a \in K^*$ and $i \neq j$. Recall from linear algebra that these elementary row operations corresponds to multiplication on the left by the elementary matrices of the form $I_n + aE_{ij}$ with $a \in K^*$ and $i \neq j$. Notice also that $(I_n + aE_{ij})^{-1} = I_n - aE_{ij}$ which is a transvection of the very same form. Therefore, again by linear algebra, $I_n = E_k \cdots E_2 E_1 A$ where each E_i is of the form $I_n + aE_{ij}$ with $a \in K^*$ and $i \neq j$. But then $A = E_1^{-1} E_2^{-1} \cdots E_k^{-1}$ where each E_i^{-1} is of the form $I_n + aE_{ij}$ with $a \in K^*$ and $i \neq j$.

The proof for $GL_n(K)$ is identical, but we need to note that using only elementary row operations of the form $aR_j + R_i$ with $a \in K^*$ and $i \neq j$, we can only reduce an element of $GL_n(K)$ down to $\mathrm{diag}(1, 1, \ldots, b)$ with $b \in K^*$.

We now have enough machinery to classify the normal subgroups of $GL_n(K)$ and thereby show that almost all the groups $PSL_n(K)$ are simple groups. Here now is the main result from which this follows.

Theorem 3.5 *If $GL_n(K) \neq GL_2(2), GL_2(3)$ and $H \leq GL_n(K)$ such that for all $A \in SL_n(K)$ and $B \in H$ we have $A^{-1}BA \in H$, then either $SL_n(K) \leq H$ or $H \leq Z(GL_n(K))$.*

We shall delay the proof of this theorem for a bit and first look at some of the consequences. The reader should check that the above result is false when $GL_n(K) = GL_2(2)$ or $GL_2(3)$. The first two corollaries are immediate and require no proof.

Corollary 3.3 *If $GL_n(K) \neq GL_2(2), GL_2(3)$ and $H \lhd GL_n(K)$, then $SL_n(K) \leq H$ or $H \leq Z(GL_n(K))$.*

Corollary 3.4 *If $GL_n(K) \neq GL_2(2), GL_2(3)$ and $H \lhd SL_n(K)$, then either $H = SL_n(K)$ or $H \leq Z(SL_n(K))$.*

Corollary 3.5 *If $GL_n(K) \neq GL_2(2), GL_2(3)$, then $GL_n(K)' = SL_n(K)$.*

Proof 3.9 *Since $GL_n(K)/SL_n(K) \cong K^*$ is abelian, we have that $GL_n(K)' \leq SL_n(K)$. So then $GL_n(K)' \lhd SL_n(K)$ and thus, by Corollary 3.4, if we can show there is an element of $GL_n(K)'$ not in $Z(SL_n(K))$, then we will be done. Take, for instance the commutator $[I_n + E_{12}, I_n + E_{21}]$ which is not a scalar matrix.*

Corollary 3.6 *If $GL_n(K) \neq GL_2(2), GL_2(3)$, then $PSL_n(K)$ is simple.*

Proof 3.10 *Let $\overline{N} \lhd PSL_n(K)$ and write $\overline{N} = N/Z(SL_n(K))$, where $N \lhd SL_n(K)$. Then by Corollary 3.4, either $N = SL_n(K)$ or $N \leq Z(SL_n(K))$ and so either $\overline{N} = PSL_n(K)$ or $\overline{N}$ is the trivial subgroup.*

Remark 3.3 *Let's look at some of the smallest groups of the form $PSL_n(K)$.*

1. *$PSL_2(2)$ and $PSL_2(3)$ are not simple. One can see this by either pointing out that their corresponding orders are 6 and 12 and there are no simple groups of these orders. Another way to see this is to show that $PGL_2(2) \cong S_3$ and $PGL_2(3) \cong A_4$ both of which are not simple groups.*

2. *$PSL_2(4)$ and $PSL_2(5)$ both have order 60 and one can show that the only simple group of order 60 is A_5. Hence, $PSL_2(4)$ and $PSL_2(5)$ do not give rise to any new simple groups.*

3. *$PSL_2(7)$ has order 168 which is not the order of any alternating group. Therefore, $PSL_2(7)$ is the smallest new simple group of this type.*

4. *$PSL_3(4)$ has order 20,160= $8!/2 = |A_8|$, however $PSL_3(4)$ is not isomorphic to A_8. This can be shown by looking at the element $(1\ 2\ 3\ 4\ 5)(6\ 7\ 8) \in A_8$ of order 15 and proving that $PSL_3(4)$ has no element of order 15. Hence, there are two non-isomorphic non-abelian simple groups of order 20,160.*

We now prove Theorem 3.5.

Proof 3.11 *Certainly, every $H \leq Z(GL_n(K))$ has the property that for all $A \in SL_n(K)$ and $B \in H$ we have $A^{-1}BA \in H$, so we may assume that H has an element outside of $Z(GL_n(K))$. There are two cases for this proof which are each proved in very different manners – one highly computational and the other not as much. These two cases arise due to the fact that transvections are conjugate in $SL_n(K)$ only when $n > 2$.*

Case 1: $n = 2$

First, we need to show that if H contains a transvection of the form $I_2 + aE_{ij}$ where $a \in K^*$ and $i \neq j$, then $SL_2(K) \leq H$. So suppose $I_2 + aE_{ij} \in H$ where $a \in K^*$ and $i \neq j$. We shall show then that every transvection of the form $I_2 + aE_{ij}$ where $a \in K^*$ and $i \neq j$ is in H and hence, by Theorem 3.4, we will be done with the case of $n = 2$. First note that it is enough to show that every transvection of the form $I_n + aE_{12} \in H$ where $a \in K^*$, since by assumption we would then have

$$\begin{bmatrix} 1 & 0 \\ -a & 1 \end{bmatrix} = \begin{bmatrix} 0 & -1 \\ 1 & 0 \end{bmatrix}^{-1} \begin{bmatrix} 1 & a \\ 0 & 1 \end{bmatrix} \begin{bmatrix} 0 & -1 \\ 1 & 0 \end{bmatrix} \in H.$$

Therefore, let's assume some matrix of the form $\begin{bmatrix} 1 & a \\ 0 & 1 \end{bmatrix} \in H$ with $a \in K^*$. Notice then that for any $b \in K^*$, we have

$$\begin{bmatrix} 1 & ab^2 \\ 0 & 1 \end{bmatrix} = \begin{bmatrix} b^{-1} & 0 \\ 0 & b \end{bmatrix}^{-1} \begin{bmatrix} 1 & a \\ 0 & 1 \end{bmatrix} \begin{bmatrix} b^{-1} & 0 \\ 0 & b \end{bmatrix} \in H.$$

Hence, for all $b, c \in K^*$, we have

$$\begin{bmatrix} 1 & ab^2 \\ 0 & 1 \end{bmatrix} \begin{bmatrix} 1 & ac^2 \\ 0 & 1 \end{bmatrix}^{-1} = \begin{bmatrix} 1 & a(b^2 - c^2) \\ 0 & 1 \end{bmatrix}.$$

Now in the case that $char(K) \neq 2$, the equation $a(b^2 - c^2) = d$ has a solution in b and c for any choice of $d \in K^*$. Indeed, the solution is $b = 2^{-1}(a^{-1}d + 1)$ and $c = 2^{-1}(a^{-1}d - 1)$. Note this does not work for $char(K) = 2$ for then $2 = 0$ which has no inverse. So we are reduced to proving this case of $n = 2$ when $char(K) = 2$.

In the case that $char(K) = 2$, first note that K^* must contain an element d with the property that $d^4 \neq 1$, for otherwise $|K| = 3$ or 5 (since K^* is cyclic). If we set $b = ad^2$, then by the work above

$$\begin{bmatrix} 1 & b \\ 0 & 1 \end{bmatrix} \in H \text{ by conjugating } \begin{bmatrix} 1 & a \\ 0 & 1 \end{bmatrix} \text{ by } \begin{bmatrix} d^{-1} & 0 \\ 0 & d \end{bmatrix}.$$

If we set

$$c = a^{-1}(1 + d^{-2}) = a[a^{-2}(1 + d^{-2})] = a[a^{-2} + (a^{-1}d)^2] = a(a^{-1} + a^{-1}d)^2,$$

then again by the work above

$$\begin{bmatrix} 1 & c \\ 0 & 1 \end{bmatrix} \in H \text{ by conjugating } \begin{bmatrix} 1 & a \\ 0 & 1 \end{bmatrix} \text{ by } \begin{bmatrix} (a^{-1} + a^{-1}d)^{-1} & 0 \\ 0 & a^{-1} + a^{-1}d \end{bmatrix}.$$

Before we continue, note that it is easy compute (exercise) that $abc = a + b$, $bc = 1 + d^2$ and $ac = 1 + d^{-2}$ which we will need in a moment. As we saw above, we know that

$$\begin{bmatrix} 1 & 0 \\ -a & 1 \end{bmatrix} \text{ and } \begin{bmatrix} 1 & 0 \\ -b & 1 \end{bmatrix} \text{ are in } H,$$

thus also in H is

$$\begin{bmatrix} 1 & 0 \\ -a & 1 \end{bmatrix} \begin{bmatrix} 1 & c \\ 0 & 1 \end{bmatrix} \begin{bmatrix} 1 & 0 \\ -b & 1 \end{bmatrix} = \begin{bmatrix} 1-bc & c \\ abc-a-b & 1-ac \end{bmatrix} = \begin{bmatrix} -d^2 & c \\ 0 & -d^{-2} \end{bmatrix}.$$

Now take any $e \in K^*$ and set $x = e(d^{-4}-1)^{-1}$. Notice that also in H is

$$\left[\begin{bmatrix} -d^2 & c \\ 0 & -d^{-2} \end{bmatrix}, \begin{bmatrix} 1 & -x \\ 0 & 1 \end{bmatrix} \right] = \begin{bmatrix} -d^2 & c \\ 0 & -d^{-2} \end{bmatrix}^{-1} \begin{bmatrix} -d^2 & c \\ 0 & -d^{-2} \end{bmatrix}^{\begin{bmatrix} 1 & -x \\ 0 & 1 \end{bmatrix}}.$$

Furthermore, one can easily compute that

$$\left[\begin{bmatrix} -d^2 & c \\ 0 & -d^{-2} \end{bmatrix}, \begin{bmatrix} 1 & -x \\ 0 & 1 \end{bmatrix} \right] = \begin{bmatrix} 1 & e \\ 0 & 1 \end{bmatrix}.$$

Therefore, any matrix of the form $\begin{bmatrix} 1 & e \\ 0 & 1 \end{bmatrix}$ with $e \in K^*$ is in H, thus finishing the case of characteristic 2.

Recall, in the beginning of the proof that we chose $A \in H - Z(GL_n(K))$. In linear algebra, we know that A is similar to a matrix in rational canonical form. Let us first consider the possibilities for this form. A might have two linear invariant factors $t-a$, $t-b$ in which case the rational canonical form would be $\begin{bmatrix} a & 0 \\ 0 & b \end{bmatrix}$, but since $A \in SL_2(K)$ it must, in fact, be $C = \begin{bmatrix} a & 0 \\ 0 & a^{-1} \end{bmatrix}$. Furthermore, $a^2 \neq 1$, since A is not in $Z(SL_2(K))$. The other possibility is that A has a single quadratic invariant factor $t^2 - at - b$ in which case the rational canonical form would be $\begin{bmatrix} 0 & 1 \\ b & a \end{bmatrix}$, but again since $A \in SL_2(K)$ it must, in fact, be $C = \begin{bmatrix} 0 & 1 \\ -1 & a \end{bmatrix}$. Let $P \in GL_2(K)$ be the matrix by which we conjugate A to put it in rational canonical form. Let's consider each case separately. First, suppose that $C = \begin{bmatrix} a & 0 \\ 0 & a^{-1} \end{bmatrix}$ and set $B = \begin{bmatrix} 1 & 1 \\ 0 & 1 \end{bmatrix}$. Consider the commutator $[C, B]$ which one can compute to be $\begin{bmatrix} 1 & 1-a^{-2} \\ 0 & 1 \end{bmatrix}$. Since $a^2 \neq 1$, this implies that $[C, B]$ is a transvection of the form $I_2 + (1-a^{-2})E_{12}$. Notice also that

$$[C, B] = [A^P, B] = A^{-P}A^{PB} = \left(A^{-1}A^{BP^{-1}} \right)^P \in H^P.$$

Now suppose that $C = \begin{bmatrix} 0 & 1 \\ -1 & a \end{bmatrix}$ and for any $b \in K^*$ set $B = \begin{bmatrix} 1 & -b^2 \\ 0 & 1 \end{bmatrix}$. Then, as above $[C^{-1}, B] = \begin{bmatrix} 1 & -b^2 \\ -b^2 & 1+b^4 \end{bmatrix}$ is in H^P for all $b \in K^*$. Conjugating $[C^{-1}, B]$

by $\begin{bmatrix} b^{-1} & -b^{-1} \\ 0 & b \end{bmatrix} \in SL_n(K)$ *yields the matrix* $\begin{bmatrix} 0 & 1 \\ -1 & 2+b^4 \end{bmatrix}$ *which must also be in* H^P – *this follows, since* H^P *has the same property as* H *does, i.e.* $A \in SL_n(K)$ *and* $B \in H^P$ *we have* $A^{-1}BA \in H^P$, *since* $B = D^P$ *for some* $D \in H$ *and so*

$$A^{-1}BA = \left(\left(A^{P^{-1}} \right)^{-1} DA^{P^{-1}} \right)^P \in H^P.$$

Then for all $b \in K^*$, H^P *contains the matrices*

$$\begin{bmatrix} 0 & 1 \\ -1 & 2+b^4 \end{bmatrix}^{-1} \begin{bmatrix} 0 & 1 \\ -1 & 2+1^4 \end{bmatrix} = \begin{bmatrix} 1 & b^4-1 \\ 0 & 1 \end{bmatrix}.$$

Therefore, H^P *will have a transvection unless* $b^4 = 1$ *for all* $b \in K^*$, *but as we saw above this could only occur if* $|K| = 3$ *or* 5. *So we need another strategy for finding a transvection in* $SL_2(5)$ *(the group* $SL_2(3)$ *is ruled out by assumption). From our above work we know that* $\begin{bmatrix} 0 & 1 \\ -1 & 3 \end{bmatrix} \in H^P$ *and also the commutator of*

$\begin{bmatrix} 0 & 1 \\ -1 & a \end{bmatrix}^{-1}$ *and* $\begin{bmatrix} 1 & -2 \\ 0 & 1 \end{bmatrix}$ *which equals* $\begin{bmatrix} 1 & -2 \\ -2 & 0 \end{bmatrix}$ *when* $q = 5$. *Conjugate this result by* $\begin{bmatrix} 2 & -1 \\ -2 & -1 \end{bmatrix} \in SL_2(5)$ *to get* $\begin{bmatrix} 0 & 1 \\ -1 & 1 \end{bmatrix} \in H^P$. *And lastly, note then that*

$$\begin{bmatrix} 0 & 1 \\ -1 & 3 \end{bmatrix}^{-1} \begin{bmatrix} 0 & 1 \\ -1 & 1 \end{bmatrix} = \begin{bmatrix} 1 & 2 \\ 0 & 1 \end{bmatrix},$$

which is a transvection in H^P. *Thus, in every case,* H^P *has a transvection of the form* $I_2 + aE_{12}$ *where* $a \in K^*$. *Since, as we saw above,* H^P *has the same property as* H *does, i.e.* $A \in SL_n(K)$ *and* $B \in H^P$ *we have* $A^{-1}BA \in H^P$, *by our earlier work,* $SL_2(K) \leq H^P$ *and so* $SL_n(K) = SL_n(K)^{P^{-1}} \leq H$. *This completes the first case.*

Case 2: $n > 2$

Recall again our choice of $A \in H - Z(GL_n(K))$. *By Remark 3.2.4, there must be a transvection* $T \in SL_n(K)$ *which does not commute with* A. *Set* $B = [A,T] \in GL_n(K)' = SL_n(K)$. *Note also that* $B = A^{-1}A^T \in H$ *and by assumption,* $B \neq I_n$.

Claim 3.5 H *contains a transvection*

If B *were a transvection we would be done, so we may assume that* B *is not a tranvection. Let* U *be the* $n-1$ *dimensional subspace of* K^n *fixed pointwise by the transvection* T^{-A} *and* W *is the* $n-1$ *dimensional subspace of* K^n *fixed pointwise by the transvection* T, *then* $B = T^{-A}T$ *fixes pointwise* $U \cap W$, *a subspace of* K^n *of dimension at least* $n-2$. *Now, if* $dim(U \cap W) = n-1$, *then* $B \in H$ *is a transvection, contrary to our assumption. Hence, it must be the case that* $dim(U \cap W) = n-2$. *Set* U' *equal to the kernel of* $B - I_n$ *and* W' *the image of* $B - I_n$, *and so* $dim(U') + dim(W') = n$. *Since* B *is not a transvection, it must be the case that* $dim(U') \leq n-2$. *Since*

$dim(U \cap W) = n - 2$, *this implies that* $dim(U') = n - 2$ *and so* $dim(W') = 2$. *Let* W'' *be a subspace of* K^n *of dimension* $n - 1$ *containing* W' *(here we use the fact that* $n > 2$*). Take any* $w \in W''$ *and write* $Bw = (B - I_n)w + w \in W' + W'' \subseteq W''$. *Therefore,* B *fixes setwise* W''. *Let* C *be a transvection which fixes* W'' *pointwise. Then* C^{-B} *is also a transvection fixing* W'' *pointwise. Now* $[B, C] = C^{-B}C$ *fixes* W'' *pointwise as well, but could be the identity. Note that* $[B, C] = B^{-1}B^C \in H$, *so if* $[B, C]$ *were a transvection we would be done. So we may assume that* $[B, C] = I_n$ *for all tranvections* C *which fix* W'' *pointwise (we will show this is not possible). Since* B *is not a tranvection, there must be a* $w \in W''$ *such that* $Bw \neq w$ *and yet* $Bw \in W''$. *Fix a* $v \in K^n - W''$ *and select the transvection* D *which fixes* W'' *pointwise with the additional property that* $Dv = w + v$. *By our assumption,* $[B, D] = I_n$ *and so* $BD = DB$. *In particular,* $BD(v) = DB(v)$. *Since* $(B - I_n)v \in W' \subseteq W''$ *this implies that* $(B - I_n)v = w'$ *for some* $w' \in W''$ *and so* $Bv = v + w'$. *Therefore,* $DB(v) = D(v + w') = w + v + w'$ *while* $BD(v) = B(w + v) = Bw + v + w'$ *and so* $Bw = w$, *a contradiction.*

Having proved this claim, since all transvections are conjugate in $SL_n(K)$ *it follows that* H *contains all the transvections. Since* $SL_n(K)$ *is generated by the transvections we then have that* $SL_n(K) \leq H$.

EXERCISES

1 Verify that any matrix of the form $I_n + aE_{ij}$ where $a \in K^*$ and $i \neq j$ is a transvection, where E_{ij} is a matrix filled with zeros except that the ij-th entry is 1.

2 Verify in Lemma 3.2 that $a_1 u_1 + \cdots + a_{n-1} u_{n-1} \in U$ is linearly independent from $u_1, \ldots, u_{i-1}, u_{i+1}, \ldots, u_{n-1}$.

3 Verify in Lemma 3.2 that $D^{-1}(I_n + E_{12})D = I_n + E_{12}$.

4 Verify that the result of Theorem 3.5 is false when $GL_n(K) = GL_2(2)$ or $GL_2(3)$.

5 In the proof of Theorem 3.5 verify that $abc = a + b$, $bc = 1 + d^2$ and $ac = 1 + d^{-2}$

CHAPTER 4

Group Action

I N THIS CHAPTER, we present a powerful idea in group theory which is group action. The consequences of group action are many and the applications are incredibly useful as we shall see in the later sections in this chapter. Indeed, if someone wants to see concrete applications in group theory, then group action certainly fits the bill. In Section 4.1, we introduce the notion of group action. In Sections 4.2 and 4.3, we see a very nice application of group action in the field of combinatorics. Section 4.4 presents some deep theoretical consequences which follow from group action. The important Sylow Theorems in Section 4.5 follow from group action. We then use the Sylow Theorems to classify finite groups of particular sizes in Section 4.6, which in turn assists us in classifying completely finite abelian groups in Section 4.7.

4.1 GROUP ACTION ON A SET

In this section, we present a way in which a group with all its structure may interact with a set which has no structure whatsoever. This notion will lead us to many deep results in the theory of groups.

Definition 4.1 *Let $(G, *)$ be a group and X any set. We say G **acts on** X if there is a binary operation $\cdot$ from $G \times X$ to X having the following properties:*

1. *For all $g, h \in G$ and all $x \in X$ we have $g \cdot (h \cdot x) = (g * h) \cdot x$.*

2. *For all $x \in X$ we have $1 \cdot x = x$.*

*One also says that G defines a **group action** on X or that X is a G-**set**.*

Example 4.1 *Here, we present several examples of group action.*

1. *Let $G = S_n$ with $* = \circ$ (composition) and $X = \{1, 2, 3, \ldots, n\}$. We can let G act on X as follows: for $\sigma \in G$ and $i \in X$, define the group action $\sigma \cdot i = \sigma(i)$. For instance, in S_4, if σ is a $90°$ rotation of a square, then $\sigma \cdot 2 = \sigma(2) = 3$. One needs to verify that we have indeed defined a group action.*

   First take $\sigma, \tau \in G$ and $i \in X$. Then

   $$\sigma \cdot (\tau \cdot i) = \sigma(\tau(i)) = (\sigma \circ \tau)(i) = (\sigma \circ \tau) \cdot i.$$

DOI: 10.1201/9781003335283-4

Second, for the identity permutation ι and $i \in X$, we have

$$\iota \cdot i = \iota(i) = i.$$

2. *Let $G = GL_n(\mathbb{R})$ and $X = \mathbb{R}^n$. Then G acts on X via matrix multiplication, i.e. for $A \in GL_n(\mathbb{R})$ and $\vec{v} \in \mathbb{R}^n$ (viewed as a column vector), then $A \cdot \vec{v} = A\vec{v}$ where the operation is matrix multiplication. The reader can easily verify that this indeed defines a group action.*

3. *Let $G = S_n$ and $A = \{a_1, a_2, \ldots, a_n\}$ be any set with n elements. Then for any $k = 1, 2, 3, \ldots$ we can have G act on A^k as follows:*

$$\sigma \cdot (a_{i_1}, a_{i_1}, \ldots, a_{i_k}) = (a_{\sigma(i_1)}, a_{\sigma(i_1)}, \ldots, a_{\sigma(i_k)}).$$

Remark 4.1 *One may view a group action (and sometimes this is given as the definition) as a homomorphism $\pi : G \to Sym(X)$. Indeed, given that G acts on X, then we can define $\pi(g)$ to be the permutation which sends each x to $g \cdot x$. Likewise, given a homomorphism from G to $Sym(X)$, we can define a group action as $g \cdot x = [\pi(g)](x)$.*

Example 4.2 *We now present four important group actions which we shall use in later discussions. The reader should verify that all four examples do indeed define group actions.*

1. *Let a group G act on itself (i.e. $X = G$) by left multiplication, i.e. for $a \in G$ and $g \in X$ define $a \cdot g = ag$ using the group operation.*

2. *Let a group G act on itself by conjugation, i.e. for $a \in G$ and $g \in X$ define $a \cdot g = aga^{-1}$ using the group operation.*

3. *Let G be a group and $H \leq G$. Set $X = \{gH : g \in G\}$, the collection of left cosets of G modulo H. Let G act on X by left multiplication, i.e. for $a \in G$ and $gH \in X$ define $a \cdot (gH) = (ag)H$ (note that one needs to verify that the action is well-defined).*

4. *Let G be a group and X be the collection of subgroups G. Let G act on X by conjugation, i.e. for $a \in G$ and $H \in X$ define $a \cdot H = aHa^{-1}$ (note that one should check that $aHa^{-1} \leq G$).*

We now define several important structures associated with a group action.

Definition 4.2 *Let G be a group acting on a set X with $g \in G$ and $x \in X$.*

1. *The **stabilizer** of x, written $G_x = \{g \in G : gx = x\}$, i.e. the group elements which fix a particular x.*

2. *The **fixator** of g, written $X_g = \{x \in X : gx = x\}$, i.e. the set elements fixed by a particular g.*

3. The **orbit** of x, written $Gx = \{gx \;:\; g \in G\}$, i.e. the elements of the set that can be realized by allowing all of the group to act on a fixed x.

One can easily show that G_x is a subgroup of G. Note that both X_g and Gx are subsets of X.

Example 4.3 *Let us return now to our four important examples above and compute the structures we just defined.*

1. G_g *is the trivial subgroup,* $X_g = \emptyset$ *unless* $g = 1$ *in which case* $X_1 = G$, *and* $Gg = G$.

2. $G_g = C_G(g)$ *(recall the centralizer of* g *in* G*), and so is* $X_g = C_G(g)$*, while* $Gg = g^G = \{aga^{-1} \;:\; a \in G\}$ *called the* **conjugacy class** *of* g *in* G.

3. $G_{gH} = gHg^{-1}$ *the collection of group elements which are conjugate to an element of* H *via the element* g*, and* $G(aH) = X$*. Note that in particular* $G_H = H$.

4. G_H *are the set of all elements in* G *that satisfy the normal property for* H*. The notation for this set is* $N_G(H)$*, i.e.*

$$N_G(H) = \{g \in G \;:\; g^{-1}hg \in H \text{ for all } h \in H\},$$

and is called the **normalizer** *of* H *in* G*. One can show that* $H \lhd N_G(H) \le G$*. The orbit* $GH = H^G = \{a^{-1}Ha \;:\; a \in G\}$*.*

It is useful to note that orbits can be defined as the equivalence classes of a particular equivalence relation on the set X the group G acts upon. Define the following relation on X: $x \sim y$ iff there is a $g \in G$ such that $gx = y$. This defines an equivalence relation on X with equivalence classes being precisely the orbits of G acting on X. Indeed, $\sim$ is reflexive since for any $x \in X$ we have $1x = x$ (definition of group action). We have symmetry, since if $x \sim y$, then there is a $g \in G$ with $gx = y$, but then using the definition of group action this can be rewritten as $g^{-1}y = x$ and so $y \sim x$. For transitivity, if $x \sim y$ and $y \sim z$, then there are $g, h \in G$ with $gx = y$ and $hy = z$. But then $(hg)x = z$ using the definition of group action, and thus $x \sim z$. If we take any $x \in X$ and compute the equivalence class

$$[x] = \{y \in X \;:\; y \sim x\} = \{y \in X \;:\; \exists g \in G, \, y = gx\}$$

$$= \{gx \;:\; g \in G\} = Gx.$$

One use of this observation is the immediate result that any two orbits of a group action are either disjoint or coincide (since equivalence classes have this property).

We now begin our discussion on counting results in this setting.

Theorem 4.1 *Let G be a group acting on a set X. Then for a given $x \in X$ there is a one-to-one correspondence between the elements in the orbit Gx and the cosets of G modulo G_x. In particular, if G or X is finite, then $|Gx| = [G : G_x]$. If both G and X are finite, then we have $|G| = |G_x||Gx|$ and so the size of an orbit divides the size of G.*

Proof 4.1 *We simply define a map* $f : G/G_x \to Gx$ *by* $f(gG_x) = gx$. *First note that* f *certainly maps onto* Gx *by its very definition. Second,* f *is both well-defined and one-to-one, since*

$$gG_x = hG_x \quad \text{iff} \quad h^{-1}g \in G_x \quad \text{iff} \quad h^{-1}gx = x \quad \text{iff} \quad gx = hx \quad \text{iff} \quad f(gG_x) = f(hG_x).$$

Thus, f *is a bijection which proves the result.*

Corollary 4.1 *Let* G *be a finite group with* $H \leq G$ *and* $g \in G$, *then*

1. *The size of the conjugacy class of* g *equals* $[G : C_G(g)]$.

2. *The number of the conjugacy classes of* H *equals* $[G : N_G(H)]$.

Proof 4.2 *For the first statement, consider the group action in Example 4.2.2. The result follows immediately from Theorem 4.1 and our computations of* Gx *and* G_x *in this setting.*

For the second statement, consider the group action in Example 4.2.4. The result follows immediately from Theorem 4.1 and our computations of Gx *and* G_x *in this setting.*

Example 4.4 *Consider the cycle types (conjugacy classes) of elements in* S_4.

CycleType	Number
$(*)(*)(*)(*)$	1
$(*)(*)(**)$	$\frac{4!}{2! \cdot 2} = 6$
$(**)(**)$	$\frac{4!}{2! \cdot 2 \cdot 2} = 3$
$(*)(***)$	$\frac{4!}{3} = 8$
$(****)$	$\frac{4!}{4} = 6$

Notice how the sizes of the conjugacy classes divide the order of the group S_4, *since conjugacy classes are orbits. Let's take it a step further. It's easy to show (and is left as an exercise) that a subgroup of a group is normal iff the subgroup is a union of conjugacy classes. Equipped with this fact we can now show that the only subgroup of* S_4 *of order 12 is* A_4.

Proposition 4.1 *If* $H \leq S_4$ *and* $|H| = 12$, *then* $H = A_4$.

Proof 4.3 *First note that any subgroup of order 12 in* S_4 *is normal, since it has index 2 in* S_4. *Therefore, by Exercise 10, the subgroup must be a union of conjugacy classes. Of course, since it is a subgroup it must contain the singleton conjugacy class* $\{1\}$ *which leaves 11 more elements to select. The sizes of the other four conjugacy* *classes are 6, 3, 8 and 6. The only way to make eleven is using 3 and 8. Hence, there is only one way to make a subgroup of order 12, and so it must be A_4 (note also that the cycle types in the conjugacy classes of size 3 and 8 are indeed even permutations).*

Example 4.5 *Let $G = S_n$ and $\tau = (1\ 2)$. We will count the number of conjugates of τ in two different ways.*

First, we will count them directly. Note for any $\sigma \in S_n$ that $\sigma \tau \sigma^{-1} = (\sigma(1)\ \sigma(2))$ (see Exercise 9 in Section 2.4) and so the conjugates of τ consist of all the transpositions in S_n. The number of transpositions in S_n equals $n(n-1)/2$, since $(k\ l) = (l\ k)$. Therefore, the number of conjugates of τ is $n(n-1)/2$.

Now we will count the number of conjugates of τ using Corollary 4.1.1. We will have to count $C_G(\tau)$ in order to do this. Note that $\sigma \tau = \tau \sigma$ iff $\sigma \tau \sigma^{-1} = \tau$ iff $(\sigma(1)\ \sigma(2)) = (1\ 2)$ iff either $\sigma(1) = 1$ and $\sigma(2) = 2$ or $\sigma(1) = 2$ and $\sigma(2) = 1$. Hence, $|C_G(\tau)| = (n-2)! + (n-2)! = 2(n-2)!$. Now, by Corollary 4.1, the number of conjugates of τ is

$$[G : C_G(\tau)] = \frac{|G|}{|C_G(\tau)|} = \frac{n!}{2(n-2)!} = \frac{n(n-1)}{2}.$$

One final observation is that Example 4.2.1 and 3 do not yield much information when we apply Theorem 4.1 to each of them. For Example 1, we get $|G| = [G : 1]$ which is trivial and Example 3 yields $|G/H| = [G : H]$ which is simply the definition of index.

EXERCISES

1 Verify group action for Example 4.1, parts 2 and 3.

2 Prove the alternate definition discussed in Remark 4.1 is equivalent to the one given in this text.

3 Verify group action for each action defined in Example 4.2.

4 For Example 4.2.3, verify that the action is well-defined.

5 For Example 4.2.4, verify that $aHa^{-1} \le G$.

6 Let $G = S_4$ and set X equal to the set of all transpositions in S_4. Set $x = (2,4) \in X$. Let G act on X as follows: For $\sigma \in G$ and $(i,j) \in X$ define $\sigma(i,j) = (\sigma(i), \sigma(j))$

    a. Verify that the above definition does indeed define a group action.

    b. List the elements of G_x and thus compute $|G_x|$.

    c. List the elements of Gx and thus compute $|Gx|$.

    d. Now compute $|Gx|$ using the Proposition which states that $|Gx| = [G : G_x]$.

7 Verify for G acting on a set X and $x \in X$ that G_x is a subgroup of G.

8 Verify all the statements for each of the four examples given in Example 4.3 regarding stabilizers, fixators and orbits.

9 Prove that for $H \leq G$ we have $H \lhd N_G(H) \leq G$ (there are two things to prove here).

10 Prove that a subgroup of a group is normal iff the subgroup is a union of conjugacy classes (note that one direction was proved in Exercise 15 of Section 2.9).

11 Prove that an element g in a group G has a conjugacy class of size 1 iff $g \in Z(G)$.

12 Prove that in an abelian group all conjugacy classes are of size 1.

13 Give an example of two non-isomorphic groups of the same size with the same number of conjugacy classes of any given size. Must the groups be abelian for this to be possible? (explain)

14 Consider a finite group G acting on a set X.

   a. For $g \in G$ and $x \in X$ show that $G_{gx} = g^{-1}G_x g$.

   b. For $g \in G$ and $x \in X$ show that $|G_x| = |g^{-1}G_x g|$.

   c. For $x, y \in X$ show that if $Gx = Gy$, then $|G_x| = |G_y|$.

15 Let G be a group and $H \leq G$. Consider the action of G on G/H by left multiplication (i.e. $a(gH) = (ag)H$).

   a. Express this action in terms of a homomorphism ϕ (see Remark 4.1).

   b. Show that $ker\phi \leq H$.

   c. Assume that $[G : H] = 2$. Apply the First Isomorphism Theorem on part a. to show that $[G : ker\phi] \leq 2$.

   d. Assume that $[G : H] = 2$. Use parts b. & c. to show that $[G : ker\phi] = 2$.

   e. Assume that $[G : H] = 2$. Use parts b. & d. to show that $H = ker\phi$.

   f. Using part e., Show that if $H \leq G$ and $[G : H] = 2$, then H is normal in G.

16 A group G acts transitively on a set X if for any $x, y \in X$, there exists a $g \in G$ such that $gx = y$. A group G acts doubly transitive on a set X if for every x_1, y_1, x_2, y_2 with $x_1 \neq y_1$ and $x_2 \neq y_2$, there exists a $g \in G$ such that $gx_1 = x_2$ and $gy_1 = y_2$.

   Show that G acts doubly transitively on a set X iff for any $x \in G$, G_x acts transitively on $X - \{x\}$ and G acts transitively on X.

4.2 BURNSIDE'S LEMMA

In this section, we introduce a result, called Burnside's Lemma, which has surprising consequences in combinatorics allowing us to count various things with ease. Although its proof is quite straightforward, it yields a nice formula for counting orbits. Before we prove this result we need some more terminology.

Definition 4.3 *Let G be a group acting on a set X. The* **characteristic function** *associated with this action is a function $f : G \times X \to \{0,1\}$ defined by*

$$f(g,x) = \begin{cases} 1, & gx = x \\ \\ 0, & gx \neq x \end{cases}$$

Lemma 4.1 (Burnside's Lemma) *If G is a finite group acting on a finite set X, then the number of orbits of X equals*

$$\frac{1}{|G|} \sum_{g \in G} |X_g|.$$

Proof 4.4 *Notice that*

$$\sum_{g \in G} |X_g| = \sum_{g \in G} \left(\sum_{x \in X} f(g,x) \right) = \sum_{x \in X} \left(\sum_{g \in G} f(g,x) \right)$$

$$= \sum_{x \in X} |G_x| = \sum_{x \in X} \frac{|G|}{|Gx|} = |G| \left(\sum_{x \in X} \frac{1}{|Gx|} \right).$$

Furthermore, consider a typical orbit $Gx = \{x_1, \ldots, x_r\}$ and notice that

$$\sum_{i=1}^{r} \frac{1}{|Gx_i|} = \sum_{i=1}^{r} \frac{1}{|Gx|} = \frac{r}{|Gx|} = 1.$$

Therefore,

$$\sum_{g \in G} |X_g| = |G| \left(\sum_{x \in X} \frac{1}{|Gx|} \right) = |G| \times (\text{The number of orbits of } X).$$

Example 4.6 *We now give some nice counting arguments which use Burnside's Lemma.*

1. *Consider the letters a, a, b, b, b, c, c and suppose we wish to count the number of distinct (nonsense) words we can produce using all seven letters (these are called* **arrangements**). *For instance, one such word could be acabcbb. Consider $G = S_7$ acting on the set X of permutations – arrangements – of the seven letters in the natural way (i.e. for instance, if σ is the 7-cycle (1 2 3 4 5 6 7), then $\sigma(acabcbb) = bacabcb$). Let x be the word aabbbcc. First note that the size of G_x equals 2!3!2!, since any $\sigma \in G_x$ may permute the first two letters, the*

second three letters and the last two letters of x and still fix x. Second, the size of Gx equals

$$[G : G_x] = \frac{|G|}{|G_x|} = \frac{7!}{2!3!2!} = \begin{pmatrix} 7 \\ 2, 3, 2 \end{pmatrix} = 210.$$

Now, it is intuitively clear that the orbit of x equals the entire set X and so the number of arrangements of a, a, b, b, b, c, c equals the size of this orbit, namely 210. In other words, this group action has only one orbit, so we need not employ Burnside's Lemma to show this fact. However, with a lot of patience one could actually compute the number of arrangements. Let us try this computation with a smaller example.

We will count the number of arrangements of the letters a, a, b. In this case $G = S_3$ which consists of

$$1, (1\ 2\ 3), (1\ 3\ 2), (2\ 3), (1\ 3), (1\ 2).$$

One can check that their corresponding fixators are

$$X_1 = X, \qquad X_{(1\ 2\ 3)} = \emptyset, \qquad X_{(1\ 3\ 2)} = \emptyset,$$

$$X_{(2\ 3)} = \{baa\}, \qquad X_{(1\ 3)} = \{aba\}, \qquad X_{(1\ 2)} = \{aab\}.$$

Therefore, since there is one orbit,

$$\frac{1}{|G|} \sum_{g \in G} |X_g| = \frac{1}{3!}(|X| + 0 + 0 + 1 + 1 + 1) = 1.$$

Hence, $|X| = 3$, which is indeed the case, namely aab, aba and baa.

Assuming we know there is one orbit, one can easily extend this argument to produce a general formula for counting arrangements, namely suppose we consider producing arrangements using the n letters

$$\underbrace{a_1, a_1, \ldots, a_1}_{n_1\ times}, \underbrace{a_2, a_2, \ldots, a_2}_{n_2\ times}, \ldots, \underbrace{a_r, a_r, \ldots, a_r}_{n_r\ times},$$

where $n = n_1 + n_2 + \cdots + n_r$. Then the number of distinct arrangements of those letters is

$$\begin{pmatrix} n \\ n_1, n_2, \ldots, n_r \end{pmatrix} = \frac{n!}{n_1! n_2! \cdots n_r!}.$$

2. *Consider a circle divided into 6 equal sectors as in Figure 4.1. Suppose we can color a sector either black or white and we wish to count the number of distinct ways of coloring the circle in the sense that two colorings of the circle are distinct if you cannot get from one to the other by rotating the circle. One says distinct up to rotation. To do this we label the sectors with numbers.*

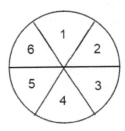

Figure 4.1 Symmetries of the triangle.

We consider the subgroup H of S_6 consisting of the six rotations of the circle $0°$, $60°$, $120°$, $180°$, $240°$ and $300°$. These elements of S_6 are

$$\sigma_0 = 1, \qquad \sigma_1 = (1\ 2\ 3\ 4\ 5\ 6), \qquad \sigma_2 = (1\ 3\ 5)(2\ 4\ 6),$$

$$\sigma_3 = (1\ 4)(2\ 5)(3\ 6), \qquad \sigma_4 = (1\ 5\ 3)(2\ 6\ 4), \qquad \sigma_5 = (1\ 6\ 5\ 4\ 3\ 2).$$

*Let X be the collection of all possible ways to color sectors $1, 2, 3, 4, 5$ and 6. The size of X is $2^6 = 64$. Let H act on X in the natural way by rotating the circle the appropriate number of degrees, i.e. σ_i will rotate the circle $(60i)°$. Notice that for a given coloring of the circle, the orbit of this coloring is precisely the set of colorings which are considered **not** distinct. Therefore, if we wish to count the number of distinct colorings of the circle, then we need only count the number of orbits of this action. Burnside's Lemma does exactly this thing. We first need to compute the six fixators. Certainly, the fixator of of identity is the entire set X. The fixator of σ_1 consists of the circle with all white sectors and the circle with all black sectors. The fixator of σ_2 consists of the circle with all white sectors, the circle with all black sectors, the circle with even numbered sectors being black, and the circle with odd numbered sectors being black. The fixator of σ_3 consists of the circle with all white sectors, the circle with all black sectors, the three circles with two opposite sectors being black and the three circles with two opposite sectors being white. Finally $X_{\sigma_4} = X_{\sigma_2}$ and $X_{\sigma_5} = X_{\sigma_1}$. Therefore, using Burnside's Lemma, the number of orbits equals*

$$\frac{1}{|H|}\left(|X_{\sigma_0}| + |X_{\sigma_1}| + |X_{\sigma_2}| + |X_{\sigma_3}| + |X_{\sigma_4}| + |X_{\sigma_5}|\right) = \frac{1}{6}(64+2+4+8+4+2) = 14.$$

The reader may wish to list the 14 distinct colorings.

3. *Consider the same setup as Example 2, but in addition assume that the coloring of a circle shows through to the back of the circle. If we wish to count the number of distinct colorings this means we are counting the number of distinct colorings up to rotation and reflection. There are exactly six rigid reflections of the circle, namely the reflections across the three diameters marked on the circle and the reflections across the three diameters which bisect opposite sectors. Name these six reflections to be $\sigma_6, \ldots, \sigma_{11}$ and note that $D_6 = \{\sigma_0, \sigma_1, \ldots, \sigma_{11}\}$. The reader should check that the first three reflections have fixators of size 8 while the second*

three have fixators of size 16. Therefore, using Burnside's Lemma, the number of orbits equals

$$\frac{1}{|K|} \sum_{i=0}^{11} |X_{\sigma_i}| = \frac{1}{12}(64 + 2 + 4 + 8 + 4 + 2 + 8 + 8 + 8 + 16 + 16 + 16) = 13.$$

Perhaps surprising, this additional restriction on distinctness barely reduces their number. The reader should look at the list of distinct colorings compiled in Example 2 and decide which two of the 14 colorings are now being equated.

EXERCISES

1 List the 14 distinct colorings in Example 4.6.2

2 Compute by hand the additional fixators in Example 4.6.3.

3 Which two of the 14 colorings in Example 4.6.2 are now being equated in Example 4.6.3?

4 Count the number of unique dominos first using a simple combinatorial argument, then by using group action.

4.3 POLYA'S FORMULA

We first define Polya's Formula and then we will apply it to counting distinct colorings. It arises from Burnside's Lemma for counting orbits. Polya's Formula is derived from the cycle types in the group acting on the colorings. Each cycle type found in the group corresponds to a monomial in Polya's Formula.

Example 4.7 *If a permutation in S_{12} has cycle type $(*)(*)(**)(**)(**)(****)$ the corresponding monomial in Polya's Formula is $x_1^2 x_2^3 x_4$.*

In general, if a permutation has cycle type which includes n cycles each of length m, then x_m^n is included in the monomial in Polya's Formula corresponding to that cycle type. The coefficient of this monomial will be the number of permutations in the group acting on the colorings of that particular cycle type.

Definition 4.4 *Let G be a subgroup of S_n acting on a set of colorings. Polya's Formula is a polynomial in unknowns $x_1, x_2, \ldots, x_n$ of the form*

$$\frac{1}{|G|} \sum_{\sigma \in G} x_1^{e_1} x_2^{e_2} \cdots x_n^{e_n},$$

where in the formula above, the σ has e_i cycles of length i in its cycle type (for $i = 1, 2, \ldots, n$).

Example 4.8 *Consider the example from the previous section for coloring the circle with six sectors.*

1. If our group is the six rotations, then the cycle decompositions, cycle types and Polya monomials are listed below:

Permutation	Cyclic Decomposition	Cycle Type	Polya Monomial
0°	(1)(2)(3)(4)(5)(6)	(*)(*)(*)(*)(*)(*)	x_1^6
60°	(1 2 3 4 5 6)	(* * * * **)	x_6
120°	(1 3 5)(2 4 6)	(* * *)(* * *)	x_3^2
180°	(1 4)(2 5)(3 6)	(**)(**)(**)	x_2^3
240°	(1 5 3)(2 6 4)	(* * *)(* * *)	x_3^2
300°	(1 6 5 4 3 2)	(* * * * **)	x_6

Therefore, Polya's Formula is

$$P(x_1, x_2, x_3, x_6) = \frac{1}{6}(x_1^6 + 2x_6 + 2x_3^2 + x_2^3)$$

2. If our group is the six rotations and six reflections, then we add the following six rows to the table:

Permutation	Cyclic Decomposition	Cycle Type	Polya Monomial
μ_1	(1)(4)(2 6)(3 5)	(*)(*)(**)(**)	$x_1^2 x_2^2$
μ_2	(2)(5)(1 3)(4 6)	(*)(*)(**)(**)	$x_1^2 x_2^2$
μ_3	(3)(6)(2 4)(1 5)	(*)(*)(**)(**)	$x_1^2 x_2^2$
μ_4	(1 2)(3 6)(4 5)	(**)(**)(**)	x_2^3
μ_5	(2 3)(1 4)(5 6)	(**)(**)(**)	x_2^3
μ_6	(3 4)(2 5)(1 6)	(**)(**)(**)	x_2^3

Therefore, Polya's Formula is

$$P(x_1, x_2, x_3, x_6) = \frac{1}{12}(x_1^6 + 2x_6 + 2x_3^2 + 4x_2^3 + 3x_1^2 x_2^2).$$

As you probably noticed, Polya's Formula looks very similar to Burnside's Lemma for counting orbits. In fact, each monomial corresponds to a fixator. Polya's Formula has two primary uses: To count the number of distinct colorings and to produce the inventory of unique colorings. First, to count the number of distinct colorings simply evaluate all the x_i by the number of colors used. This makes sense since all the numbers in a cycle must be colored with the same color to remain fixed, thus there are as many ways to color the numbers in that cycle as there are colors.

Example 4.9 *Consider again Example 4.8.*

1. *If we are looking for the distinct colorings up to rotation, we have seen that Polya's Formula is $P(x_1, x_2, x_3, x_6) = \frac{1}{6}(x_1^6 + 2x_6 + 2x_3^2 + x_2^3)$. So the number of distinct colorings with two colors is $P(2, 2, 2, 2) = \frac{1}{6}(2^6 + 2 \cdot 2 + 2 \cdot 2^2 + 2^3) = 14$, which got us to our answer much quicker than Burnside's Lemma. In fact, now we can easily compute distinct colorings with three colors to be $P(3, 3, 3, 3) = \frac{1}{6}(3^6 + 2 \cdot 3 + 2 \cdot 3^2 + 3^3) = 130$.*

2. *If we are looking for the distinct colorings up to rotation and reflection, we have seen that Polya's Formula is $P(x_1, x_2, x_3, x_6) = \frac{1}{12}(x_1^6 + 2x_6 + 2x_3^2 + 4x_2^3 + 3x_1^2 x_2^2)$. So the number of distinct colorings with two colors is $P(2, 2, 2, 2) = \frac{1}{12}(2^6 + 2 \cdot 2 + 2 \cdot 2^2 + 4 \cdot 2^3 + 3 \cdot 2^2 \cdot 2^2) = 13$. Again, we can easily compute distinct colorings with three colors to be $P(3, 3, 3, 3) = \frac{1}{12}(3^6 + 2 \cdot 3 + 2 \cdot 3^2 + 4 \cdot 3^3 + 3 \cdot 3^2 \cdot 3^2) = 92$.*

Now let's address the second use of Polya's Formula, namely to produce the inventory of all distinct colorings. Polya's Formula will be used as a *generating function* to list all the possible colorings. How it works is as follows: Let $P(x_1, x_2, \ldots, x_n)$ be Polya's Formula – a multi-variate polynomial in the unknowns $x_1, x_2, \ldots, x_n$. Suppose our colors are $c_1, c_1, \ldots, c_m$. If we evaluate each x_k at $\sum_{i=1}^m c_i^k$, we will get a polynomial in $c_1, c_2, \ldots, c_m$ which will describe explicitly the full inventory of distinct colorings. Indeed, replacing x_k by $\sum_{i=1}^m c_i^k$ is saying we must color all the elements in a k-cycle the same color, having m colors to choose from. Having done so, the coefficient of $c_1^{e_1} c_2^{e_2} \cdots c_n^{e_n}$ in $P(\sum_{i=1}^m c_i, \sum_{i=1}^m c_i^2, \ldots, \sum_{i=1}^m c_i^n)$ corresponds to the number of ways to color using e_1 colors of c_1, e_2 colors of c_2, $\ldots$, e_m colors of c_m.

Example 4.10 *Let's return to Example 4.8.*

1. *For distinct colorings of the six sectors up to rotation, we derived Polya's Formula $P(x_1, x_2, x_3, x_6) = \frac{1}{6}(x_1^6 + 2x_6 + 2x_3^2 + x_2^3)$. Suppose first we are coloring using black and white. To get the full inventory of colorings we evaluate*

$$P(b+w, b^2+w^2, b^3+w^3, b^6+w^6) = \frac{1}{6}((b+w)^6 + 2(b^6+w^6) + 2(b^3+w^3)^2 + (b^2+w^2)^3)$$

$$= w^6 + bw^5 + 3b^2w^4 + 4b^3w^3 + 3b^4w^2 + b^5w + b^6.$$

What this is telling us is that there is one way to color them all white, one way to color using one black and five white, three ways to color using two black and four white, etc. Notice we still have to decide what the colorings are, but at least *we know how many we are looking for of each type. Notice also that if we set $b = 1$ and $w = 1$ in the resulting polynomial, we once again get the number of distinct colorings. Using three colors b, w, r we begin to see the true power of Polya's Formula:*

$$P(b + w + r, b^2 + w^2 + r^2, b^3 + w^3 + r^3, b^6 + w^6 + r^6)$$

$$= \frac{1}{6}((b + w + r)^6 + 2(b^6 + w^6 + r^6) + 2(b^3 + w^3 + r^3)^2 + (b^2 + w^2 + r^2)^3)$$

$$= w^6 + rw^5 + bw^5 + 3r^2w^4 + 5brw^4 + 3b^2w^4 + 4r^3w^3 + 10br^2w^3 + 10b^2rw^3 + 4b^3w^3$$

$$+ 3r^4w^2 + 10br^3w^2 + 16b^2r^2w^2 + 10b^3rw^2 + 3b^4w^2 + r^5w + 5br^4w + 10b^2r^3w$$

$$+ 10b^3r^2w + 5b^4rw + b^5w + r^6 + br^5 + 3b^2r^4 + 4b^3r^3 + 3b^4r^2 + b^5r + b^6.$$

2. *Let's repeat the process for colorings distinct up to rotation and reflection. We derived Polya's Formula in this case to be*

$$P(x_1, x_2, x_3, x_6) = \frac{1}{12}(x_1^6 + 2x_6 + 2x_3^2 + 4x_2^3 + 3x_1^2x_2^2).$$

Thus, for two colors we evaluate

$$P(b + w, b^2 + w^2, b^3 + w^3, b^6 + w^6)$$

$$= \frac{1}{12}((b + w)^6 + 2(b^6 + w^6) + 2(b^3 + w^3)^2 + 4(b^2 + w^2)^3 + 3(b + w)^2(b^2 + w^2)^2)$$

$$= w^6 + bw^5 + 3b^2w^4 + 3b^3w^3 + 3b^4w^2 + b^5w + b^6.$$

Comparing the inventory in the last example, focusing on the coefficients of b^3w^3 we see that it is here that the number of colorings was reduced by one. Let's try three colors:

$$P(b + w + r, b^2 + w^2 + r^2, b^3 + w^3 + r^3, b^6 + w^6 + r^6)$$

$$= \frac{1}{12}((b + w + r)^6 + 2(b^6 + w^6 + r^6) + 2(b^3 + w^3 + r^3)^2$$

$$+ 4(b^2 + w^2 + r^2)^3 + 3(b + w + r)^2(b^2 + w^2 + r^2)^2)$$

$$= w^6 + rw^5 + bw^5 + 3r^2w^4 + 3brw^4 + 3b^2w^4 + 3r^3w^3 + 6br^2w^3 + 6b^2rw^3 + 3b^3w^3$$

$$+ 3r^4w^2 + 6br^3w^2 + 11b^2r^2w^2 + 6b^3rw^2 + 3b^4w^2 + r^5w + 3br^4w + 6b^2r^3w$$

$$+ 6b^3r^2w + 3b^4rw + b^5w + r^6 + br^5 + 3b^2r^4 + 3b^3r^3 + 3b^4r^2 + b^5r + b^6.$$

Example 4.11 *This example deals with counting distinct graphs. If two vertices in the graph are connected by an edge, we will consider that edge colored black. If they are not connected by an edge, we will consider that (nonexistent) edge painted white.*

Let's count the number of graphs with five vertices. Before we can do this we need to count the number of edges it can have, but this is not difficult.

$$\text{The number of edges will be } \binom{5}{2} = 10.$$

Therefore, the number of graphs possible is $2^{10} = 1024$. In general,

$$\text{the number of graphs with } n \text{ vertices is } 2^{\binom{n}{2}}.$$

Now some of these graphs with five vertices are not distinct. For instance, the graph containing only the edge connecting vertices 1 and 2 is the same as the graph containing only the edge connecting vertices 3 and 4. What we really want to count is the number of distinct (or in this case we say **non-isomorphic**) graphs. What then do we mean by distinct in this case? What we mean is there is no permutation of the vertices which preserves all the edge connections, i.e. fixes the coloration. So it boils down to counting something we already know how to do.

Let's start with an easier set of graphs, namely one with four vertices. So our set X is the collection of all graphs with four vertices (there are $2^6 = 64$). Now the entire permutation group $G = S_4$ is acting on the set vertices of the graphs in X. We require a table of all the possible cycle types in S_4 and the number of each type (we did a similar thing in Section 3.4 using S_5).

Cycle Type	Number
$(*)(*)(*)(*)$	1
$(*)(*)(**)$	$\frac{4!}{2! \cdot 2} = 6$
$(*)(***)$	$\frac{4!}{3} = 8$
$(**)(**)$	$\frac{4!}{2! \cdot 2 \cdot 2} = 3$
$(****)$	$\frac{4!}{4} = 6$

Since X is really the collection of colorations of edges, we need to find the corresponding permutation of the edges in order to create Polya's Formula. We shall denote the collection of edges by $\overline{12}, \overline{13}, \overline{14}, \overline{23}, \overline{24}, \overline{34}$, where $\overline{mn}$ means the edge connecting the vertices numbered m and n. Let's add one more columns to our table using a generic *permutation of each vertex cycle type.*

Vertex Cycle Type	Number	Edge Cycle Type
$(1)(2)(3)(4)$	1	$(\overline{12})(\overline{13})(\overline{14})(\overline{23})(\overline{24})(\overline{34})$
$(1)(2)(3\ 4)$	6	$(\overline{12})(\overline{13}\ \overline{14})(\overline{23}\ \overline{24})(\overline{34})$
$(1)(2\ 3\ 4)$	8	$(\overline{12}\ \overline{13}\ \overline{14})(\overline{23}\ \overline{24}\ \overline{34})$
$(1\ 2)(3\ 4)$	3	$(\overline{12})(\overline{13}\ \overline{24})(\overline{14}\ \overline{23})(\overline{34})$
$(1\ 2\ 3\ 4)$	6	$(\overline{12}\ \overline{23}\ \overline{34}\ \overline{14})(\overline{13}\ \overline{24})$

Therefore, Polya's Formula is

$$\frac{1}{24}(x_1^6 + 9x_1^2 x_2^2 + 8x_3^2 + 6x_2 x_4),$$

and the number of distinct graphs with four vertices is

$$\frac{1}{24}(2^6 + 9 \cdot 2^2 \cdot 2^2 + 8 \cdot 2^2 + 6 \cdot 2 \cdot 2) = 11.$$

We can exhibit the inventory of distinct graphs with four vertices by evaluating

$$\frac{1}{24}((b+w)^6 + 9(b+w)^2(b^2+w^2)^2 + 8(b^3+w^3)^2 + 6(b^2+w^2)(b^4+w^4)),$$

but since we only care about existing edges we can replace w by 1 and evaluate

$$\frac{1}{24}((b+1)^6 + 9(b+1)^2(b^2+1^2)^2 + 8(b^3+1^3)^2 + 6(b^2+1^2)(b^4+1^4))$$

$$= 1 + b + 2b^2 + 3b^3 + 2b^4 + b^5 + b^6.$$

In Figure 4.2, we display the distinct graphs.

EXERCISES

1 Investigate the case of graphs with five vertices as we did with four in Example 4.11.

2 We wish to paint the roof of a house (see Figure 4.3). There are four sections of roof each of which can be painted in one of two colors: sienna and ochre. Our goal is to investigate the number of distinct colorings up to (two) rotations and (two) reflections.

   a. Use Burnside's Lemma to count the number of distinct colorings.

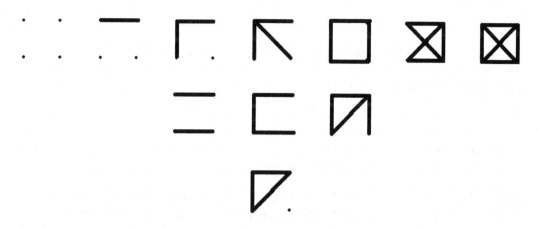

Figure 4.2 The distinct graphs with four vertices.

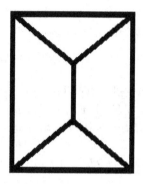

Figure 4.3 The roof of a house.

   b. Create the Polya Polynomial associated with this problem.

   c. Use part b. to count the number of distinct colorings.

   d. Use part b. to list the inventory of colorings.

   e. Exhibit an example of each of the distinct colorings.

4.4   SOME CONSEQUENCES OF GROUP ACTION

In this section, we prove some important results in group theory using group action.

    First, we re-prove Cayley's Theorem using group action. Notice below how easily the result falls out using group action.

Theorem 4.2 (Cayley's Theorem) *If $|G| = n < \infty$, then G is isomorphic to a subgroup of S_n.*

Proof 4.5 *Let G act on G by left multiplication. This induces a homomorphism $\pi : G \to Sym(G)$. The kernel of this map is trivial, since if for some $a \in G$ we have $\pi(a) = 1$, then $[\pi(a)](g) = g$ for all $g \in G$. In particular, $[\pi(a)](a) = a$ or $a^2 = a$,* *which implies $a = 1$. Thus by FTH,*

$$G \cong G/1 = G/\ker \pi \cong \pi(G) \leq Sym(G) \cong S_n.$$

Thus, G is isomorphic to a subgroup of S_n.

The next result generalizes the result that a subgroup of index 2 must be normal.

Theorem 4.3 (Ore's Theorem) *Let $H \leq G$ a group. If $[G : H] = p$ and p is the smallest prime dividing the order of G, then $H \lhd G$.*

Proof 4.6 *Let G act on G/H by left multiplication. This group action gives rise to a homomorphism $\pi : G \to Sym(G/H)$. The kernel of this map is contained in H, since if $\pi(g) = 1$, then in particular, $[\pi(g)](H) = H$ or $gH = H$ which implies $g \in H$. We show, in fact, that $\ker \pi = H$ (and so $H \lhd G$). Suppose to the contrary that $\ker \pi \neq H$. By FTH, $G/\ker \pi \cong \pi(G) \leq Sym(G/H) \cong S_p$ and so $[G : \ker \pi] = |\pi(G)|$ which divides $|Sym(G/H)| = |S_p| = p!$. We also know that*

$$[G : \ker \pi] = [G : H][H : \ker \pi] = p[H : \ker \pi],$$

and so p divides $[G : \ker \pi]$. Write $p! = [G : \ker \pi]m$ for some integer m, then multiply m on both sides of $[G : \ker \pi] = p[H : \ker \pi]$ to get $p! = [G : \ker \pi]m = p[H : \ker \pi]m$. Cancelling a p on both sides of the last equation gives that $[H : \ker \pi] \neq 1$ divides $(p - 1)!$. Therefore, there exists a prime $q < p$ which divides $[H : \ker \pi]$, and thus divides $|H|$, and thus divides $|G|$, which is a contradiction.

Remark 4.2 *We point out that in general if $H \leq G$ and $[G : H] = p$ a prime, then it need not be the case that $H \lhd G$. Consider the group $G = S_3$ and $H = \langle \mu_1 \rangle$ where $\mu = (2\ 3)$. Notice that $[G : H] = |G|/|H| = 6/2 = 3$ a prime, yet H is not normal in G. Indeed, take $\rho = (1\ 2\ 3) \in G$ and compute*

$$\rho^{-1}\mu\rho = (3\ 2\ 1)(2\ 3)(1\ 2\ 3) = (1\ 2) \notin H.$$

Our goal now is to derive using group action an important combinatorial group equation called the **class equation**.

Theorem 4.4 *Let G be any finite group. Then*

$$|G| = |Z(G)| + \sum_{i=1}^{m}[G : C_G(g_i)],$$

for some natural number m and some $g_i \in G$ with $[G : C_G(g_i)] > 1$. Note that $m = 0$ exactly when G is abelian.

Proof 4.7 *Let G act on itself by conjugation as in Example 4.2.2. Since orbits are equivalence classes on the set G and G is finite, we have*

$$G = \bigsqcup_{i=1}^{r} Gg_i,$$

for some $g_i \in G$ and some positive integer r. Now some of these orbits Gg_i may contain only a single element and we've seen that $|Gg_i| = 1$ iff $g_i \in Z(G)$. Therefore, all the singleton orbits of this action may be combined to obtain $Z(G)$. If G is abelian, then all the orbits are singletons and $G = Z(G)$. Otherwise, write the union as

$$G = Z(G) \sqcup \bigsqcup_{i=1}^{m} Gg_i.$$

for some $g_i \in G$ with $|Gg_i| > 1$ and some positive integer m. This yields the equation

$$|G| = |Z(G)| + \sum_{i=1}^{m} |Gg_i| = |Z(G)| + \sum_{i=1}^{m} [G : C_G(g_i)],$$

for some $g_i \in G$ with $[G : C_G(g_i)] > 1$ and some positive integer m.

Example 4.12 *We will illustrate the class equation with an example. Consider the dihedral group $D_4 = \{1, \rho_1, \rho_2, \rho_3, \mu_1, \mu_2, \mu_3, \mu_4\}$, where*

$$\rho_1 = (\,1\,2\,3\,4\,), \quad \rho_2 = (\,1\,3\,)(\,2\,4\,), \quad \rho_3 = (\,1\,4\,3\,2\,),$$

$$\mu_1 = (\,2\,4\,), \quad \mu_2 = (\,1\,2\,)(\,3\,4\,), \quad \mu_3 = (\,1\,3\,), \quad \mu_4 = (\,1\,4\,)(\,2\,3\,).$$

Let G act on G by conjugation. We first compute all the distinct orbits of this action (recall that orbits partition the set being acting upon into disjoint equivalence classes). The orbit of 1 is always $\{1\}$, since for all $g \in G$ we have $g1g^{-1} = 1$. We now compute the orbit of ρ_1:

$$\{1\rho_1 1^{-1}, \rho_1 \rho_1 \rho_1^{-1}, \rho_2 \rho_1 \rho_2^{-1}, \rho_3 \rho_1 \rho_3^{-1}, \mu_1 \rho_1 \mu_1^{-1}, \mu_2 \rho_1 \mu_2^{-1}, \mu_3 \rho_1 \mu_3^{-1}, \mu_4 \rho_1 \mu_4^{-1}\}$$

$$= \{\rho_1, \rho_3\}.$$

The orbit of ρ_2 is

$$\{1\rho_2 1^{-1}, \rho_1 \rho_2 \rho_1^{-1}, \rho_2 \rho_2 \rho_2^{-1}, \rho_3 \rho_2 \rho_3^{-1}, \mu_1 \rho_2 \mu_1^{-1}, \mu_2 \rho_2 \mu_2^{-1}, \mu_3 \rho_2 \mu_3^{-1}, \mu_4 \rho_2 \mu_4^{-1}\}$$

$$= \{\rho_2\}.$$

The orbit of μ_1 is

$$\{1\mu_1 1^{-1}, \rho_1 \mu_1 \rho_1^{-1}, \rho_2 \mu_1 \rho_2^{-1}, \rho_3 \mu_1 \rho_3^{-1}, \mu_1 \mu_1 \mu_1^{-1}, \mu_2 \mu_1 \mu_2^{-1}, \mu_3 \mu_1 \mu_3^{-1}, \mu_4 \mu_1 \mu_4^{-1}\}$$

$$= \{\mu_1, \mu_3\}.$$

Finally, the orbit of μ_2 is

$$\{1\mu_2 1^{-1}, \rho_1 \mu_2 \rho_1^{-1}, \rho_2 \mu_2 \rho_2^{-1}, \rho_3 \mu_2 \rho_3^{-1}, \mu_1 \mu_2 \mu_1^{-1}, \mu_2 \mu_2 \mu_2^{-1}, \mu_3 \mu_2 \mu_3^{-1}, \mu_4 \mu_2 \mu_4^{-1}\}$$

$$= \{\mu_2, \mu_4\}.$$

Hence, the partition of D_4 into disjoint orbits is

$$D_4 = \{1\} \sqcup \{\rho_2\} \sqcup \{\rho_1, \rho_3\} \sqcup \{\mu_1, \mu_3\} \sqcup \{\mu_2, \mu_4\}.$$

As in the proof of the Class Equation, the singleton orbits combine to form the center of the group and so we see that $Z(D_4) = \{1, \rho_2\}$.

We now give several consequences which can be proved using the class equation.

Corollary 4.2 *If G is a group with $|G| = p^n$ for some prime p and positive integer n, then p divides $Z(G)$, so in particular $Z(G) \neq 1$.*

Proof 4.8 *By the class equation, we have*

$$p^n = |Z(G)| + \sum_{i=1}^{m}[G : C_G(g_i)],$$

for some $g_i \in G$ with $[G : C_G(g_i)] > 1$ and some positive integer m. Note that if the sum were empty, then $G = Z(G)$ and the result follows trivially. So assume now that the sum has terms. Since for each i, $|G| = [G : C_G(g_i)]|C_G(g_i)|$ this implies that $[G : C_G(g_i)]$ divides $|G|$. Now $|G| = p^n$ and $[G : C_G(g_i)] > 1$, so it must be that $[G : C_G(g_i)] = p^k$ for some $1 \leq k \leq n$. Regardless of what k is, we can conclude that each $[G : C_G(g_i)]$ is a multiple of p and therefore $\sum_{i=1}^{m}[G : C_G(g_i)]$ is a multiple of p. So we can rewrite the class equation as $p^n = |Z(G)| + Lp$, for some integer L. Therefore, $|Z(G)| = p(p^{n-1} - L)$ ande we can conclude that p divides $|Z(G)|$.

Corollary 4.3 *If G is a group with $|G| = p^2$ for some prime p, then G is abelian.*

Proof 4.9 *By Corollary 4.2, we have p divides $|Z(G)|$ and since $|Z(G)|$ divides $|G|$ it must be that either $|Z(G)| = p$ or p^2. But $|Z(G)|$ cannot be p, for if it were, then we would have $|G/Z(G)| = |G|/|Z(G)| = p^2/p = p$. This would imply that $G/Z(G) \cong \mathbb{Z}_p$ a cyclic group which by Lemma 2.12 implies that G is abelian, contradicting the fact that $Z(G) \neq G$. Hence, $|Z(G)|$ must be p^2 and so G must be abelian.*

We point out that there is another way to obtain this last result without the use of the class equation (which we will see later). We now prove Cauchy's Lemma using the class equation.

Theorem 4.5 (Cauchy's Theorem) *If G is any finite group whose order is divisible by a prime p, then G has a element of order p.*

Proof 4.10 *By induction, we assume all groups of smaller size than G have the desired property of this theorem (note that the lemma holds vacuously for the trivial group). If G is abelian, then we are done by Lemma 2.14. So we may assume G is nonabelian. By the Class Equation, we know*

$$|G| = |Z(G)| + \sum_{i=1}^{m}[G : C_G(g_i)],$$

for some positive integer m and some $g_i \in G$ with $[G : C_G(g_i)] > 1$. Note that the sum is not empty, since G is assumed not to be abelian. Note also that for each g_i we have $|G| = [G : C_G(g_i)]|C_G(g_i)|$ and since $[G : C_G(g_i)] > 1$, this makes $C_G(g_i)$ a proper subgroup of G. If p were to divide the order of one of the $C_G(g_i)$, then by induction such a $C_G(g_i)$, and therefore G, would have an element of order p thus proving the *theorem. So we may assume that for all the g_i it is not the case that p divides the order of $C_G(g_i)$. Since p is prime and $|G| = [G : C_G(g_i)]|C_G(g_i)|$, it must then be true then that for all the g_i it is the case that p divides $[G : C_G(g_i)]$. Therefore, p divides the sum*

$$\sum_{i=1}^{m} [G : C_G(g_i)].$$

Since p divides this sum and also divides $|G|$, by the class equation it must be that p divides $|Z(G)|$. Since G is assumed to be nonabelian, this makes $Z(G)$ a proper subgroup of G and so by induction has (and therefore G has) an element of order p.

There are other ways to obtain Cauchy's Lemma one of which we will present later which uses the Sylow Theorems. With the use of the lemma we prove next, we can give an alternative proof of Ore's Theorem.

Lemma 4.2 *Let G act on a set X and ϕ be the homomorphism associated with that group action. Let $Y \subseteq X$ be a set of representatives for the distinct orbits of the group action and $H = \bigcap_{x \in Y} G_x$, then*

$$ker(\phi) = \bigcap_{x \in X} G_x = \bigcap_{g \in G} H^g \leq H.$$

Proof 4.11 *The results are all contained in the following equivalent statements: $k \in ker(\phi)$ iff $\phi(k) = 1_X$ iff $\phi(k)(x) = x$ all $x \in X$ iff $k \cdot x = x$ all $x \in X$ iff $k \in G_x$ all $x \in X$ iff $k \in G_{g \cdot y}$ all $g \in G$ and all $y \in Y$ iff $k \in gG_y g^{-1}$ all $g \in G$ and all $y \in Y$ iff $k \in \bigcap_{g \in G} H^g$.*

Certainly, $\bigcap_{g \in G} H^g \leq H$, since $H = H^1$.

Corollary 4.4 *If $H \leq G$ a group with $[G : H] = n$, then there exists $K \triangleleft G$ with $K \leq H$ such that $[G : K]$ divides $n!$.*

Proof 4.12 *Let G act on G/H by left multiplication and ϕ be the associated homomorphism for the group action. Set $K = ker(\phi)$ and we shall show that K is the desired normal subgroup we seek. First, by Lemma 4.2,*

$$K = \bigcap_{gH \in G/H} G_{gH} \subseteq G_H = H.$$

Second, as in the proof of Cayley's Theorem, $\phi(G)$ is isomorphic to a subgroup of S_n which implies $|\phi(G)|$ divides $|S_n| = n!$. By the Fundamental Theorem of Homomorphisms, $G/K \simeq \phi(G)$ so that $[G : K] = |G/K| = |\phi(G)|$ which divides $n!$.

Theorem 4.6 (Ore's Theorem (revisited)) *If G is a finite group and $H \leq G$ with $[G : H] = p$ where p is the smallest prime dividing $|G|$, then $H \triangleleft G$.*

Proof 4.13 *By Corollary 4.4, there is a $K \leq H$ with $K \triangleleft G$ and $[G : K]$ divides $p!$. Set $m = [G : K]$ and note that m divides $|G|$. Notice that for any prime q dividing m we have q divides $|G|$ and q divides $p!$. But then, since p is the smallest prime dividing*

$|G|$, it must be that $q = p$. Therefore, m must be a power of p, say $m = p^k$. Now since p^k divides $p!$ it must be that $k = 1$ and so $m = p$. Thus, $[G : K] = p = [G : H]$ and

$$|K| = \frac{|G|}{[G : K]} = \frac{|G|}{[G : H]} = |H|.$$

Since $K \leq H$ and $|K| = |H|$ we have $H = K \lhd G$.

EXERCISES

1 As in Example 4.12, determine the class equation for each of the following groups:

   a. Q_8

   b. D_5

2 Let G be a group with $|G| = p^n$ for p prime and n a positive integer.

   a. Show that if $H \leq G$ and $|H| = p^{n-1}$, then $H \lhd G$.

   b. Show that if $H \leq G$ and $H \neq G$, then there exists $g \in G$ H such that $gHg^{-1} = H$.

4.5 SYLOW THEORY

For the purposes of this section we shall always assume that G is a finite group. Sylow Theory is an extremely important component in the study and classification of finite groups. There are basically three main results in this area called (naturally) the First, Second and Third Sylow Theorem. Before we can get to these results we need to define terms and prove a few preliminary results.

Definition 4.5 *For a given prime p and H a subgroup of a group G,*

   *1. H is a p-**subgroup** if $|H| = p^k$ for some natural number k.*

   *2. H is a p-**Sylow** subgroup if $|H| = p^n$ for some natural number n and p^{n+1} does not divide $|G|$. In other words, H is a maximal p-subgroup.*

   *3. G is a p-**group** if $|G| = p^n$ for some natural number n.*

We will now derive a more general class equation for use in the Sylow Theorems and as a way to give an alternate and quite elegant proof of Cauchy's Theorem.

Theorem 4.7 (General Class Equation) *If G acts on a finite set X, then*

$$|X| = |Y| + \sum_{i=1}^{m} [G : G_{x_i}],$$

where $[G : G_{x_i}] > 1$ for some $x_i \in X$ and positive integer m and $Y = \{x \in X : |Gx| = 1\}$ (again, there may be no terms in the sum). If, in addition, G is a p-group, then $|X| \equiv |Y| \pmod{p}$.

Proof 4.14 *The derivation of this more general class equation is identical to the work done to derive the original class equation, so there is no need to reproduce that argument here. We will verify the statement that $|X| \equiv |Y| \pmod{p}$ in the case that G is a p-group. Considering the general class equation, if there were no terms in the sum, then the statement follows trivially. Otherwise, notice that for each i, $|G| = [G : G_{x_i}]|G_{x_i}|$. Set $|G| = p^n$ so that $[G : G_{x_i}]$ divides p^n which implies that p must divide $[G : G_{x_i}] > 1$ for all i. Hence, p divides the sum, $\sum_{i=1}^{m}[G : G_{x_i}]$, which by the general class equation implies that p divides the difference $|X| - |Y|$, i.e. $|X| \equiv |Y| \pmod{p}$.*

Theorem 4.8 (Cauchy's Theorem) *If G is a group whose order is divisible by a prime p, then G must have a element of order p.*

Proof 4.15 *Let X be the set consisting of p-tuples $(g_1, g_2, \ldots, g_p)$ of elements of G with the property that $g_1 g_2 \cdots g_p = 1$. Note that X is non-empty, since $(1, 1, \ldots, 1) \in X$. In fact, the size of X must be $|G|^{p-1}$ since we can choose the first $p-1$ components of an p-tuples freely in G with the last being determined as $(g_1 g_2 \cdots g_{p-1})^{-1}$. Note also that the size of X implies in particular that p divides $|X|$. Take the p-cycle $\sigma = (1\ 2\ \cdots\ p) \in S_p$ and let the subgroup $\langle \sigma \rangle$ act on X by permuting the subscripts of any p-tuple. In other words,*

$$\sigma^k(g_1, g_2, \ldots, g_p) = (g_{\sigma^k(1)}, g_{\sigma^k(2)}, \ldots, g_{\sigma^k(p)}).$$

We leave it to the reader to check (by induction on k) that $\sigma^k(g_1, g_2, \ldots, g_p) \in X$ and that we indeed have a group action on a set. By Theorem 4.7, since $\langle \sigma \rangle$ is a p-group we have that $|X| \equiv |Y|\pmod{p}$. Thus, since p divides $|X|$, it follows that p also divides $|Y|$. Let us look more closely at the set Y. Recall, that Y consists of elements of X which have orbits each consisting of one element. Note that Y is non-empty, since $(1, 1, \ldots, 1) \in Y$. Take one such $(g_1, g_2, \ldots, g_p) \in Y$. Then

$$\langle \sigma \rangle(g_1, g_2, \ldots, g_p) = \{(g_1, g_2, \ldots, g_p)\},$$

so in particular, $\sigma(g_1, g_2, \ldots, g_p) = (g_1, g_2, \ldots, g_p)$ or $(g_{\sigma(1)}, g_{\sigma(2)}, \ldots, g_{\sigma(p)}) = (g_1, g_2, \ldots, g_p)$ or $(g_2, g_3, \ldots, g_p, g_1) = (g_1, g_2, \ldots, g_p)$. But then $g_1 = g_2 = \cdots = g_p$. Recall the property of elements of X, that $g_1 g_2 \cdots g_p = 1$ and this yields if we set $g = g_1 = g_2 = \cdots = g_p$ that $g^p = 1$. Finally, note that since p divides $|Y|$ we know we can find a $g \neq 1$ with this property $g^p = 1$ and so we have proved the theorem.

We are now ready to prove the three Sylow Theorems.

Theorem 4.9 (First Sylow Theorem) *Let G be a finite subgroup such that p^k divides $|G|$ with $k \geq 0$. Then G has a subgroup of order p^k. In particular, G has a p-Sylow subgroup.*

Proof 4.16 *Let G be a group of order $p^n r$ where p is a prime, n is a natural number and p does not divide r. The proof is by induction on $|G|$. When $|G| = 1$, the result holds trivially. By induction, assume that all groups of order less than the order of*

G satisfy the theorem. If G had a proper subgroup H of order $p^n s$, then by induction H (and therefore G) would have a p-subgroup of order p^k for $0 \leq k \leq n$. In the case that no such subgroup H exists, we consider the class equation,

$$|G| = |Z(G)| + \sum_{i=1}^{m}[G : C_G(g_i)],$$

for some positive integer m and some $g_i \in G$ with $[G : C_G(g_i)] > 1$. Furthermore, the subgroups $C_G(g_i)$ must have order $p^{n_i}r_i$ with $n_i < n$ and p does not divide r_i. Notice that for all i

$$p^n r = |G| = [G : C_G(g_i)]|C_G(g_i)| = [G : C_G(g_i)]p^{n_i}r_i.$$

Therefore, it must be the case that p divides $[G : C_G(g_i)]$ for all i and so by the class equation p divides $|Z(G)|$. By Cauchy's Theorem, $Z(G)$ has an element, say z, of order p. Set $H = \langle z \rangle \leq Z(G)$, and thus $H \triangleleft G$. Then G/H is a group of order $p^{n-1}r$ less than the order of G and so has a p-subgroup $K/H \leq G/H$ of order p^k for $0 \leq k \leq n-1$, where $H \leq K \leq G$ (by the Correspondence Theorem). Since $|H| = p$ and $|H|$ divides $|K|$, this imples that K is a p-subgroup of G of order p^{k+1} where $1 \leq k+1 \leq n$. The only p-subgroup we have missed is the trivial one, which certainly exists.

Theorem 4.10 (Second Sylow Theorem) *Any p-subgroup of a group G can be conjugated into any p-Sylow of G, i.e. if H is a p-subgroup of G and P is a p-Sylow of G, then there exist $g \in G$ such that $gHg^{-1} \leq P$.*

Proof 4.17 *Let H act on $X = G/P$ by left multiplication. As in the general class equation, let Y consist of those elements of X with one element orbits. First note that Y is non-empty. Indeed, Since $|G/P|$ is not divisible by p (otherwise P is not a p-Sylow) and $|X| \equiv |Y| \pmod{p}$ this implies that p does not divide $|Y|$ and so in particular, $|Y| \neq 0$. So take any $gP \in Y$ and note that for all $h \in H$ we have $hgP = gP$ or equivalently $ghg^{-1} \in P$, and so $gHg^{-1} \leq P$.*

Remark 4.3 *We make several remarks about conjugation.*

1. *Note that if P is a p-Sylow subgroup of G and $g \in G$, then gPg^{-1} is also a p-Sylow subgroup of G. To see this simply consider the bijection $f : P \to gPg^{-1}$ by $f(h) = ghg^{-1}$. More generally, if $H \leq G$, then for any $g \in G$ we have $gHg^{-1} \leq G$ and $|gHg^{-1}| = |H|$.*

2. *We remind the reader of some exponential notation introduced for group elements. For $g, h \in G$ a group, we write $h^g = ghg^{-1}$. Similarly, for $H \leq G$, we write $H^g = gHg^{-1}$.*

Corollary 4.5 *For any finite group G*

1. *Every p-subgroup of G is contained in a p-Sylow of G.*

2. *Any two p-Sylows of G are conjugate in G. Thus, for any p-Sylow subgroup of G the set $\{gPg^{-1} : g \in G\}$ consists of all the p-Sylow subgroups of G.*

3. *If the number of p-Sylow subgroups in G is equal to 1, then this unique subgroup is normal in G.*

4. *The number of p-Sylow subgroups in G equals $[G : N_G(P)]$, where P is any p-Sylow subgroup of G.*

Proof 4.18 *The first part is easy, for if H is a p-subgroup of G, then by the Second Sylow Theorem, there is a $g \in G$ such that $H^g \leq P$ where P is a p-Sylow of G. But then $H \leq P^{g^{-1}}$ a conjugate of a p-Sylow which is again a p-Sylow (see Remark 4.3).*

The second part is also easy. Take two p-Sylows of G, say P and Q. By the Second Sylow Theorem, there is a $g \in G$ such that $P^g \leq Q$. But as we pointed out P^g is a p-Sylow as well which means $|P^g| = |Q|$ and so $P^g = Q$.

To prove the third part, let P be the unique p-Sylow subgroup. For any $g \in G$, since $g^{-1}Pg$ is also a p-Sylow subgroup and there is only one, it must be that $g^{-1}Pg = P$ which proves that $P \lhd G$.

To prove the fourth part, by Corollary 4.1 and Corollary 4.5.ii, the number of p-Sylows in G equals $[G : N_G(P)]$ for any p-Sylow P of G.

Definition 4.6 *For any prime p, let n_p denote the number of p-Sylow subgroups of a given group G.*

Theorem 4.11 (Third Sylow Theorem) *If $|G| = p^n m$ with p not dividing m and $n > 0$, then n_p divides m and is congruent to 1 modulo p.*

Proof 4.19 *By the First Sylow Theorem, there is a p-Sylow $P \leq G$. Let X be the set of all p-Sylows of G and have P act on X by conjugation. This action makes sense, since every conjugate of a p-Sylow of G is again a p-Sylow of G. As usual, Y will denote the elements of X which have single element orbits. By Theorem 4.7, $|X| \equiv |Y| \pmod{p}$. We will show that $Y = \{P\}$ so that $|Y| = 1$ and thus prove half of the result.*

First note that $P \in Y$ since certainly $g^{-1}Pg = P$ for any $g \in P$. Now suppose that $Q \in Y$. Then $g^{-1}Qg = Q$ all $g \in P$ and so $P \leq N_G(Q)$. Since $P \leq N_G(Q)$ and $|P| = p^n$, then p^n divides $|N_G(Q)|$ so that $|N_G(Q)| = p^n k$ with p not dividing k. This means that P is a p-Sylow of $N_G(Q)$ as well. Using the same argument, Q is a p-Sylow of $N_G(Q)$. By Corollary 4.5, there is a $g \in N_G(Q)$ such that $g^{-1}Qg = P$. But, since $g \in N_G(Q)$, we have $P = g^{-1}Qg = Q$.

Finally, since $P \leq N_G(P)$, as in the above argument, $|N_G(P)| = p^n k$ with p not dividing k. So then

$$p^n m = |G| = [G : N_G(P)]|N_G(P)| = [G : N_G(P)]p^n k,$$

which implies that $m = [G : N_G(P)]k$ and so $[G : N_G(P)]$, which by Corollary 4.5.4 is the number of p-Sylows of G, must divide m.

EXERCISES

1 Verify the statement in Theorem 4.8 that

    a. The action defined in the proof is indeed a group action.

    b. $\sigma^k(g_1, g_2, \ldots, g_p) \in X$ for $k \geq 0$.

2 Prove the following exponential laws for $g, h, k \in G$ a group:

    a. $h^g k^g = (hk)^g$.

    b. $(h^k)^g = h^{(kg)}$.

    c. $h^{(kg)} = (h^g)^{(gk)}$.

3 Prove that any subgroup of $Z(G)$ must be normal.

4.6   CLASSIFYING FINITE GROUPS WITH SYLOW THEORY

We start this section with a general result that goes very far in classifying finite groups. We first need to prove a couple of number theoretic results.

Lemma 4.3 *Given p is a prime number and m is any positive integer such that m divides $p - 1$.*

    1. The congruence equation $x^m \equiv 1 \pmod{p}$ has m distinct solutions modulo p.

    2. If a is a solution to $x^m \equiv 1 \pmod{p}$ and k is the smallest positive integer such that $x^k \equiv 1 \pmod{p}$, then k divides m.

Proof 4.20 *Without diving prematurely into ring theory (all of this is proved later in the text), one can show given a polynomial $f(x)$ of degree m with integer coefficients that the congruence equation $f(x) \equiv 0 \pmod{p}$ has at most m distinct solutions in $\mathbb{Z}_p$. In fact, there is a criterion for when such an $f(x)$ has exactly m solutions – it occurs iff $f(x)$ divides $x^p - x$. Therefore, if m divides $p - 1$, then for some integer k we have*

$$x^{p-1} - 1 = x^{mk} - 1 = (x^m - 1)(x^{m(k-1)} + \cdots + x^m + 1),$$

and so $x^m - 1$ divides $x^{p-1} - 1$ which in turn divides $x^p - x$ and so the first statement is proved.

    To prove the second statement, notice that if a is a solution to $x^m \equiv 1 \pmod{p}$, then $a^m = 1$ in the multiplicative group $\mathbb{Z}_p^*$. To say k is the smallest positive integer such that $x^k \equiv 1 \pmod{p}$ is to say that $o(a) = k$ in $\mathbb{Z}_p^*$. Therefore, by Lemma 2.2.1, we know that k must necessarily divide m.

Theorem 4.12 *Let G be a group with $|G| = pq$ for primes p and q.*

1. *If $p = q$, then G is an abelian group and there are only two such groups (up to isomorphism), namely $\mathbb{Z}_{p^2}$ and $\mathbb{Z}_p \oplus \mathbb{Z}_p$.*

2. *If $p < q$ and p does not divide $q - 1$, then G is cyclic, i.e. $G \cong \mathbb{Z}_{pq}$.*

3. *If $p < q$ and p divides $q - 1$, then either $G \cong \mathbb{Z}_{pq}$ is cyclic or G is a unique (up to isomorphism) non-abelian group generated by x and y and satifying the relations $x^p = 1$, $y^q = 1$ and $yx = xy^k$ where $k \not\equiv 1 \ (mod \ q)$ and $k^p \equiv 1 \ (mod \ q)$ with $k \in \mathbb{N}$.*

Proof 4.21 *If $p = q$, then we have already proved in Corollary 4.3 that G is abelian and the fact that there are only two such groups, namely $\mathbb{Z}_{p^2}$ and $\mathbb{Z}_p \oplus \mathbb{Z}_p$, follows from the classification of finite abelian groups. Assume from now on that $p < q$. By the First Sylow Theorem, there exists a p-Sylow, say P, and a q-Sylow, say Q. We point out that $P \cap Q = 1$, for the order of the subgroup $P \cap Q$ of both P and Q would have to divide both p and q and so necessarily must have order 1. Furthermore, since P and Q are each of prime order they must be cyclic. Set $P = \langle x \rangle$ and $Q = \langle y \rangle$ for some $x, y \in G$. Since p is the smallest prime dividing $|G|$ and $[G : Q] = p$, by Ore's Theorem, $Q \lhd G$. By the Third Sylow Theorem, n_p divides q and is congruent to 1 modulo p. This implies that $n_p = 1$ or q. These two cases correspond to statements 2 and 3 of this theorem. For notice if $n_p = q$, then $q \equiv 1 \ (mod \ p)$ and so p divides $q - 1$.*

Therefore, to prove statement 2, we must assume that $n_p = 1$ and so $P \lhd G$, by Corollary 4.5. Notice that $x^{-1}y^{-1}xy \in P \cap Q = 1$ and so $x^{-1}y^{-1}xy = 1$ or equivalently $xy = yx$. This in turn implies that $o(xy) = pq$. Indeed,

$$(xy)^{pq} = (x^p)^q (y^q)^p = 1^q 1^p = 1,$$

and $xy \neq 1$ otherwise $P = Q$ and finally we show that $o(xy) \neq p$ or q (similarly). If it were that $o(xy) = p$, then

$$1 = (xy)^p = x^p y^p = y^p,$$

and so q divides p which implies $q = p$, a contradiction. Since $o(xy) = pq = |G|$, this implies that $G = \langle xy \rangle$ is cyclic, thus proving statement 2.

To prove statement 3, first note that since $P \cap Q = 1$ this implies that

$$|PQ| = |P||Q|/|P \cap Q| = pq = |G|,$$

and so $G = PQ$ and G is generated by x and y. We observe further that $Q \lhd G$ implies $x^{-1}yx \in Q$ and so $x^{-1}yx = y^k$ for some natural number $k < q$ and so $yx = xy^k$. From this last equation, one can show by induction, we get that $yx^i = x^i y^{k^i}$ for $i = 1, 2, \ldots$. In particular, $yx^p = x^p y^{k^p}$ which simplifies to $y = y^{k^p}$ or $y^{k^p - 1} = 1$. But then q divides $k^p - 1$ and so $k^p \equiv 1 \ (mod \ q)$. We may assume that $k \not\equiv 1 \ (mod \ q)$, for otherwise $yx = xy$ and G is the cyclic group already discussed above. Since p is a prime and referring to Lemma 4.3 it must be that $k = 1$ and so the p distinct solutions to $x^p \equiv 1 \pmod{q}$ must be $1, k, k^2, \ldots, k^{p-1}$. For each $s \in \{1, k, k^2, \ldots, k^{p-1}\}$, one can show that x and y^s generate G with $(y^s)^q = 1$ and $y^s x = x(y^s)^k$. Furthermore, one can check that the homomorphism which takes x to x and y to y^s is, in fact, an isomorphism. Therefore, these seemingly p different groups are, in fact, the same up to isomorphism.

Example 4.13 *Here, we present instances of classifying groups of a given order pq using the theorem just proved.*

1. *Groups of order $4 = 2 \cdot 2$ must be abelian and there are only two such groups, namely $\mathbb{Z}_4$ and $\mathbb{Z}_2 \oplus \mathbb{Z}_2$.*

2. *For groups of order $6 = 2 \cdot 3$ we have 2 divides $3 - 1$ and so there is a cyclic group of order 6, namely $\mathbb{Z}_6$ and exactly one non-abelian group of order 6 which must be the familiar S_3.*

3. *Groups of order 9 as we saw for order 4 are $\mathbb{Z}_9$ and $\mathbb{Z}_3 \oplus \mathbb{Z}_3$.*

4. *For groups of order $10 = 2 \cdot 5$ we have 2 divides $5 - 1$ and so there is the cyclic group $\mathbb{Z}_{10}$ and the non-abelian dihedral group D_5. In fact, this situation can be generalized as follows: If $|G| = 2p$ for an odd prime p, then either G is isomorphic to $\mathbb{Z}_{2p}$ or G is isomorphic to the dihedral group D_p, since 2 divides $p - 1$.*

5. *For groups of order $15 = 3 \cdot 5$ we have 3 does not divide $5 - 1$ and so there is only the cyclic group $\mathbb{Z}_{15}$.*

6. *For groups of order $21 = 3 \cdot 7$ we have 3 divides $7 - 1$ and so there is the cyclic group $\mathbb{Z}_{21}$ and a non-abelian group generated by x and y and satisfying the relations $x^7 = 1$, $y^3 = 1$ and $xy = y^k x$ where $k \not\equiv 1 \pmod{7}$ and $k^3 \equiv 1 \pmod{7}$. Here is a concrete way to describe this non-abelian group:*

$$G = \left\{ \begin{bmatrix} 1 & a \\ 0 & b \end{bmatrix} : a \in \mathbb{Z}_7, \text{ but } b = 1, 2, 4 \right\},$$

where the operation is matrix multiplication modulo 7.

Example 4.14 *Here is another sampling of finite group classification arguments using the Sylow Theorems. In these examples our goal is to prove that G cannot be simple.*

1. *Any group of prime power $n > 1$ order cannot be simple (if $n = 1$, then $G \cong \mathbb{Z}_p$ which is simple). There are two ways to see this. One way to see this is to point out that the center of the group is a non-trivial normal subgroup; now it could equal the whole group, but then G would be abelian and thus non-simple. Another way is to use the Sylow Theorems. By the First Sylow Theorem, if $|G| = p^n$ for a prime p and positive integer $n > 1$, then G has a p-subgroup, say H, of order p^{n-1}. But then $[G : H] = p$ and p is the smallest (in fact, only) prime dividing $|G|$. Therefore, by Ore's Theorem, $H \lhd G$ and so G is not simple.*

2. *If $|G| = pq$ where $p < q$ are primes and p does not divides $q - 1$, then G is not simple. Indeed, as in the proof of Theorem 4.12, the q-Sylow $Q \triangleleft G$ makes G not simple.*

3. *More generally, any group of order $p^n m$ where p is prime, $n > 0$, p does not divide m and $1 < m < p$ is not simple (orders such as $6 = 3 \cdot 2$, $10 = 5 \cdot 2$, $15 = 5 \cdot 3$, $18 = 3^2 \cdot 2$, $20 = 5 \cdot 4$). By the Sylow theorems we have that $n_p \equiv 1 \pmod{p}$ and divides m. This means $n_p = 1$ or $p+1$, etc. But since $m < p$ it is necessarily the case that $n_p = 1$ (there is not enough room in G for even two p-Sylows). This yields a unique normal p-Sylow in G, thus making G not simple.*

4. *Any group of order $45 = 3^2 \cdot 5$ is not simple, since by the Third Sylow Theorem, $n_3 \equiv 1 \pmod{3}$ and must divide 5 and so $n_3 = 1$. Therefore, by Corollary 4.5.iii, the unique 3-Sylow is normal in G, thus making G not simple.*

5. *The case that any group of order $56 = 2^3 \cdot 7$ is not simple has a different sort of counting argument. By the Sylow Theorems, $n_7 \equiv 1 \pmod{7}$ and divides 8. In the case that $n_7 = 1$, then G will have a unique 7-Sylow which by Corollary 4.5.iii must be normal in G, hence making G not simple. In the case that $n_7 = 8$, let $P_1, \ldots, P_8$ represent the eight 7-Sylows of order 7. Note that for any $1 \le i < j \le 8$ the subgroup $P_i \cap P_j$ must be trivial being a proper subgroup of P_i which has order 7. Furthermore, the six non-trivial elements of each 7-Sylow have order 7 giving us $|P_1 \cup \cdots \cup P_8| = (8)(6) = 48$ elements of G of order 7. Since G has at least one 2-Sylow of order $2^3 = 8$, this means G has an additional 7 non-trivial elements of order a power of 2. But now we have counted all the elements of G. This means there is a single 2-Sylow in G which again must be normal in G, thus making G not simple.*

6. *Any group of order $30 = 2 \cdot 3 \cdot 5$ is not simple. Using the Sylow Theorems one gets that $n_5 = 1$ or 6 and $n_3 = 1$ or 10. As in the previous example the 5-Sylows intersect each other trivially and have four non-trivial elements of order 5. If you had six 5-Sylows, then you would have 24 elements of order 5. The 3-Sylows intersect each other trivially and have two non-trivial elements of order 3. If you had ten 3-Sylows, then you would have 20 elements of order 3. Since $|G| = 30$, you cannot have both $n_5 = 6$ and $n_3 = 10$ for otherwise you would be counting 45 distinct elements in G, an obvious contradiction. Therefore, either $n_5 = 1$ or $n_3 = 1$ which in turn produces either a 5-Sylow or 3-Sylow normal in G, thus making G not simple.*

7. *No group of order 96 is simple. Let G be a group of order $96 = 3 \cdot 2^5$. By the Third Sylow Theorem, the number of 2-Sylows is congruent to 1 modulo 2 and divides 3 which implies there are either one or three 2-Sylows. If there is one 2-Sylow, then it is normal in G and so G is not simple. Should there be three 2-Sylows we show once again that G is not simple. Let P and Q be two of the 2-Sylows. Consider the product PQ. We know that*

$$96 \ge |PQ| = \frac{|P||Q|}{|P \cap Q|} = \frac{32 \cdot 32}{|P \cap Q|}.$$

Therefore, $|P \cap Q| \geq \frac{32}{3}$. Since P and Q are distinct and $|P \cap Q|$ divides $|P| = 32$ it must be that $|P \cap Q| = 16$. Set N equal to the normalizer of $P \cap Q$ in G. Since the index of $P \cap Q$ in both P and Q is two, $P \cap Q$ is normal in both P and Q and so P and Q are contained in N. Therefore, $|N| > 32$ (since both P and Q are distinct subgroups of N each of order 32). Since $|N|$ divides $|G|$ we are forced to conclude that $|N| = |G|$. But this makes $P \cap Q$ normal in G and so G is not simple.

8. *This final argument does not use the Third Sylow Theorem. Indeed, should one try to apply the Third Sylow Theorem, it would not lead to the desired conclusion. We show any group of order $224 = 2^5 \cdot 7$ is not simple. Let P be a 2-Sylow of a group G of order 224. Let G act on G/P by left multiplication. This induces a homomorphism $\pi : G \to Sym(G/P)$. Since, $|G/P| = 224/32 = 7$ we have that $Sym(G/P) \cong S_7$. We have seen already that for this particular action the normal subgroup $\ker \pi \leq P$ and so $\ker \pi$ is a proper subgroup of G. We show now that $\ker \pi$ is non-trivial which makes G not simple. If it were the case that $\ker \pi = 1$, then G would embed in $Sym(G/P)$ and so $|G| = 224$ would divide $|Sym(G/P)| = |S_7| = 7!$, but this is clearly not the case. We leave it as an exercise to show that if $|G| = p^n m$ with $p \nmid m$ and $p \nmid (m - 1)!$, then G is not simple.*

EXERCISES

1 Prove that any group of order 100 has a normal subgroup of order 25.

2 Describe the 2-Sylows of D_{10}.

3 Describe all the Sylow subgroups in S_4.

4 Classify the groups of order 33.

5 Show there is no simple group of order 148.

6 Show there is no simple group of order 48.

7 Show there is no simple group of order 36.

8 Show there is no simple group of order 225.

9 Prove that the only simple groups of order < 60 are of prime order.

10 In the proof of Theorem 4.12, verify that

   a. x and y^s generate G with $(y^s)^q = 1$ and $y^s x = x(y^s)^k$.

   b. the homomorphism which takes x to x and y to y^s is, in fact, an isomorphism.

11 Prove that if $|G| = p^n m$ with $p \nmid m$ and $p \nmid (m - 1)!$, then G is not simple.

12 Check that, equipped with the techniques and results of this section, one can show order-by-order that there are no non-abelian groups of order < 60.

4.7   FINITE ABELIAN GROUPS

In the section give the complete proof of a fact we illuded to earlier – we classify all finite abelian groups. We could easily extend this result to a larger class of groups, namely finitely generated abelian groups, but it would add too much to the complication of the proof.

Lemma 4.4 *Let G an finite abelian group.*

1. *For each prime p dividing the order of G there is a unique p-Sylow subgroup of G.*

2. *If $|G| = p^n m$ and $p \nmid m$, then the unique p-Sylow subgroup of G is*

$$P = \{g \in G \ : \ g^{p^n} = 1\}.$$

3. *G is a direct sum of all its Sylow subgroups.*

Proof 4.22 *Since G is abelian every subgroup is normal and so for each p-Sylow $P \triangleleft G$. Let Q be another p-Sylow. By the Second Sylow Theorem there exists $g \in G$ such that $Q = g^{-1} P g$, but $g^{-1} P g = P$ and so $Q = P$.*

Set $H = \{g \in G \ : \ g^{p^n} = 1\}$ which is a subgroup of G, since G is abelian. Now H is a p-subgroup of G, for suppose some prime q divides $|H|$. By Cauchy's Theorem, there exists an element $h \in H$ of order q. Since $h^{p^n} = 1$, this implies q divides p^n and so $q = p$. Hence, the order of H is a power of p. Let P be the unique p-Sylow of G. By the Second Sylow Theorem, there exists $g \in G$ such that $g^{-1} H g \subseteq P$ and so $H \subseteq g P g^{-1} = P$. For the reverse inclusion, since $|P| = p^n$ and any element of a group raised to the order of the group equals the identity, by the very definition of H we have $P \subseteq H$ and so $P = H$.

We prove the third statement by induction on the order of G. If $|G| = 1$ the result is trivial. If $|G| > 1$, suppose p divides $|G|$ and write $|G| = p^n m$ with $p \nmid m$. Set $P = \{g \in G \ : \ g^{p^n} = 1\}$, the unique p-Sylow in G and $H = \{g \in G \ : \ g^m = 1\} \leq G$.

Claim 4.1 *$G = P \oplus H$*

First, if $g \in P \cap H$, then $o(g)$ divides both p^n and m which are relatively prime and so $o(g) = 1$ which implies $g = 1$. Second, take any $g \in G$. Since $|G| = p^n m$ with $p \nmid m$ it must be that $o(g) = p^l k$ where $0 \leq l \leq n$ with $p \nmid k$ and $k \mid m$. Since $\gcd(p^l, k) = 1$ there exist integers r and s such that $p^l r + ks = 1$. Notice that $g = g^{p^l r + ks} = (g^k)^s (g^{p^l})^r \in PH$, since $(g^k)^{p^n} = (g^{p^l k})^{p^{n-l}} = 1^{p^{n-l}} = 1$ and $(g^{p^l})^m = (g^{p^l k})^{m/k} = 1^{m/k} = 1$.

By induction H, which has order $|G|/|P| < |G|$, is a direct sum of all its Sylow subgroups which in turn makes the result true for G.

The next result is a technical lemma needed for the classification of finite abelian groups, and otherwise has little value.

Lemma 4.5 *Let G be a finite abelian p-group and $g \in G$ be of maximal order in G. Set $H = \langle g \rangle$. For any $aH \in G/H$ there exists $b \in G$ such that $\langle b \rangle \cap H = 1$ and $bH = aH$.*

Proof 4.23 *Set $s = o(aH)$ in G/H so that $a^s H = (aH)^s = H$. Therefore, $a^s = g^k$ for some positive integer k. Write $k = p^n m$ with $p^n \leq o(g)$ and $p \nmid m$. Note that*

$$o(g^m) = \frac{o(g)}{\gcd(o(g), m)} = \frac{o(g)}{1} = o(g).$$

Since $(aH)^{o(a)} = a^{o(a)} H = H$ this implies that s divides $o(a)$. Therefore,

$$o(a^s) = \frac{o(a)}{\gcd(o(a), s)} = \frac{o(a)}{s}.$$

On the other hand,

$$o(a^s) = o(g^k) = o(g^{p^n m}) = o((g^m)^{p^n}) = \frac{o(g^m)}{\gcd(o(g^m), p^n)} = \frac{o(g)}{\gcd(o(g), p^n)} = \frac{o(g)}{p^n}.$$

Therefore, $o(a)p^n = o(g)s$. By assumption, $o(a) \leq o(g)$ and both are powers of the prime p (since G is a p-group), thus s is also a power of p – in fact, s divides p^n. Hence, we can write $p^n = st$ for some integer t. Set $h = g^{mt} \in H$. Note that

$$h^s = g^{mst} = g^{p^n m} = g^k = a^s.$$

Set $b = ah^{-1}$. We show that this is the b we are looking for as stated in the lemma. First, $aH = bH$, since $b = ah^{-1} \in aH$. Second, suppose that $c \in \langle b \rangle \cap H$. Then $c = b^e = a^e h^{-e} \in H$ for some integer e. Therefore, $a^e \in H$ and so $(aH)^e = a^e H = H$. But then s divides e and we can write $e = sf$ for some integer f. Hence,

$$c = b^e = b^{sf} = (b^s)^f = (a^s h^{-s})^f = (a^s a^{-s})^f = 1^f = 1.$$

Thus, the lemma is proved.

Lemma 4.6 *If G is a finite abelian p-group, then $G = \langle g \rangle \oplus K$ where $\langle g \rangle$ is a maximal cyclic group in G, i.e. there is no $a \in G$ with $\langle g \rangle \subsetneqq \langle a \rangle \subseteq G$.*

Proof 4.24 *The proof is by induction on the order of G. If $|G| = 1$ the result holds trivially. If $|G| > 1$, then $|G| = p^n$ with $n \geq 1$. Should G be cyclic, then G itself is maximal cyclic and $K = 1$. We can therefore assume G is not cyclic with $|G| = p^n$ and $n > 1$. Since G is finite, by well ordering, we can always find a $g \in G$ with $\langle g \rangle$ cyclic and of maximal size. Set $H = \langle g \rangle$. By Cauchy's Lemma we know $|H| \geq p$ and so $|G/H| < |G|$. Since G is finite, by a finite number of applications of induction we can write $G/H = \overline{H}_1 \oplus \cdots \oplus \overline{H}_n$ where each $\overline{H}_i$ is maximal cyclic in G/H. By Lemma 4.5, we can choose $a_1, \ldots, a_n \in G$ such that each $\overline{H}_i = \langle a_i H \rangle$ and $\langle a_i \rangle \cap H = 1$. Set $K = \langle a_1, \ldots, a_n \rangle$. The following claim will complete the proof:*

Claim 4.2 $G = H \oplus K$

First, if $x \in H \cap K$, then $x = a_1^{m_1} \cdots a_n^{m_n} \in H$. Therefore, $H = xH = (a_1 H)^{m_1} \cdots (a_n H)^{m_n}$. Since we have unique representation in $\overline{H}_1 \oplus \cdots \oplus \overline{H}_n$ and $H = (a_1 H)^0 \cdots (a_n H)^0$, it must be that $m_1 = \cdots = m_n = 0$ and so $x = a_1^0 \cdots a_n^0 = 1$. Second, take an $x \in G$. Then $xH \in G/H$ and so $xH = (a_1 H)^{m_1} \cdots (a_n H)^{m_n} = (a_1^{m_1} \cdots a_n^{m_n})H$. Thus, $x \in (a_1^{m_1} \cdots a_n^{m_n})H = H(a_1^{m_1} \cdots a_n^{m_n}) \in HK$.

Corollary 4.6 *If G is a finite abelian p-group, then G is a direct sum of cyclic subgroups of prime power order.*

Proof 4.25 *This follows immediately by a finite number of applications of Lemma 4.6.*

We are now ready to classify finite abelian groups. We will, in fact, give two classifications in the sense that we will give two different ways of listing the non-isomorphic abelian groups of a given finite order.

Theorem 4.13 (Classification of Finite Abelian Groups I) *If G is a non-trivial finite abelian group, then G is isomorphic to a direct sum of non-trivial cyclic subgroups of prime power order. Furthermore, this direct sum representation is unique up to isomorphism and order.*

Proof 4.26 *By Lemma 4.4, G is a direct sum of its Sylow subgroups and then by Corollary 4.6 each of these Sylow subgroups can be written as a direct sum of cyclic groups of prime power order. This proves the existence part of the theorem.*

We now show the uniqueness of this representation. Without loss of generality, we can assume G is a p-group, since by Lemma 4.4, G is a direct sum of its Sylow subgroups. We prove existence in this simpler case by induction on the order of G. If $|G| = 2$, then G is cyclic and G itself is the unique representation of itself as a product of cyclic groups of prime power order. Now assume $|G| > 2$ and suppose $G = H_1 \oplus \cdots \oplus H_n$ and $G = K_1 \oplus \cdots \oplus K_m$ where the H_i and K_i are cyclic subgroups of G of prime power order. Suppose for each i we have $H_i \cong \mathbb{Z}_{p^{a_i}}$ and $K_i \cong \mathbb{Z}_{p^{b_i}}$ and by reordering we may assume $a_1 \geq a_2 \geq \cdots \geq a_n$ and $b_1 \geq b_2 \geq \cdots \geq b_m$. Consider the map $\phi : G \to G$ by $\phi(g) = g^p$ which is a homomorphism, since G is abelian. Let's focus on the image of this map.

1. *The kernel of this map consists of elements of order p and is non-trivial by Cauchy's Lemma. By FTH, $G/\ker\phi \cong \phi(G)$ and so $|\phi(G)| = |G|/|\ker\phi| < |G|$.*

2. *$\phi(G) = H_1^p \oplus \cdots \oplus H_n^p$ where each $H_i \cong \mathbb{Z}_{p^{a_i-1}}$. We should point out that the sum remains direct, since*

$$H_i^p \cap (H_1^p + \cdots + H_{i-1}^p + H_{i+1}^p + \cdots + H_n^p)$$
$$\subseteq H_i \cap (H_1 + \cdots H_{i-1} + H_{i+1} + \cdots + H_n) = 1.$$

For each $H_i \cong \mathbb{Z}_p$ note that $H_i^p = 1$. If k equals the number of H_i isomorphic to $\mathbb{Z}_p$, then $\phi(G) = H_1^p \oplus \cdots \oplus H_{n-k}^p$. Of course a similar thing happens to the representation of $G = K_1 \oplus \cdots \oplus K_m$ yielding $\phi(G) = K_1^p \oplus \cdots \oplus K_{m-l}^p$ where l equals the number of K_i isomorphic to $\mathbb{Z}_p$.

Hence, we may invoke induction on $\phi(G)$ to get $a_i - 1 = b_i - 1$ for $i = 1, 2, \ldots, n-k$ and $n - k = m - l$. Thus, it remains to show that $k = l$. To see this, note that

$$|H_1||H_2| \cdots |H_n| = |H_1 \oplus H_2 \oplus \cdots \oplus H_n| = |G|$$

$$= |K_1 \oplus K_2 \oplus \cdots \oplus K_m| = |K_1||K_2| \cdots |K_m|.$$

By cancelling terms we now know are equal we get $|H_{n-k+1}| \cdots |H_n| = |K_{m-l+1}| \cdots |K_m|$ and since the remaining subgroups are all isomorphic to $\mathbb{Z}_p$, we have $p^k = p^l$ and so $k = l$.

Example 4.15 *Let G be an abelian group of order $200 = 2^3 \cdot 5^2$. Therefore, by Lemma 4.4, G has a unique 2-Sylow P of order 8 and a unique 5-Sylow Q of order 25 with $G = P \oplus Q$. Now by Theorem 4.13, P is isomorphic to either $\mathbb{Z}_8$, $\mathbb{Z}_4 \oplus \mathbb{Z}_2$, or $\mathbb{Z}_2 \oplus \mathbb{Z}_2 \oplus \mathbb{Z}_2$. Similarly, Q is isomorphic to either $\mathbb{Z}_{25}$ or $\mathbb{Z}_5 \oplus \mathbb{Z}_5$. Therefore, the complete list of non-isomorphic abelian groups of order 200 is*

$$\mathbb{Z}_8 \oplus \mathbb{Z}_{25}, \quad \mathbb{Z}_8 \oplus \mathbb{Z}_5 \oplus \mathbb{Z}_5, \quad \mathbb{Z}_4 \oplus \mathbb{Z}_2 \oplus \mathbb{Z}_{25}, \quad \mathbb{Z}_4 \oplus \mathbb{Z}_2 \oplus \mathbb{Z}_5 \oplus \mathbb{Z}_5$$

$$\mathbb{Z}_2 \oplus \mathbb{Z}_2 \oplus \mathbb{Z}_2 \oplus \mathbb{Z}_{25}, \quad \mathbb{Z}_2 \oplus \mathbb{Z}_2 \oplus \mathbb{Z}_2 \oplus \mathbb{Z}_5 \oplus \mathbb{Z}_5.$$

Theorem 4.14 (Classification of Finite Abelian GroupsII) *If G is a non-trivial finite abelian group, then G is isomorphic to a direct sum of non-trivial cyclic subgroups, $G = H_1 \oplus H_2 \oplus \cdots \oplus H_n$, with $|H_i|$ dividing $|H_{i-1}|$ for $i = 2, 3, \ldots, n$.*

Proof 4.27 *The proof is by induction on the number of distinct prime divisors of G. For the base case, G is then a p-group and, by Corollary 4.6, can be written as a direct sum of cyclic groups of prime power order. Now simply rearrange the cyclic subgroups into decreasing order. The subgroups successively divide, since they all have order a power of the same prime.*

Now assume more than one prime divides the order of the group. Suppose p is a prime dividing $|G|$ and P is the p-Sylow subgroup of G. As in the proof of Lemma 4.4, we can write $G = P \oplus H$. By induction, we can represent $H = H_1 \oplus \cdots \oplus H_n$ where each H_i is cyclic and $|H_i|$ divides $|H_{i-1}|$ for $i = 2, 3, \ldots, n$. By Corollary 4.6, we may write $P = K_1 \oplus \cdots \oplus K_m$ where each K_i is a cyclic group of order a power of p. Suppose for each i we have $H_i \cong \mathbb{Z}_{n_i}$ and $K_i \cong \mathbb{Z}_{p^{a_i}}$ and by reordering the K_i we have $a_1 \geq a_2 \geq \cdots \geq a_m$. Note that for each $i < \min(m, n)$, since $\gcd(n_i, p^{a_i}) = 1$, Exercise 6 in Section 2.5, we know $H_i + K_i = H_i \oplus K_i$, remains cyclic (of order $n_i p^{a_i}$) and $|H_i \oplus K_i|$ still divides $|H_{i+1} \oplus K_{i+1}|$. Therefore, pair off in successive order as many H_i and K_i as possible to get the desired representation.

Example 4.16 *The proof just presented gives the algorithm for producing the desired representation. Let's return to abelian groups G of order 200. Recall that G has a 2-Sylow P of order 8 and a 5-Sylow Q of order 25. As before P is isomorphic to either $\mathbb{Z}_8$, $\mathbb{Z}_4 \oplus \mathbb{Z}_2$, or $\mathbb{Z}_2 \oplus \mathbb{Z}_2 \oplus \mathbb{Z}_2$ and Q is isomorphic to either $\mathbb{Z}_{25}$ or $\mathbb{Z}_5 \oplus \mathbb{Z}_5$. Then the complete list of non-isomorphic abelian groups of order 200 is*

$$\mathbb{Z}_8 \oplus \mathbb{Z}_{25} \cong \mathbb{Z}_{200}$$

$$\mathbb{Z}_8 \oplus (\mathbb{Z}_5 \oplus \mathbb{Z}_5) = (\mathbb{Z}_8 \oplus \mathbb{Z}_5) \oplus \mathbb{Z}_5 \cong \mathbb{Z}_{40} \oplus \mathbb{Z}_5$$

$$(\mathbb{Z}_4 \oplus \mathbb{Z}_2) \oplus \mathbb{Z}_{25} = (\mathbb{Z}_4 \oplus \mathbb{Z}_{25}) \oplus \mathbb{Z}_2 \cong \mathbb{Z}_{100} \oplus \mathbb{Z}_2$$

$$(\mathbb{Z}_4 \oplus \mathbb{Z}_2) \oplus (\mathbb{Z}_5 \oplus \mathbb{Z}_5) = (\mathbb{Z}_4 \oplus \mathbb{Z}_5) \oplus (\mathbb{Z}_2 \oplus \mathbb{Z}_5) \cong \mathbb{Z}_{20} \oplus \mathbb{Z}_{10}$$

$$(\mathbb{Z}_2 \oplus \mathbb{Z}_2 \oplus \mathbb{Z}_2) \oplus \mathbb{Z}_{25} = (\mathbb{Z}_2 \oplus \mathbb{Z}_{25}) \oplus \mathbb{Z}_2 \oplus \mathbb{Z}_2 \cong \mathbb{Z}_{50} \oplus \mathbb{Z}_2 \oplus \mathbb{Z}_2$$

$$(\mathbb{Z}_2 \oplus \mathbb{Z}_2 \oplus \mathbb{Z}_2) \oplus (\mathbb{Z}_5 \oplus \mathbb{Z}_5) = (\mathbb{Z}_2 \oplus \mathbb{Z}_5) \oplus (\mathbb{Z}_2 \oplus \mathbb{Z}_5) \oplus \mathbb{Z}_2 \cong \mathbb{Z}_{10} \oplus \mathbb{Z}_{10} \oplus \mathbb{Z}_2$$

Corollary 4.7 *If G is a finite abelian group and $d > 0$ divides the order of G, then G has a subgroup of order d.*

EXERCISES

1 Classify the abelian groups of the given order first using Theorem 4.13 and then using Theorem 4.14

   a. Abelian groups of order 35.

   b. Abelian groups of order 20.

   c. Abelian groups of order 36.

   d. Abelian groups of order 72.

   e. Abelian groups of order 216.

   f. Abelian groups of order 30.

2 Prove that if G is an abelian group of order square-free (i.e. a product of distinct primes), then G is cyclic.

3 Prove Corollary 4.7.

CHAPTER 5

Group Presentation and Representations

I N THIS CHAPTER, we introduce two topics in group theory which can each fill an entire textbook. Therefore, one can view this chapter an giving exposure to this two topics with a light overview. In Section 5.1, we introduce the notion of a *free group*, which is a group generated by a set of elements, as a prelude to Section 5.2, *group presentations*, in which the group generators have additional rules for how they relate to each other. The free groups are *free* of these additional rules. Finally, in Section 5.3, we introduce *group representations*, in which we represent group elements by matrices.

5.1 FREE GROUPS

One can think of free groups as groups defined with the minimal amount of structure, i.e. the group axioms themselves. We start with the definition of free groups.

Definition 5.1 *A group F is* **free** *if there is a subset X of F satisfying the following property: For any group G and function $f : X \to G$ there is a unique homomorphism extending f to F, i.e. there is a homomorphism $\phi : F \to G$ such that $\phi(x) = f(x)$ for all $x \in X$.*

In this case, X is called a **basis** *for F and we say that F is* **free on** X.

To get a feel for this definition, consider the setting of a vector space V. Recall the result that if X is a collection of basis vectors for V and we assign each $x \in X$ to a vector in another vector space W, say $f(x)$, then there is a unique vector space homomorphism (typically called a *linear transformation*) which sends each $x \in X$ to the same place that f did. Therefore, in this setting every vector space would be free. Such is not the case for groups as we shall see.

Example 5.1 *The additive group $\mathbb{Z}$ is a free group on the singleton set $X = \{1\}$ or $X = \{-1\}$. Indeed, if we have a function $f : \{1\} \to G$, then we can define a homomorphism $\phi : \mathbb{Z} \to G$ by $\phi(n) = f(1)^n$.*

DOI: 10.1201/9781003335283-5

The following result says there is a unique (up to isomorphism) free group on a set of a fixed cardinality.

Theorem 5.1 *If F is a free group on X and F' is a free group on X' and $|X| = |X'|$, then $F \cong F'$.*

Proof 5.1 *Since $|X| = |X'|$, there is a bijection $g : X \to X'$. Consider the inclusion maps $f : X \to F$ and $f' : X' \to F'$. Set $h = (f')^{-1} \circ g \circ f$ a map from X to F' and $h' = f^{-1} \circ g^{-1} \circ f'$ a map from X' to F. Since F and F' are free on X and X' (respectively), there are homomorphisms ϕ and ϕ' extending h to F and h' to F' (respectively). Notice that $\phi' \circ \phi = 1_F$, since the restriction*

$$(\phi' \circ \phi) \upharpoonright X = f^{-1} \circ g^{-1} \circ f' \circ (f')^{-1} \circ g \circ f = 1_X,$$

and since 1_F is another homomorphism extending 1_X and F is free on X, by uniqueness, $\phi' \circ \phi = 1_F$. Similarly, $\phi \circ \phi' = 1_{F'}$ and so ϕ is an isomorphism from F to F' and thus $F \cong F'$.

Theorem 5.2 *The following are true:*

1. *If F is free on X, then X generates F, i.e. $F = \langle X \rangle$.*

2. *F is a free group on X iff $F = \langle X \rangle$ and for any group G and function $f : X \to G$ there is a homomorphism extending f to F.*

3. *For every non-empty set X there is a group F such that F is free on X.*

Proof 5.2 *For the first statement, consider the inclusion map $f : X \to \langle X \rangle$. Since F is free on X we can extend this map to a homomorphism $\phi : F \to \langle X \rangle$. Since $\langle X \rangle \leq F$ we can define an inclusion monomorphism $\psi : \langle X \rangle \to F$ and it follows that $(\psi \circ \phi) \upharpoonright X = 1_X$ and so $\psi \circ \phi$ extends the map $1_X : X \to F$. Since 1_F also extends 1_X, by uniqueness it follows that $\psi \circ \phi = 1_F$. Therefore, the inclusion map ψ is also surjective and thus $\langle X \rangle = F$.*

To prove the second statement, we already have one direction by what we just proved. Therefore, assume that $F = \langle X \rangle$ and for any group G and function $f : X \to G$ there is a homomorphism $\phi : F \to G$ extending f to F. It suffices to show that this homomorphism is uniquely defined. To see this, notice that an element of F has the form $x_1^{e_1} \cdots x_n^{e_n}$ where the x_i's are in X (perhaps repeats) and $n, e_1, \ldots, e_n$ are positive integers. Since ϕ is a homomorphism we know that

$$\phi(x_1^{e_1} \cdots x_n^{e_n}) = \phi(x_1)^{e_1} \cdots \phi(x_n)^{e_n} = f(x_1)^{e_1} \cdots f(x_n)^{e_n}.$$

Hence, the homomorphism ϕ is completely determined by f and therefore must be a unique extension.

To prove the third statement, for our set X we define the corresponding set of new symbols $X' = \{x' \mid x \in X\}$ clearly of the same cardinality as X (X' will represent the corresponding inverses of the elements of X). Set $Y = X \cup X' \cup \{e\}$, where e is a *symbol not appearing in X or X' (and will denote the identity element). Now define the following sets:*

$$S = \{y_1 y_2 \cdots \mid each\ y_i \in Y\},$$

$$W = \{y_1 y_2 \cdots \mid each\ y_i \in Y\ and\ \exists n \in \mathbb{Z}^+,\ y_i = e\ for\ i \geq n\}.$$

*Let $\mathcal{R}$ be the elements in $\mathcal{W}$ for which no x and its corresponding x' ever appear adjacent to each other, and e never precedes an element of $X \cup X'$. The set S represents infinite strings or **words** in Y, $\mathcal{W}$ represents finite words in Y and $\mathcal{R}$ is called **reduced words** in Y. A typical element in $\mathcal{R}$ can be represented as $y_1^{e_1} \cdots y_k^{e_k}$, where adjacent y_i and y_{i+1} are both distinct in $X \cup X'$ and not of the form x and x' (for some $x \in X$) and represents an abbreviation for*

$$\underbrace{y_1 \cdots y_1}_{e_1\ times} \cdots \underbrace{y_k \cdots y_k}_{e_k\ times} eee \cdots .$$

Set $1 = ee \cdots$. One can easily check that $F = \mathcal{R}$ is a group via the operation concatenation and simplification to a reduced word via the following identities: for any $x \in X$ we have $xx' = e$ and $xe = x$ (and so x' is indeed the inverse of x and e is the identity element in F). It remains to show that F is free on X. To do this we employ the result just proved in part 2. First, F is certainly generated by $\langle X \rangle$ by its very definition. Second, for any map $f : X \to G$ define the homomorphism $\phi : F \to G$ by

$$\phi(y_1^{e_1} \cdots y_k^{e_k}) = \phi(y_1)^{e_1} \cdots \phi(y_k)^{e_k},$$

where $\phi(x') = \phi(x)^{-1}$ for any $x \in X$.

Remark 5.1 *Several remarks are in order.*

1. *Having proved Theorem 5.2.3, we can see where the terminology free comes from in the sense that the set X freely generates F without putting any conditions on X, such as $xy = yx$, for instance.*

2. *Without loss of generality, by Theorem 5.1, we can always assume if necessary that our free group is the one we just constructed in Theorem 5.2.3.*

3. *Notice that the construction of the free group F in the proof of Theorem 5.2.3 dictates that the identity element of F is not in X.*

Definition 5.2 *If F is a free group on a set X, then the **rank** of F is equal to the cardinality of X.*

Remark 5.2 *We make some further remarks.*

1. *If F is a free group of rank at least 2, then F is not abelian. Suppose, to the contrary, it were abelian and take $x, y \in X$. Since F is abelian we have $xy = yx$ or $x^{-1}y^{-1}xy = 1$, i.e. $x^{-1}y^{-1}xy$ reduces to 1. However, $x^{-1}y^{-1}xy$ is clearly a reduced word, since $x \neq y^{-1}$ – a contradiction (using a similar argument one can show the stronger result that such a group has a trivial center). Note this*

also proves that the free groups which are abelian are exactly the free groups generated by a single element which are infinite cyclic and therefore isomorphic to $\mathbb{Z}$.

2. *No element of a free group can have finite order except of course 1. Suppose, to the contrary, that some reduced word $x_1 \cdots x_n \in \langle X \rangle = F$ had finite order. Then for some positive integer k we would have $(x_1 \cdots x_n)^k = 1$, i.e. in particular the word*

$$\underbrace{(x_1 \cdots x_n) \cdots (x_1 \cdots x_n)}_{k \ times} \quad reduces \ to \ 1.$$

But the only way that could occur is if $x_n = x_1^{-1}$, $x_{n-1}^{-1} = x_2$, ... etc. If n is even, then $x_{n/2}^{-1} = x_{(n/2)+1}$ and $x_1 \cdots x_n$ reduces to 1 contrary to our assumption. If n is odd, then $x_{(n+1)/2}$ is its own inverse which is impossible according to how F was constructed in Theorem 5.2.3. Note that this result in turn shows the generators of a free group necessarily have infinite order, and thus every free group is infinite.

3. *Not every group is free. As we have seen in a remark above, any abelian non-cyclic group cannot be free (in fact, any group with a non-trivial center cannot be free).*

4. *A subgroup of a free group does not necessarily have the same rank as the entire group. In fact, as we shall see in the following example, the group may have finite rank while the subgroup has infinite rank. Consider the free group of rank 2 generated by $X = \{x, y\}$ and define the subgroup $H = \langle Y \rangle$ where $Y = \{x^i y x^{-i} \mid i = 1, 2, \ldots\}$. One can show that H is free on Y where Y is clearly an infinite set.*

5. *A free group can only have a single rank, for suppose that F were free on both X and Y with $|X| > |Y|$, then choose $x \in X - Y$. Since F is generated by Y, we can express $x = y_1^{e_1} \cdots y_n^{e_n}$ (reduced) for some $y_1, \ldots, y_n \in Y$ and $e_1, \ldots, e_n$ integers. But then $x^{-1} y_1^{e_1} \cdots y_n^{e_n} = 1$ or reduces to 1, which is not possible.*

EXERCISES

1 Verify that ϕ defined in Example 5.1 is indeed a homomorphism.

2 Verify in the proof of Theorem 5.2.3 that $F = \mathcal{R}$ is a group via the operation concatenation and simplification to a reduced word via the following identities: for any $x \in X$ we have $xx' = e$ and $xe = x$.

3 Referring to Remark 5.2.4, verify that H is free on Y.

4 Let F be a free group on $X = \{x_1, x_2\}$. Set $y_1 = x_1^2$, $y_2 = x_2^2$, $y_3 = x_1 x_2$ and $Y = \{y_1, y_2, y_3\}$. Prove that $G = \langle Y \rangle$ is free on Y.

5.2 GROUP PRESENTATIONS

In this section, we take a free group a step further by putting restrictions on the generators thus making the them no longer independent or free. These restrictions will be called **relations** and the generators together these relations between the generators will constitute the presentation of a group. In a sense, we exhibit the *essential* or *pure* description of the group. We now give the formal definition.

Definition 5.3 *Let F be free group on a set X and R a subset of F. Then $\langle X \mid R \rangle$ is called a* **group presentation**.

Example 5.2 *Here are some group presentations which we shall look at in detail during this section:*

1. *Let $X = \{x\}$ and $R = \{x^n\}$, for some positive integer n. Then certainly the free group on X is isomorphic to $\mathbb{Z}$ (see proof of Theorem 5.2.3). We shall show that the group presentation $\langle x \mid x^n \rangle$ represents a cyclic group of order n.*

2. *Let $X = \{x, y\}$ and $R = \{x^2, y^2, x^{-1}y^{-1}xy\}$. We shall show that the group presentation $\langle x, y \mid x^2, y^2, x^{-1}y^{-1}xy \rangle$ represents the Klein-4 group.*

3. *Let $X = \{x, y\}$ and $R = \{x^{-1}y^{-1}xy\}$. This group presentation represents a* **free abelian group**.

4. *Let $X = \{x, y\}$ and $R = \{x^4, x^2y^{-2}, xyxy^{-1}\}$. We shall show that this is the generators and relations for the quaternions.*

5. *Let $X = \{x_1, x_2, \ldots, x_n\}$ and*

   $$R = \{x_i^2,\ (x_jx_{j+1})^3,\ (x_kx_l)^2 \mid i, k, l = 1, \ldots, n-1;\ j = 1, \ldots, n-2;\ l < k-1\}.$$

   This group presentation represents, S_n, the symmetric group on n elements.

6. *Let $X = \{x, y\}$ and $R = \{x^2, y^n, xyxy\}$. This group presentation represents the dihedral group D_n.*

7. *The Generalized Quaternions have group presentation $\langle x, y \mid yxyx^{-1}, xyxy^{-1} \rangle$.*

Now we give the formal definition of what it means for a group to be represented by a group presentation. First note that for any subset R of a group G there is a notion of a smallest normal subgroup N_R of G containing R. Indeed, this result holds and is left as an exercise for the reader.

Lemma 5.1 *Let R be a subset of a group G.*

1. *There exists a smallest normal subgroup N_R of G containing R in the sense that if $N \lhd G$ and $N \supseteq R$, then $N_R \subseteq N$.*

2. *The N_R in the first statement is the intersection of all normal subgroups of G containing R.*

3. The N_R in the first statement equals

$$\{r_1^{\pm g_1} \cdots r_n^{\pm g_n} \mid r_i \in R, \ g_i \in G, \ n \in \mathbb{Z}^+\}.$$

Hence, N_R is sometimes called the **normal subgroup of** G **generated by** R.

Definition 5.4 Let F be a free group on a set X and R a subset of F and let N_R be the normal subgroup of F generated by R. A group G **has presentation** $\langle X \mid R \rangle$ – or G **is defined by generators** X **and relations** R – or $\langle X \mid R \rangle$ **is a group presentation of** G – if $G \cong F/N_R$.

Notice that in the group F/N_R the relations are being equated with the identity, so that a group having such a presentation has additional constraints, namely that each element in the relations is now equal to 1. For this reason we define the phrase that a group G **satisfies the relations** R if there is an epimorphism $\phi : F \to G$ such that $N_R \leq ker(\phi)$. We now list some useful results that will help us link groups to their presentations.

Lemma 5.2 Consider a group presentation with generators X and relations R. Let F be the free group associated with X.

1. If G has presentation $\langle X \mid R \rangle$, then G satisfies the relations R.

2. The group $F/N_R = \langle xN_R \ : \ x \in X \rangle$

3. If G is a group with $f : X \to G$ a map satisfying the property that for all $r = x_1^{e_1} \cdots x_n^{e_n} \in R$ we have $f(x_1)^{e_1} \cdots f(x_n)^{e_n} = 1$, then there exists a unique homomorphism $\Phi : F/N_R \to G$ such that $\Phi(xN_R) = f(x)$ for all $x \in X$.

4. If a finite group G satisfies the relations R, then there is an epimorphism $\psi : F/N_R \to G$.

5. If a finite group G satisfies the relations R and $|G| \leq |F/N_R|$, then G has presentation $\langle X \mid R \rangle$.

Proof 5.3 For the first statement, we are given that there is an isomorphism from $\psi : F/N_R \to G$. Consider the quotient homomorphism $\nu : F \to F/N_R$. Then the composition $\psi \circ \nu : F \to G$ is an epimorphism whose kernel is precisely the normal subgroup N_R (exercise) and thus contains N_R.

The second statement is immediate, since $F = \langle X \rangle$. For the third statement, since F is free on X we can extend f to a unique homomorphism $\phi : F \to G$. By our assumptions on f we see that $\phi(r) = 1$ for all $r \in R$ and so $N_R \leq ker(\phi)$. Since $N_R \leq ker(\phi)$, we have a well-defined homomorphism $\chi : F/N_R \to F/ker(\phi)$ by $\chi(xN_R) = xker(\phi)$ (see Exercise 3 of Section 2.10). By the Fundamental Theorem of Homomorphisms, there is a monomorphism $\psi : F/ker(\phi) \to G$ by $\psi(gN_R) = \phi(g)$. Then $\Phi = \psi \circ \chi$ is the desired homomorphism uniquely determined since ϕ was uniquely determined.

For the fourth statement, since G satisfies the relations R, there is an epimorphism $\phi : F \to G$ such that $N_R \leq ker(\phi)$. Therefore, using the Third Isomorphism Theorem, we have

$$F/ker(\phi) \cong (F/N_R)/(ker(\phi)/N_R).$$

Call this isomorphism χ mapping $(F/N_R)/(ker(\phi)/N_R)$ to $F/ker(\phi)$. By FTH, there is an isomophism $\Phi : F/ker(\phi) \to G$. Consider the quotient epimorphism $\nu : F/N_R \to (F/N_R)/(ker(\phi)/N_R)$ by $\nu(xN_R) = (xN_R)(ker(\phi)/N_R)$. Then the composition $\Phi \circ \chi \circ \nu$ is the desired epimorphism.

$$F/N_R \xrightarrow{\nu} (F/N_R)/(ker(\phi)/N_R) \xrightarrow{\chi} F/ker(\phi) \xrightarrow{\Phi} G.$$

For the fifth statement, since $|G| \leq |F/N_R|$, the epimorphism in the fourth statement is now an isomorphism thus proving the result.

Example 5.3 *Lemma 5.2 provides us with the tools necessary to associate groups with their presentations.*

1. *Let $X = \{x, y\}$ and $R = \{x^2, y^2, x^{-1}y^{-1}xy\}$. For brevity, set $N = N_R$ and $G = F/N_R$. First note that since $x^{-1}y^{-1}xy \in N$, this implies $xyN = yxN$. Furthermore, the fact that $G = \langle xN, yN \rangle$ implies G has at most 4 elements. Indeed, $(xN)^2 = x^2N = N$, $(yN)^2 = y^2N = N$ and $(xN)(yN) = (xy)N = (yx)N = (yN)(xN)$ so that the only elements in G that can be distinct are N, xN, yN and $(xy)N$. Let $H = \mathbb{Z}_2 \times \mathbb{Z}_2$ be the Klein-4 group and define the map $f : X \to H$ by $f(x) = (1, 0)$ and $f(y) = (0, 1)$. Since F is free on X there is a homomorphism $\phi : F \to H$ extending f. Since f maps onto the generators of H, ϕ is, in fact, an epimorphism. Furthermore, it is easy to see by how f is defined, that $R \subseteq ker(\phi)$ and so the smallest normal subgroup containing R, $N_R \leq ker(\phi)$. Hence, H satisfies the relations R and $|H| = 4 \geq |G|$ which implies by Lemma 5.2.5 that $H \cong G$ and so the group presentation $\langle x, y \mid x^2, y^2, x^{-1}y^{-1}xy \rangle$ represents the Klein-4 group.*

2. *Let $X = \{x\}$ and $R = \{x^n\}$, for some positive integer n. Again, set $N = N_R$ and $G = F/N_R$. As in the previous example, we note that $G = \langle xN \rangle$ and $(xN)^n = N$ so that G has at most n elements. Let $H = \mathbb{Z}_n$ and define the map $f : X \to H$ by $f(x) = 1$ which extends to a homomorphism $\phi : F \to H$. Again, ϕ is an epimorphism with $R \subseteq ker(\phi)$ which proves H satisfies the relations R and since $|H| = n \geq |G|$ we have $H \cong G$.*

3. *Let $X = \{x, y\}$ and $R = \{x^4, x^2y^{-2}, xyxy^{-1}\}$. Again, set $N = N_R$ and $G = F/N_R$. Recall (see Exercise 2c. in Section 2.7) that one way to represent the quaternions is as follows: Let*

$$A = \begin{bmatrix} 0 & 1 \\ -1 & 0 \end{bmatrix} \qquad and \qquad B = \begin{bmatrix} 0 & i \\ i & 0 \end{bmatrix}.$$

Then the quaternions, $H = \langle A, B \rangle$, the subgroup of $M_{22}(\mathbb{C})$ (under matrix multiplication) generated by A and B. One can easily check that $A^4 = I$, $A^2 = B^2$ and $ABAB^{-1} = I$. Define the map $f : \{x, y\} \to H$ by $f(x) = A$ and $f(y) = B$ which extends to a homomorphism $\phi : F \to H$. Again, ϕ is an epimorphism with $R \subseteq ker(\phi)$ which proves H satisfies the relations R. The difficult part of the example is to show that G has at most 8 elements (if we can, then $G \cong H$ as it did in the examples above). First note that $\langle xN \rangle$ is a normal subgroup of G of order ≤ 4. The bound on the order follows, since $(xN)^4 = x^4 N = N$. The fact that $\langle xN \rangle \lhd G$ follows from two observations: G is generated by xN and yN and

$$(yN)^{-1}(xN)^k(yN) = (y^{-1}x^k y)N = (y^{-1}x^{k-1}xyxy^{-1}yx^{-1})N$$

$$= (y^{-1}x^{k-1})N(xyxy^{-1})N(yx^{-1})N = (y^{-1}x^{k-1})N(yx^{-1})N$$

$$= (y^{-1}x^{k-1}yx^{-1})N = \cdots = (y^{-1}yx^{-k})N = x^{-k}N = (xN)^{-k} \in \langle xN \rangle.$$

Second, we show that $G/\langle xN \rangle$ has at most two cosets (this will complete the result, for then $|G| = [G : \langle xN \rangle]|\langle xN \rangle| \leq (2)(4) = 8$). We show that $G/\langle xN \rangle$ has at most the two cosets $\langle xN \rangle$ and $(yN)\langle xN \rangle$. This follows from the following two observations: First, these cosets generate $G/\langle xN \rangle$, since xN and yN generate G (note that $\langle xN \rangle = (xN)\langle xN \rangle$). Secondly, observe the following products of the generators:

$$\langle xN \rangle \langle xN \rangle = \langle xN \rangle, \; since \; (xN)^2 = x^2 N \in \langle xN \rangle.$$

$$(yN)\langle xN \rangle \langle xN \rangle = (yN)\langle xN \rangle, \; since \; (yN)(xN)(xN) = (yN)(x^2 N) \in (yN)\langle xN \rangle.$$

$$\langle xN \rangle (yN)\langle xN \rangle = (yN)\langle xN \rangle, \; since \; (xN)(yN)(xN) = (xyx)N$$
$$= (yN) \in (yN)\langle xN \rangle.$$

$$\langle yN \rangle \langle yN \rangle = \langle xN \rangle, \; since \; (yN)^2 = y^2 N = (x^{-2}x^2)N(y^2 N) = (xN)^{-2} \in \langle xN \rangle.$$

Remark 5.3 *Presentations for a fixed group are not unique. For instance, one can show that $\mathbb{Z}_n$ is defined by the generators $X = \langle x, y \rangle$ with relations $R = \{x^n, x^{-1}y\}$.*

Example 5.4 *Using Sylow Theory and Group Presentation we are now in a position to classify groups of certain small orders. To illustrate this we will classify groups of order 6 and 8.*

1. *Let G be a group of order 6. Using Sylow Theory one can show that G has a cyclic 2-Sylow of order 2 – call it $P = \langle x \rangle$ with $o(x) = 2$; and a normal cyclic 3-Sylow of order 3 – call it $Q = \langle y \rangle$ with $o(y) = 3$. There are two possibilities: either x and y commute or they don't. If they do, then as in the proof of Theorem 4.12, the order of xy is 6 and $G \cong \mathbb{Z}_6$. Let's consider the case when $xy \neq yx$. One can check that the six elements of G can be represented as $1, x, y, y^2, xy, xy^2$ and as a result yx must equal xy^2, or equivalently $xyxy = 1$. But this is the presentation which uniquely defines S_3 and thus $G \cong S_3$.*

2. *Let G be a group of order 8. If G is an abelian group, then the classification is known. Indeed, we know that G is isomorphic to either $\mathbb{Z}_8$, $\mathbb{Z}_4 \times \mathbb{Z}_2$ or $\mathbb{Z}_2 \times \mathbb{Z}_2 \times \mathbb{Z}_2$. So we may assume that G is non-abelian. Now G has an element of order 4, otherwise $g^2 = 1$ for all $g \in G$ and so G would be abelian. Set $x \in G$ with $o(x) = 4$ and define $H = \langle x \rangle$. Since $[G : H] = 2$ we know $H \lhd G$. Let $G/H = \{H, yH\}$ for some $y \notin H$. Since $G/H \cong \mathbb{Z}_2$ we have $(yH)^2 = H$ and so $y^2 \in H$. Now $y^2 \neq x$, for otherwise $o(y) = 8$ making $G \cong \mathbb{Z}_8$ abelian. Likewise $y^2 \neq x^3$, since $o(x^3) = 4$. Hence, either $y^2 = 1$ or $y^2 = x^2$. Now since $H \lhd G$ we know $yxy^{-1} \in H$, and since $o(yxy^{-1}) = o(x) = 4$, then either $yxy^{-1} = x$ or $yxy^{-1} = x^3$. We show $yxy^{-1} \neq x$, for otherwise $xy = yx$. But, since $G/H = \{H, yH\}$ and $H = \langle x \rangle$ this implies $G = \langle y, x \rangle$ and so G would be abelian. Thus, $yxy^{-1} = x^3$. So we have two group presentations:*

$$G_1 = \langle\, x, y \mid x^4 = 1,\ y^2 = 1,\ xyxy = 1\, \rangle,$$

$$G_2 = \langle\, x, y \mid x^4 = 1,\ y^2 x^{-2} = 1,\ xyxy^{-1} = 1\, \rangle.$$

We've seen these group presentations earlier. Indeed, $G_1 \cong D_4$ and $G_2 \cong Q_8$.

EXERCISES

1 Show that S_3 has presentation $\langle\, x, y \mid x^2 = 1,\ y^3 = 1,\ (xy)^2 = 1\, \rangle$.

2 Show that A_4 has presentation $\langle\, x, y \mid x^2 = 1,\ y^3 = 1,\ (xy)^3 = 1\, \rangle$.

3 Prove all parts of Lemma 5.1.

4 In the proof of Lemma 5.2.1, verify that $\psi \circ \nu : F \to G$ is an epimorphism whose kernel is precisely the normal subgroup N_R.

5.3 GROUP REPRESENTATION

In this section, we give a very naive introduction to group representation. The topic of group representation is an immense topic which is far too big to fit in a single section let alone an entire chapter. Basically we show how one can represent groups as a collection of matrices over the complex numbers. More specifically, we will be looking at *linear finite dimensional* representations of finite groups.

Definition 5.5 *Let G be a finite group. A* **linear representation** *of G is a group homomorphism $\phi : G \to GL_n(\mathbb{C})$ where n is called the* **degree** *of the representation. A representation is* **faithful** *if ϕ is a group monomorphism.*

Example 5.5 *Here, we list several examples of (faithful) group representations, some of which the reader has already encountered in the text.*

1. *For any finite group G the trivial map $\phi(g) = I_n$ for all $g \in G$ is the* **trivial** *representation of G of degree n. This representation is faithful only for the trivial group.*

2. *A degree 1 representation of a group G is a group homomorphism $\phi : G \to \mathbb{C}^*$ and since G is of finite order this implies that $\phi(g)$ must be a root of unity for all $g \in G$, and therefore the image lies on the unit circle in the complex plane. In particular, if G is a cyclic group of order n, then $\phi(g)$ is an nth root of unity for all $g \in G$.*

For instance, if $G = \langle g \rangle$ is cyclic of order 4, then ϕ maps onto $\{\pm 1, \pm i\} = \langle i \rangle$. Now ϕ maps the identity to 1 and ϕ is determined by where it sends g, i.e. $\phi(g) = -1, i$ or $-i$, so there are three representations of G.

3. *If $G = \{1, g, h, k\}$ is the Klein-4 group, then the following is a degree 2 representation of G: $\phi : G \to GL_2(\mathbb{C})$ by*

$$\phi(1) = \begin{bmatrix} 1 & 0 \\ 0 & 1 \end{bmatrix}, \quad \phi(g) = \begin{bmatrix} -1 & 0 \\ 0 & -1 \end{bmatrix}, \quad \phi(h) = \begin{bmatrix} 0 & 1 \\ 1 & 0 \end{bmatrix}, \quad \phi(k) = \begin{bmatrix} 0 & -1 \\ -1 & 0 \end{bmatrix}.$$

4. *If $G = \{\pm 1, \pm i, \pm j, \pm k\}$, the quarternions, then the following is a degree 2 representation of G: $\phi : G \to GL_2(\mathbb{C})$ by*

$$\phi(i) = \begin{bmatrix} 0 & 1 \\ -1 & 0 \end{bmatrix} \quad \text{and} \quad \phi(j) = \begin{bmatrix} 0 & i \\ i & 0 \end{bmatrix}.$$

Note that Q_8 has presentation $\langle x, y \mid x^4 = 1, x^2 = y^2, xyx = y \rangle$ and one can check that $x = \phi(i)$ and $y = \phi(j)$ satisfy these relations.

5. *Another degree two representation of the quaternions is the following: $\phi : G \to GL_2(\mathbb{C})$ by*

$$\phi(i) = \begin{bmatrix} i & 0 \\ 0 & -i \end{bmatrix} \quad \text{and} \quad \phi(j) = \begin{bmatrix} 0 & 1 \\ -1 & 0 \end{bmatrix}.$$

Again, one can check that $x = \phi(i)$ and $y = \phi(j)$ satisfy the relations $x^4 = 1, x^2 = y^2, xyx = y$.

6. *A degree 2 representation of a cyclic group is easy to understand in terms of a rotation group. Let $G = \langle g \rangle$ with $o(g) = n$. Define the linear representation $\phi : G \to GL_2(\mathbb{C})$ by*

$$\phi(g^k) = \begin{bmatrix} \cos\left(\frac{2\pi k}{n}\right) & -\sin\left(\frac{2\pi k}{n}\right) \\ \sin\left(\frac{2\pi k}{n}\right) & \cos\left(\frac{2\pi k}{n}\right) \end{bmatrix}.$$

7. *The dihedral group made up of rotations and reflections of an n-gon has a similar representation. The rotations are represented just as we did for the previous example of a cyclic group and the reflections are sent to*

$$\begin{bmatrix} \cos\left(\frac{2\pi k}{n}\right) & \sin\left(\frac{2\pi k}{n}\right) \\ \sin\left(\frac{2\pi k}{n}\right) & -\cos\left(\frac{2\pi k}{n}\right) \end{bmatrix},$$

where the matrix above represents the reflection across a line that makes an angle of $\pi k/n$ with the x-axis.

8. *A degree four representation of the quaternions is the following:* $\phi : G \rightarrow GL_4(\mathbb{R})$ *by*

$$\phi(i) = \begin{bmatrix} 0 & -1 & 0 & 0 \\ 1 & 0 & 0 & 0 \\ 0 & 0 & 0 & -1 \\ 0 & 0 & 1 & 0 \end{bmatrix} \quad and \quad \phi(j) = \begin{bmatrix} 0 & 0 & -1 & 0 \\ 0 & 0 & 0 & 1 \\ 1 & 0 & 0 & 0 \\ 0 & -1 & 0 & 0 \end{bmatrix}.$$

Again, one can check that $x = \phi(i)$ and $y = \phi(j)$ satisfy the relations $x^4 = 1, x^2 = y^2, xyx = y$.

9. *The group S_n has a representation of degree n defined as follows: Let $e_1, e_2, \ldots e_n$ be the standard basis for $\mathbb{C}$. We shall view elements of $GL_n(\mathbb{C})$ as linear transformations and define the group representation $\phi : S_n \rightarrow GL_n(\mathbb{C})$ in terms of this basis. In other words, for $\sigma \in S_n$ the image $\phi(\sigma)$ will be defined as $\phi(\sigma)e_i = e_{\sigma(i)}$. One can show that ϕ is indeed a group homomorphism. Such a representation is an example of a **permutation** representation. Let's take the specific example of S_3 and let's represent the permutation $\sigma = (1\ 2\ 3)$. Therefore,*

$$[\phi(\sigma)](e_1) = e_2, \quad [\phi(\sigma)](e_2) = e_3, \quad [\phi(\sigma)](e_3) = e_1.$$

Hence,

$$\phi(\sigma) = \begin{bmatrix} 0 & 0 & 1 \\ 1 & 0 & 0 \\ 0 & 1 & 0 \end{bmatrix}.$$

*These are the so-called **permutation** matrices.*

10. *A group G acting on a finite set $X = \{x_1, x_2, \ldots, x_n\}$ induces a permutation representation, since G can be viewed as a collection of functions of the set X. Indeed, for each $g \in G$, define the map $f_g : X \rightarrow X$ as follows: $f_g(x) = gx$, where gx is the action of g on x. In the previous example, these functions were bijections, but they need not necessarily be so. As in the previous example we view elements of $GL_n(\mathbb{C})$ as linear transformations and define the group representation $\phi : G \rightarrow GL_n(\mathbb{C})$ in terms of the standard basis. If we index our standard basis by the elements of X, then the group representation of G is $\phi : G \rightarrow GL_n(\mathbb{C})$ and is given by $\phi(g)e_{x_i} = e_{gx_i}$. This representation will be faithful iff the action is faithful (a group action is faithful if $gx = hx$ implies $g = h$).*

Remark 5.4 *We make several remarks related to the examples given above.*

1. *From our examples it is clear that group representation is not unique for a given group, even faithful group representations. Take for example the representation of a cyclic group with a degree 1, then a degree 2 representation.*

2. *Group representation is not unique even if we fix the degree. Indeed, if $\phi : G \to GL_n(\mathbb{C})$ is a group representation of G of degree n, then for any invertible matrix $A \in GL_n(\mathbb{C})$, the map $\psi : G \to GL_n(\mathbb{C})$ by $\psi(g) = A^{-1}\phi(g)A$ is also a group representation of G of degree n. Hence, there are, in fact, infinitely many different representations of a fixed degree for a given group G.*

3. *Since every finite group of order n can be embedded in S_n (Cayley's Theorem) and S_n has a faithful permutation representation, this implies that every finite group has a faithful linear representation.*

EXERCISES

1 Verify for Examples 5.5.4,.5,.8 that $x = \phi(i)$ and $y = \phi(j)$ satisfy the relations $x^4 = 1, x^2 = y^2, xyx = y$.

2 As was done in Example 5.5.9, compute the representation of $\sigma = (1\ 3)$.

3 Using Remark 5.4.3, compute a permutation representation of the Klein-4 group.

CHAPTER 6

Solvable and Nilpotent Groups

I N THIS CHAPTER, we introduce two families of groups: nilpotent and solvable. One can think of these groups as generalizations of abelian groups. The main reason we cover this theory is for the application of solvable groups to answer the question if there exists a formula for finding the roots of a polynomial of degree five or more, otherwise known as *solvability by radicals*. *Solvable* groups are named as such because of this connection. In Section 6.1, we remind the reader of some important subgroups as well as introduce some additional important subgroups. In Section 6.2, we discuss certain chains of subgroups of a group which we use in Section 6.3 to define nilpotent and solvable groups.

6.1 SOME RELEVANT SUBGROUPS

There are several subgroups of a given group that come up a lot in our discussion of nilpotent and solvable groups. This section is meant to present to the reader, remind the reader in some cases, and discuss these subgroups as well as their implications.

Definition 6.1 *For a given group G the center of G, written*

$$Z(G) = \{z \in G \ : \ zg = gz \ \ for \ all \ \ g \in G\}.$$

Remark 6.1 *Here, we list some examples and results that the reader may also wish to verify.*

1. *$Z(G)$ is always an abelian normal subgroup of G.*

2. *$Z(G)$ is characteristic in G (a subgroup H is **characteristic** in a group G if for every automorphism ϕ of G, we have $\phi(H) \leq H$).*

3. *The center of $GL_n(F)$ is the collection of all scalar matrices (i.e. matrices of the form aI_n where $a \in F^*$), where $GL_n(F)$ is the group of $n \times n$ matrices over a field F with matrix multiplication.*

DOI: 10.1201/9781003335283-6

4. The center of $SL_n(F)$ is also a collection of all scalar matrices with the added restriction that $a^n = 1$, where $SL_n(F)$ is the subgroup of $GL_n(F)$ with determinant equal to 1.

5. The center of $U_3(F)$, strictly upper triangular 3×3 matrices, are the matrices of the form $\begin{bmatrix} 1 & 0 & a \\ 0 & 1 & 0 \\ 0 & 0 & 1 \end{bmatrix}$, where $a \in F$.

6. The center of the dihedral group D_4 is the subgroup containing the identity and $180°$ rotation.

7. Consider the epimorphism $\iota : G \to Inn(G)$ defined by $\iota(g) = i_g$, where i_g is the inner automorphism defined by $i_g(x) = gxg^{-1}$. Then $ker(\iota) = Z(G)$ and so $G/Z(G) \cong Inn(G)$.

8. For any group G and $g \in G$, we have that $g \in Z(C_G(g))$ where $C_G(g) = \{a \in G : ga = ag\}$. Hence, every element of a group G lies in an abelian subgroup of G.

9. If $G/Z(G)$ is cyclic, then G is abelian.

10. $Z(Sym(X)) = 1$ when $|X| \geq 3$.

The following result is not so easy to deduce and so we shall prove this in detail:

Theorem 6.1 *If G is a group with trivial center, then $Aut(G)$ has a trivial center as well.*

Proof 6.1 *Let's call a homomorphism which commutes with all the elements of $Inn(G)$ a **normal** homomorphism.*

First, we show that if ϕ is a normal automorphism, then there is a homomorphism $\psi : G \to Z(G)$ such that $\phi(g) = \psi(g)^{-1}g$ for all $g \in G$.

To show this, first assume that ϕ is a normal automorphism. Define the map on G, $\psi(g) = g\phi(g)^{-1}$. Notice that $\phi(g) = \psi(g)^{-1}g$, as desired. Furthermore, ψ maps into $Z(G)$. To see this, first notice that for all $g \in G$, $\phi i_g(g) = i_g \phi(g)$ which implies that $\phi(g) = g\phi(g)g^{-1}$, and so $g\phi(g)^{-1} = \phi(g)^{-1}g$. Second, notice that for all $g, x \in G$, $\phi i_{g^{-1}}(xg^{-1}) = i_{g^{-1}}\phi(xg^{-1})$, and so $\phi(g^{-1}xg^{-1}g) = g^{-1}\phi(xg^{-1})g$, and so $\phi(g)^{-1}\phi(x) = g^{-1}\phi(x)\phi(g)^{-1}g$, which implies that $g\phi(g)^{-1}\phi(x) = \phi(x)\phi(g)^{-1}g$. But then, by above, $g\phi(g)^{-1}\phi(x) = \phi(x)g\phi(g)^{-1}$. Since ϕ maps onto G, we can conclude that for all $h \in G$, $g\phi(g)^{-1}h = hg\phi(g)^{-1}$ which can be rewritten as $\psi(g)h = h\psi(g)$. This shows that ψ maps into $Z(G)$. Having shown this, we can now prove that ψ is a homomorphism. Indeed, for all $g, h \in G$,

$$\psi(gh) = gh\phi(gh)^{-1} = gh\phi(h)^{-1}\phi(g)^{-1} = g\psi(h)\phi(g)^{-1} = g\phi(g)^{-1}\psi(h) = \psi(g)\phi(h).$$

Now assume that $Z(G)$ is trivial. We first show that the only normal automorphism of G is the identity automorphism. Let ϕ be a normal automorphism of G. By *the work above, we know that $\psi(g) = g\phi(g)^{-1}$ defines a homomorphism from G into $Z(G) = 1$, by assumption. Therefore, for all $g \in G$, we have $1 = \psi(g) = g\phi(g)^{-1}$ and so $\phi(g) = g$ for all $g \in G$, i.e. ϕ is the identity automorphism. With this result in mind we can now show that $Z(Aut(G))$ is trivial. Take any $\phi \in Z(Aut(G))$, so in particular ϕ is a normal automorphism, and so by the work above ϕ is the identity automorphism.*

We leave it to the reader to show that, in fact, the map defined by $\phi(g) = \psi(g)^{-1}g$, where $\psi : G \to Z(G)$ is a homomorphism, is necessarily a normal endomorphism.

We remind the reader of another important subgroup of a group.

Definition 6.2 *Let X be a non-empty subset of a group G. The **subgroup generated by** X, written $\langle X \rangle$ is the collection of all finite products of elements of X and their inverses. The set X is called the **generating set** of $\langle X \rangle$.*

If $X = \{g_1, g_2, \ldots, g_n\}$ a finite set, then we write $\langle g_1, g_2, \ldots, g_n \rangle$ for $\langle X \rangle$. Note that if $X = \{g\}$, then $\langle X \rangle$ is simply the cyclic subgroup generated by g. One needs to check, of course, that $\langle X \rangle$ is indeed a subgroup of G. Furthermore, one can show that $\langle X \rangle$ is the smallest subgroup of G containing the set X.

Example 6.1 *Here, we list several examples groups and their generators.*

1. *The Klein-4 group $V = \{e, a, b, c\}$ is generated by $X = \{a, b\}$.*

2. *The quaternions are generated by the set $X = \{i, j\}$.*

3. *The dihedral group D_4 (rotations and reflections of a square) is generated by any single rotation and any single reflection.*

Before we go on to our next important subgroup (which will appear in the next section) we remind the reader of some notation.

Definition 6.3 *For $g, h \in G$ a group, the **conjugate of h by g in G**, written $h^g = g^{-1}hg$. The **conjugacy class** of $H \le G$, written $H^g = \{h^g : h \in H\}$.*

We leave it to the reader to verify the following simple statements which attest to the fact that conjugation is a lot like exponentiation:

Lemma 6.1 *Let G be any group with $H \le G$. Then*

1. *For $h, g, k \in G$, $(hk)^g = h^g k^g$*

2. *For $h, g, k \in G$, $(h^k)^g = h^{kg}$*

3. *For $h, g, k \in G$, $(h^g)^{-1} = (h^{-1})^g$*

4. *For $g \in G$, $H^g \le G$.*

5. For $h_1, \ldots, h_n, g \in G$, $(h_1 \cdots h_n)^g = h_1^g \cdots h_n^g$.

6. For $h_1, \ldots, h_n, g \in G$, $((g^{h_1})^{h_2} \cdots)^{h_n} = g^{h_1 \cdots h_n}$.

We continue our discussion of several important subgroups of a given group.

Definition 6.4 *Let G be a group and $g, h \in G$. The **commutator of** g **and** h, written*
$$[g, h] = g^{-1}h^{-1}gh = g^{-1}g^h.$$

*More generally, if $g_1, g_2, \ldots, g_n \in G$ then the **commutator of** $g_1, g_2, \ldots, g_n$, written*
$$[g_1, g_2, \ldots, g_n] = [\,[g_1, g_2, \ldots, g_{n-1}], g_n]\quad \text{(defined recursively).}$$

The following are some simple properties of commutators which the reader may wish to verify:

Theorem 6.2 *If $g, h, k \in G$ a group, then*

1. $[g, h]^{-1} = [h, g]$

2. $[g, h]^k = [g^k, h^k]$

3. $[gh, k] = [g, k]^h [h, k]$

4. $[g, hk] = [g, k][g, h]^k$

5. $[g, h^{-1}] = ([g, h]^{h^{-1}})^{-1}$

6. $[g^{-1}, h] = ([g, h]^{g^{-1}})^{-1}$

7. $[g, h^{-1}, k]^h [h, k^{-1}, g]^k [k, g^{-1}, h]^g = 1$

The next result we shall prove in detail. First, we need some additional notation. For $g, h, k \in G$ define $g^{h+k} = g^h g^k$.

Theorem 6.3 *Let $g, h \in G$ be a group and $n \in \mathbb{Z}^+$. Then*

1. $[g^n, h] = [g, h]^{g^{n-1} + g^{n-2} + \cdots + g + 1}$

2. *If $[g, h] \in Z(\langle g, h \rangle)$, then $[g^n, h] = [g, h]$.*

Proof 6.2 *We prove the first statement by induction and the use of Theorem 6.2.3. The base case, $n = 1$, is immediate. For $n > 1$,*
$$[g^n, h] = [g^{n-1}g, h] = [g^{n-1}, h]^g [g, h] = \left([g, h]^{g^{n-2} + g^{n-3} + \cdots + g + 1}\right)^g [g, h] =$$
$$[g, h]^{g^{n-1} + g^{n-2} + \cdots + g}[g, h] = [g, h]^{g^{n-1} + g^{n-2} + \cdots + g + 1}.$$

For the second statement, if $[g, h] \in Z(\langle g, h \rangle)$, then in particular $[g, h]$ commutes with any power of g, and so
$$[g, h]^{g^{n-1} + g^{n-2} + \cdots + g + 1} = [g, h]^{g^{n-1}}[g, h]^{g^{n-2}} \cdots [g, h] = [g, h][g, h] \cdots [g, h] = [g, h]^n.$$

The result then follows from the first statement.

Definition 6.5 *Let G be a group and X and Y be two non-empty subsets of G. The* **commutator subgroup of X and Y,** *written*

$$[X, Y] = \langle\ [x, y]\ :\ x \in X, y \in Y\ \rangle,$$

the subgroup generated by the commutators $[x, y]$. More generally for $X_1, X_2, \ldots, X_n$ non-empty subsets of G, the **commutator of $X_1, X_2, \ldots, X_n$,** *written*

$$[X_1, X_2, \ldots, X_n] = [\ [X_1, X_2, \ldots, X_{n-1}], X_n]\quad (\text{defined recursively}).$$

Remark 6.2 *For the next result, we remind the reader of some definitions and results we saw earlier in the text.*

1. *For X and Y subsets of a group G, the set XY is the collection of all products of an element of X and an element of Y.*

2. *If $H, K \leq G$ a group, then the HK and KH are subgroups of G exactly when $HK = KH$, which is true in the special case that H and K are both normal subgroups of G.*

Theorem 6.4 *If H, K and L are normal subgroups of a group G, then $[HK, L] = [H, L][K, L]$.*

Proof 6.3 *First note that since H, K and L are normal in G, so are $[HK, L]$, $[H, L]$ and $[K, L]$. For instance, for $h \in H$, $l \in L$ and $g \in G$, we have $[h, l]^g = [h^g, l^g] \in [H, L]$ which shows that $[H, L] \triangleleft G$. By the remarks above, this in turn implies that $[H, L][K, L] \leq G$.*

Hence, to show that $[HK, L] \subseteq [H, L][K, L]$, it is enough to show that each $[hk, l] \in [HK, L]$ is also in $[H, L][K, L]$. To see this, note that

$$[hk, l] = [h, l]^k[k, l] = [h^k, l^k][k, l] \in [H, L][K, L].$$

Next, to show that $[H, L][K, L] \subseteq [HK, L]$, first note that since $[H, L], [K, L] \triangleleft G$, we have that $[H, L][K, L] = [K, L][H, L]$ and $[H, L], [K, L] \leq G$. Therefore, it's enough to show a single product $[h, l][h, l']$ is in $[HK, L]$ where $h \in H$ and $l, l' \in L$. Now to see this, note that for any $k \in K$, using Theorem 6.2.3,

$$[h, l][h, l'] = k[h, l]^k[h, l']^k k^{-1} = ([hk, l][hk, l'])^{k^{-1}},$$

and this last expression is in $[HK, L]$ being a normal subgroup of G.

Definition 6.6 *For a group G, the* **derived subgroup of G** *(or sometimes called the* **commutator subgroup of G**), *written $G' = [G, G]$.*

Using Theorem 6.2.1, one sees that G' is, in fact, the collection of all finite products of commutators in G (no need for their inverses). Another easy fact to verify is that G' is characteristic in G, and hence normal in G. Here are some additional properties which we prove in detail:

Theorem 6.5 *Let G and K be groups and H, L be subgroups of G. Then*

1. *If $\phi : G \to K$ a homomorphism, then $\phi(G') \leq K'$*

2. *$G' \leq H$ iff $H \triangleleft G$ and G/H is abelian.*

3. *G' is the smallest normal subgroup of G which will form an abelian factor group.*

4. *If $L/H \leq Z(G/H)$, then $[G, L] \leq H$.*

Proof 6.4 *For the first statement, take any $[g_1, h_1] \cdots [g_n, h_n] \in G'$. By properties of a homomorphism,*

$$\phi([g_1, h_1] \cdots [g_n, h_n]) = \phi([g_1, h_1]) \cdots \phi([g_n, h_n]) = [\phi(g_1), \phi(h_1)] \cdots [\phi(g_n), \phi(h_n)] \in K'.$$

For the second statement, first assume that $G' \leq H$. Then for all $g \in G$ and $h \in H$, we have

$$g^{-1}hg = h(h^{-1}g^{-1}hg) = h[h, g] \in H,$$

and so $H \triangleleft G$. To see that G/H is abelian, notice that for $g_1, g_2 \in G$,

$$g_1 H g_2 H = g_2 H g_1 H g_1^{-1} H g_2^{-1} H g_1 H g_2 H = g_2 H g_1 H [g_1, g_2] H = g_2 H g_1 H.$$

Now assume that $H \triangleleft G$ and G/H is abelian. Then for all $g_1, g_2 \in G$, we have

$$g_1^{-1} H g_2^{-1} H g_1 H g_2 H = g_1^{-1} H g_1 H g_2^{-1} H g_2 H = H,$$

and so $[g_1, g_2] H = H$. Hence, $[g_1, g_2] \in H$ and thus by the remark after the definition of G', it follows that $G' \leq H$.

The third statement follows immediately from the second statement, since it says in particular that any normal subgroup H of a group G with G/H abelian necessarily contains the derived group G'.

For the fourth statement, since $L/H \leq Z(G/H)$ we have that $gHlH = lHgH$ for all $g \in G$ and $l \in L$. This can be rewritten as $[g, l] H = H$ and so $[g, l] \in H$.

EXERCISES

1 Prove the following results listed in Remark 6.1:

   a. $Z(G)$ is always an abelian normal subgroup of G.

   b. $Z(G)$ is characteristic in G (a subgroup H is characteristic in a group G if for every automorphism ϕ of G, we have $\phi(H) \leq H$).

   c. The center of $U_3(F)$, strictly upper triangular 3×3 matrices, are the matrices of the form $\begin{bmatrix} 1 & 0 & a \\ 0 & 1 & 0 \\ 0 & 0 & 1 \end{bmatrix}$, where $a \in F$.

   d. The center of the dihedral group D_4 is the subgroup containing the identity and $180°$ rotation.

e. Consider the epimorphism $\iota : G \to Inn(G)$ defined by $\iota(g) = i_g$, where i_g is the inner automorphism defined by $i_g(x) = gxg^{-1}$. Then $ker(\iota) = Z(G)$ and so $G/Z(G) \cong Inn(G)$.

f. For any group G and $g \in G$, we have that $g \in Z(C_G(g))$ where $C_G(g) = \{a \in G : ga = ag\}$ (hence, every element of a group G lies in an abelian subgroup of G).

g. $Z(Sym(X)) = 1$ when $|X| \geq 3$

2 Prove that if $H \leq K \leq L \leq G$, H is characteristic in K and K is characteristic in L, then H is characteristic in L.

3 Show that in the proof of Theorem 6.1 the map defined by $\phi(g) = \psi(g)^{-1}g$, where $\psi : G \to Z(G)$ is a homomorphism, is necessarily a normal endomorphism.

4 Prove Lemma 6.1

5 Prove Theorem 6.2

6 Explain why in the proof of Theorem 6.4 in order to prove $[H,L][K,L] \subseteq [HK,L]$, it's enough to show a single product $[h,l][h,l']$ is in $[HK,L]$ where $h \in H$ and $l,l' \in L$.

7 Prove that G' is characteristic in G, for any group G.

6.2   SERIES OF GROUPS

We now consider chains of subgroups within a group, called **series** in order to eventually define solvable and nilpotent groups. First, we need to distinguish between several kinds of series and then we will narrow our focus to particular examples of series.

Definition 6.7 *Let $H_0 \subseteq H_1 \subseteq \cdots \subseteq H_n$ be subgroups of a group G with $H_0 = 1$ and $H_n = G$.*

1. *A* **subnormal series** *has the property that $H_i \lhd H_{i+1}$ for all $i = 0, 1, \ldots, n-1$.*

2. *A* **normal series** *has the property that $H_i \lhd G$ for all $i = 0, 1, \ldots, n$.*

3. *An* **abelian** *series is a subnormal series with the additional property that H_{i+1}/H_i is abelian for all $i = 0, 1, \ldots, n-1$.*

4. *A* **central** *series is a normal series with the additional property that $H_{i+1}/H_i \subseteq Z(G/H_i)$ for all $i = 0, 1, \ldots, n-1$.*

The H_i are called the **terms** *in the series and n is called the* **length** *of the series.*

Remark 6.3 *We make some simple observations.*

1. *Clearly, every normal series is also subnormal (but not vice-versa).*

2. $K \lhd H \lhd G$ *does not necessarily imply* $K \lhd G$, *however if* K *is a characteristic in* $H \lhd G$, *then* $K \lhd G$.

Example 6.2 *Here, we list some examples of series.*

1. *The following is a normal series in the abelian group* $(\mathbb{Z}, +)$ *for any positive integers* a *and* n:
$$\{0\} \le a^n \mathbb{Z} \le a^{n-1} \mathbb{Z} \le \cdots a\mathbb{Z} \le \mathbb{Z}.$$
For instance, if we take $a = 2$ *and* $n = 4$ *we get the normal series*
$$\{0\} \le 16\mathbb{Z} \le 8\mathbb{Z} \le 4\mathbb{Z} \le 2\mathbb{Z} \le \mathbb{Z}.$$
These series are certainly abelian and central as well.

2. *The following is a subnormal series in the group* D_4, *the dihedral group:*
$$\{\rho_0\} \le \{\rho_0, \mu_1\} \le \{\rho_0, \rho_2, \mu_1, \mu_2\} \le D_4.$$

We generate certain series which will be used in defining nilpotent and solvable groups, but first, we need to define some additional subgroups.

Definition 6.8 *Let* G *be a group, then*

1. *For any natural number* n, *the* n-**th derived subgroup**, *written* $G^{(n)}$, *is defined recursively as follows:* $G^{(0)} = G$, $G^{(1)} = G'$ *and* $G^{(n+1)} = (G^{(n)})'$.

2. *For any natural number* n, G^n *is defined recursively as follows:* $G^0 = G$, $G^1 = G' = [G, G]$ *and* $G^{n+1} = [G, G^n]$.

3. *For any natural number* n, *the* n-**th center** *of* G, *written* $Z_n(G)$ *is defined recursively as follows:* $Z_0(G) = 1$, $Z_1(G) = Z(G)$ *and* $Z_{n+1}(G) = \{g \in G \mid [G, g] \le Z_n(G)\}$.

Remark 6.4 *It should be clear to the reader that* $G^{(n+1)} \lhd G^{(n)}$ *and that* $G^{(n)}/G^{(n+1)}$ *is abelian. Here are some other facts about the* n-*th derived subgroup.*

Theorem 6.6 *For any group* G *and* m *and* n *natural numbers,*

1. $G^{(n)}$ *is characteristic in* G *(and so* $G^{(n)} \lhd G$).

2. $(G^{(m)})^{(n)} = G^{(mn)}$.

3. *If* $H \le G$, *then* $H^{(n)} \le G^{(n)}$.

Proof 6.5 *We prove the first statement by induction. For the* $n = 0$ *case* $G^{(0)} = G$ *which is certainly characteristic in itself. For* $n > 0$, *since* $G^{(n)} = (G^{(n-1)})'$ *is characteristic in* $G^{(n-1)}$ *which, by induction, is in turn characteristic in* G, *it follows that* $G^{(n)}$ *is characteristic in* G *(see Exercise 2 in Section 6.1).*

The remaining statements are left to the reader to prove.

Definition 6.9 *The series,* $G = G^{(0)} \geq G^{(1)} \geq G^{(2)} \geq \cdots$ *is called the* **derived series**

Note that if a derived series terminates in 1 it is then an abelian normal series.

Example 6.3 S_n *has a derived series terminating in 1 when* $n < 5$ *(left as an exercise).*

We need a lemma before we can prove some facts about G^n. The proof of this lemma will be omitted since it would require us to first introduce several definitions and to prove several results which will take us too far off the beaten path.

Lemma 6.2 *Let* H, K, L *be subgroups of* G. *If two of the subgroups* $[H, K, L]$, $[K, L, H]$, $[L, H, K]$ *are contained in a normal subgroup of* G, *then so is the third.*

Here, we list some facts concerning G^n.

Theorem 6.7 *For any group* G *and* n *a natural number,*

1. G^n *is characteristic in* G *(and so* $G^n \lhd G$).

2. $G^{n+1} \leq G^n$.

3. $[G^m, G^n] \leq G^{m+n}$.

4. $(G^m)^n = G^{mn}$.

Proof 6.6 *We prove the second and third statement and leave the first and fourth as exercises. To prove the second statement, since* $G^{n-1} \leq G$, *it is enough to show a single commutator* $[g, h] \in G^n$ *(where* $g \in G$ *and* $h \in G^{n-1}$) *is also in* G^{n-1}. *Now* $[g, h] = h^{-g}h \in G^{n-1}$, *since, by the first statement,* $G^{n-1} \lhd G$.
    We prove the third statement by induction on m. *The case* $m = 0$ *follows immediately from the second statement. For* $m > 0$, *by Lemma 6.2,* $[G^m, G^n] = [G^{m-1}, G, G^n]$ *is contained in the product* $[G, G^n, G^{m-1}][G^n, G^{m-1}, G]$ $(G^n \lhd G$ *implies that* $[G, G^n, G^{m-1}][G^n, G^{m-1}, G]$ *is a normal subgroup of* G). *Now, by induction, we know that* $[G, G^n, G^{m-1}][G^n, G^{m-1}, G] \leq G^{m+n+1}$.

Definition 6.10 *The series,* $G = G^0 \geq G^1 \geq G^2 \geq \cdots$ *is called the* **lower central series**

Note that if a lower central series terminates in 1 it is then (as we shall see) a central normal series.

Example 6.4 *If* $G = B_3(F)$, *upper triangular matrices over a field* F, *then* $G = G^0 \geq G^1 = G' \geq G^2 = 1$ *forms a lower central series.*

Here are some facts concerning $Z_n(G)$:

Theorem 6.8 *For any group* G *and* n *a natural number,*

1. $[G, Z_{n-1}(G)] \leq Z_n(G)$.

2. $Z_m(G/Z_n(G)) = Z_{m+n}(G)/Z_n(G)$.

3. $Z_n(G)$ is characteristic in G (and so $Z_n(G) \lhd G$).

4. $Z_n(G) \leq Z_{n+1}(G)$.

5. $[G^m, Z_n(G)] \leq Z_{n-m}(G)$, if $n \geq m$.

Proof 6.7 *We prove the second statement by induction on m. The $m = 0$ case follows immediately from the definition (convince yourself). For $m > 0$, by induction,*

$$Z_m(G/Z_n(G)) = \{g \in G/Z_n(G) \mid [G/Z_n(G), g] \leq Z_{m-1}(G/Z_n(G))\}$$

$$= \{g \in G/Z_n(G) \mid [G/Z_n(G), g] \leq Z_{m+n-1}(G)/Z_n(G)\}$$

$$= \{g \in G \mid [G, g] \leq Z_{m+n-1}(G)\}/Z_n(G) = Z_{m+n}(G)/Z_n(G).$$

We leave the rest of the theorem as an exercise.

Definition 6.11 *The series $1 = Z_0(G) \leq Z_1(G) \leq Z_2(G) \leq \cdots$ is called the **upper central series***

Note that if it terminates in G it is then a central normal series.

Example 6.5 *We give several examples of an upper central series.*

1. *If $G = D_4$, then $1 = Z_0(G) \leq Z_1(G) = \{\rho_0, \ \rho_2\} \leq Z_2(G) = D_4$ forms an upper central series.*

2. *If $G = U_3(F)$, then $1 = Z_0(G) \leq Z_1(G) \leq Z_2(G) = G$ forms an upper central series.*

3. *$G = B_3(F)$ does not form a upper central series, since $Z_1(G), Z_2(G), \ldots$ are each the collection of non-trivial scalar matrices (left as an exercise).*

The next result connects the structures G^n and $Z_n(G)$ and is the basis for the terminology *upper* and *lower*.

Theorem 6.9 *For and group G, if for some natural number n we have $G^n = 1$, then for $i = 0, 1, \ldots, n$, $G^{n-i} \leq Z_i(G)$.*

Proof 6.8 *We prove this by induction on i. For $i = 0$ the statement to prove becomes $1 \leq 1$ which is evidently true. For $0 < i \leq n$, notice that $[G, G^{n-i}] = G^{n-i+1} = G^{n-(i-1)}$ which by induction is contained in $Z_{i-1}(G)$. But then by the definition of $Z_i(G)$ it follows that $G^{n-i} \leq Z_i(G)$.*

EXERCISES

1 Give an example showing that $K \triangleleft H \triangleleft G$ does not necessarily imply $K \triangleleft G$.

2 Prove that if K is a characteristic in $H \triangleleft G$, then $K \triangleleft G$.

3 Verify that Example 6.2.2 is an abelian series.

4 Verify that $G^{(n+1)} \triangleleft G^{(n)}$ and that $G^{(n)}/G^{(n+1)}$ is abelian.

5 Prove Theorem 6.6.2 & .3.

6 Verify that S_n has a derived series terminating in 1 when $n < 5$.

7 Prove Theorem 6.7.1 & .4.

8 Prove Theorem 6.8.1,.3,.4 & .5 (note that the proof of part 4 follows a similar pattern to the proof of part 3 in Theorem 6.7).

9 Verify for the group $G = B_3(F)$ that $Z_1(G), Z_2(G), \ldots$ are each the collection of non-trivial scalar matrices.

6.3 SOLVABLE AND NILPOTENT GROUPS

We have now reached the goal of this chapter, namely the study of solvable and nilpotent groups. These two classes of groups are a generalization of abelian groups in the sense that both can be constructed from abelian groups via a finite number of extensions.

Definition 6.12 *A group G is called* **solvable** *if it has an abelian series; recall that this means there is a subnormal series $1 = G_0 \leq G_1 \leq \cdots \leq G_n = G$ with each G_{i+1}/G_i abelian. The length n for the shortest such series is called the* **solvability class** *of G (or G is called a* **solvable group of class** n). *A solvable group of class 2 is called a* **metabelian** *group.*

Certainly, a solvable group of class 1 is an abelian group. The following result reveals the tenacity of solvable groups.

Theorem 6.10 *The following are true:*

1. *Every subgroup of a solvable group is solvable.*

2. *The image of any homomorphism with solvable domain is solvable.*

3. *Every factor group of a solvable group is solvable.*

Proof 6.9 *Let G be a solvable group. Thus G has a subnormal series $1 = G_0 \leq G_1 \leq \cdots \leq G_n = G$ with each G_{i+1}/G_i abelian. To prove the first statement, consider the series*

$$1 = H \cap G_0 \leq H \cap G_1 \leq \cdots \leq H \cap G_n = H.$$

One can easily verify that the series is subnormal and by the Second Isomorphism Theorem, each

$$H \cap G_{i+1}/H \cap G_i \cong (H \cap G_{i+1})G_i/G_i \le G_{i+1}/G_i.$$

In other words, $H \cap G_{i+1}/H \cap G_1$ is isomorphic to a subgroup of an abelian group and is therefore abelian as well. Hence, $1 = H \cap G_0 \le H \cap G_1 \le \cdots \le H \cap G_n = H$ is an abelian series and so H is solvable.

To prove the second statement, consider the series

$$1 = \phi(1) = \phi(G_0) \le \phi(G_1) \le \cdots \le \phi(G_n) = \phi(G).$$

Certainly, the subnormal property is preserved by the homomorphism and by the First and Third Isomorphism Theorem each quotient

$$\phi(G_{i+1})/\phi(G_i) \cong (G_{i+1}/ker(\phi))/(G_i/ker(\phi)) \cong G_{i+1}/G_i,$$

which is assumed to be abelian. Hence, $1 = \phi(1) = \phi(G_0) \le \phi(G_1) \le \cdots \le \phi(G_n) = \phi(G)$ is an abelian series and so $\phi(G)$ is solvable.

To prove the third statement, consider the series

$$\{N\} = G_0 \le G_1 N/N \le \cdots \le G_n N/N = G/N.$$

By the Third Isomorphism Theorem (part 1), each $G_i N/N \lhd G_{i+1} N/N$, since $G_i N$ and N are both normal in $G_{i+1}N$ (check). By the Third Isomorphism Theorem (part 2),

$$(G_{i+1}N/N)/(G_i N/N) \cong G_{i+1}N/G_i N \cong G_{i+1}/G_i,$$

which we know to be abelian. Hence, $\{N\} = G_0 \le G_1 N/N \le \cdots \le G_n N/N = G/N$ is an abelian series for G/N and so G/N is solvable.

Definition 6.13 *A group G is called **solvable** if it has an abelian series; recall that this means there is a subnormal series $1 = G_0 \le G_1 \le \cdots \le G_n = G$ with each G_{i+1}/G_i abelian. The length n for the shortest such series is called the **solvability class** of G (or G is called a **solvable group of class** n). A solvable group of class 2 is called a **metabelian** group.*

Definition 6.14 *A group G is called **nilpotent** if it has a normal series $1 = G_0 \le G_1 \le \cdots \le G_n = G$ with each $G_{i+1}/G_i \le Z(G/G_i)$. The length n for the shortest such series is called the **nilpotency class** of G (or G is called a **nilpotent group of class** n).*

Certainly, a nilpotent group of class 1 is an abelian group. Nilpotent groups have a similar tenacious result. Since the proof is very similar in nature to the proof just given, we leave it to the reader as an exercise to provide the proof.

Theorem 6.11 *The following are true:*

1. *Every subgroup of a nilpotent group is nilpotent.*

2. *The image of any homomorphism with nilpotent domain is nilpotent.*

Remark 6.5 *We need to point out that not every solvable group is nilpotent. Take for example the group $G = S_3$ which is solvable (since it has a derived series – see Exercise 6 of Section 6.2), yet is certainly not nilpotent. Since, $Z(G) = 1$, then for all i, by Thereom 6.8.2, $Z(G/G_i) = Z(G)/Z_i(G) = 1$. Therefore, for all i, we cannot possible have $G_{i+1}/G_i \leq Z(G/G_i) = 1$. Yet it is certainly the case that every nilpotent group is solvable (convince yourself). In fact, we will see later that if a group is nilpotent of class n, then it is also solvable of class $\leq n$.*

Lemma 6.3 *If G is a p-group, then G is nilpotent.*

Proof 6.10 *We prove this result by induction. Assume every p-group of order less than the order of G is nilpotent. Since $Z(G) \neq 1$, this implies $G/Z(G)$ is p-group of smaller order than G and therefore is nilpotent. Consider the canonical homomorphism $\nu : G \to G/Z(G)$. The the preimages of the central series of $G/Z(G)$ form a central series for G, thus making G nilpotent.*

We wish now to connect the notions of solvability and nilpotency with some of the specific series we defined in the previous lesson. This will make the study of these classes of groups more concrete in the sense that we can prove more results about these groups, in particular about their solvable and nilpotent classes.

Theorem 6.12 *For a group G,*

1. *If G has an abelian series $1 = G_0 \leq G_1 \leq \cdots \leq G_n = G$ then for all $i = 0, 1, \ldots, n$ we have $G^{(i)} \leq G_{n-i}$.*

2. *G is solvable of class n iff $G^{(n)} = 1$ and $G^{(n-1)} \neq 1$, for some n.*

3. *If G has a central series $1 = G_0 \leq G_1 \leq \cdots \leq G_n = G$ then for all $i = 0, 1, \ldots, n$ we have $G_i \leq Z_i(G)$.*

4. *G is nilpotent of class n iff $Z_n(G) = G$ and $Z_{n-1}(G) \neq G$, for some n.*

Proof 6.11 *We prove the first statement by induction. For $i = 0$, the statement is self evident. For $i > 0$, by induction, $G^{(i)} = (G^{(i-1)})' \leq (G_{n-(i-1)})' \leq G_{n-i}$, since $G_{n-(i-1)}/G_{n-i}$ is abelian (see Theorem 10.9.2).*
To prove the second statement, first assume that G is solvable. So then G has an abelian series $1 = G_0 \leq G_1 \leq \cdots \leq G_n = G$ and let this be the shortest such. By part 1, it follows then that $G^{(n)} \leq G_0 = 1$. Furthermore, $G^{(n-1)} \neq 1$, for otherwise the derived series would be shorter than our assumed shortest abelian series witnessing the solvability of G. Now assume that $G^{(n)} = 1$ and $G^{(n-1)} \neq 1$, for some n. Then the derived series $G = G^{(0)} \geq G^{(1)} \geq G^{(2)} \geq \cdots \geq G^{(n)} = 1$ is an abelian series witnessing the solvability of G. By the direction just proved and the fact that $G^{(n-1)} \neq 1$, it must be the case that the derived series is the shortest abelian series for G (convince yourself) and hence n is the solvability class for G.
We leave the third and fourth statements to the reader to prove in an analogous manner to the first and second.

Example 6.6 *From the examples presented in the last section, we know then that*

1. S_n *is solvable for* $n < 5$.

2. *We will show later in the text that for primes* p, q, r *if* G *is a group of order* p^m *or* $p^m q$ *or* $p^2 q^2$ *or* pqr, *then* G *is solvable.*

3. *Burnside proved a more general fact that if* G *has order* $p^m q^n$, *then* G *is solvable. The most general fact written in an amazing article of incredible length by Feit and Thomas is that any group of odd order is necessarily solvable.*

4. *The first four smallest non-solvable groups have orders 60, 120, 168 and 180. The non-solvable group of order* $60 = 2^2 \cdot 3 \cdot 5$ *is* A_5.

5. $B_3(F)$ *is nilpotent of class 2. In general,* $B_n(F)$ *is nilpotent of class* $n - 1$.

6. D_4 *is nilpotent of class 2.*

7. $U_3(F)$ *is nilpotent of class 2.*

Theorem 6.13 *If* G *is a nilpotent group of class* n, *then* G *is solvable of class* $\leq n$.

Proof 6.12 *We have already pointed out that nilpotency implies solvability, so it remains to show the statement regarding the class. So in addition, assuming* G *has a central series of shortest length* n *gives us an abelian series of length* n *which may or not be the shortest such. Hence, the best we can conclude is that* G *is solvable of class* $\leq n$.

These next series of results put the subgroup G^n in proper perspective.

Lemma 6.4 *If* G *is a nilpotent group of class* n, *then* $G^i \leq Z_{n-i}(G)$ *for* $i = 0, 1, \ldots, n$.

Proof 6.13 *The proof is by induction. The* $i = 0$ *case follows immediately from Theorem 6.12.4. For* $i > 0$, *using induction and Theorem 6.8.1,*

$$G^i = [G, G^{i-1}] \leq [G, Z_{n-(i-1)}(G)] = Z_{n-(i-2)}(G) \leq Z_{n-i}(G).$$

Lemma 6.5 *If* $G^n = 1$, *then* G *is nilpotent of class* $\leq n$.

Proof 6.14 *By assumption and Theorem 6.9, we know that* $G^{n-i} \leq Z_i(G)$ *for* $i = 0, 1, \ldots, n$. *So if we set* $i = n$ *we get* $G = G^0 \leq Z_n(G)$. *Hence,* $Z_n(G) = G$ *and so by Theorem 6.12.4,* G *is nilpotent of class* $\leq n$.

Theorem 6.14 *A group* G *is nilpotent of class* n *iff* $G^n = 1$ *and* $G^{n-1} \neq 1$.

Proof 6.15 *Assuming* G *is nilpotent of class* n, *by Lemma 6.4,* $G^n \leq Z_0(G) = 1$ *and so* $G^n = 1$. *Furthermore,* $G^{n-1} \neq 1$ *for otherwise, by Lemma 6.5,* G *would be nilpotent of class* $\leq n - 1$, *a contradiction.*

Assuming that $G^n = 1$ *and* $G^{n-1} \neq 1$, *by Lemma 6.4,* G *is nilpotent of class* $\leq n$. *Now* G *cannot be nilpotent of class* $i < n$, *for otherwise, by the reverse direction just proved we would have* $G^i = 1$, *contradicting the fact that* $G^{n-1} \neq 1$.

EXERCISES

1 Prove Theorem 6.11

2 Prove that every nilpotent group is solvable.

3 Prove Theorem 6.12.3 & .4.

4 Determine the solvability class of S_n for $n < 5$.

5 Verify the following:

   a. $B_3(F)$ is nilpotent of class 2.

   b. In general, $B_n(F)$ is nilpotent of class $n - 1$.

   c. D_4 is nilpotent of class 2.

   d. $U_3(F)$ is nilpotent of class 2.

II

Rings and Fields

CHAPTER 7

Ring Theory

I N THIS CHAPTER, we now study a set with **two** operations as opposed to groups which have one operation. This, of course, makes the study both more complicated and more rich not only because of the second operation added, but also because of the interaction between the two operations. As we shall see, rings have many structures and concepts similar to groups such as subring, ring homomorphism and quotient ring and ideals (as opposed to normal subgroups).

7.1 DEFINITION AND EXAMPLES

We now begin the study of rings by presenting basic definitions and lots of examples to back up these definitions.

Definition 7.1 *A* **ring** *is a set R together with two operations which we shall denote by the symbols of $+$ and $\cdot$ which satisfy the following properties:*

1. *$(R, +)$, denoted by R^+, forms an abelian group.*

2. *$(R, \cdot)$ satisfies the closure and associative properties of a group.*

3. *$(R, +, \cdot)$ satisfies the distributive property, namely for all $r, s, t \in R$ we have $r \cdot (s + t) = r \cdot s + r \cdot t$ and $(r + s) \cdot t = r \cdot t + s \cdot t$.*

A **commutative** *ring is a ring in which in addition $(R, \cdot)$ satisfies the commutative property. A ring* **with unity** *is a ring in which in addition $(R, \cdot)$ satisfies the identity property of a group.*

We will denote this unity in a ring with unity by 1 and we sometimes call this ring a ring **with 1**. One can show as we did for groups that this 1 is unique when it exists. We denote the additive identity by 0 and the additive inverse of a ring element r by the notation $-r$.

Example 7.1 *Here, we list some examples of rings. The ring properties are easily verifiable.*

1. *$\mathbb{Z}, \mathbb{Q}, \mathbb{R}$ and $\mathbb{C}$ with $+$ and $\cdot$ being the familiar addition and multiplication each form a commutative ring with unity.*

DOI: 10.1201/9781003335283-7

$+_4$	0	1	2	3
0	0	1	2	3
1	1	2	3	0
2	2	3	0	1
3	3	0	1	2

$\cdot_4$	0	1	2	3
0	0	0	0	0
1	0	1	2	3
2	0	2	0	2
3	0	3	2	1

Figure 7.1 The addition and multiplication table for $\mathbb{Z}_4$.

2. Let $\mathbb{Z}_n = \{0, 1, \ldots, n-1\}$ with addition and multiplication modulo n. Then $\mathbb{Z}$ is a commutative ring with 1. For instance, Figure 7.1 present the addition and multiplication tables for $\mathbb{Z}_4$:

3. For an integer n consider $n\mathbb{Z}$, all integer multiples of n with the same operations as $\mathbb{Z}$. Then for $n \neq 0$ we have a commutative ring without unity.

4. Let $M_n(R)$ be the collection of $n \times n$ matrices with entries from some ring R and define matrix addition and matrix multiplication in the usual linear algebraic way making use of the operations from the ring R. Then $M_n(R)$ is a ring (non commutative) with unity (when R has unity) and is called a **matrix ring**.

5. Let $R[x]$ be the collection of polynomials whose coefficients come from a ring R and define polynomial addition and multiplication in the usual algebraic way making use of the operations from the ring R. Then $R[x]$ is a ring called a **polynomial ring**.

6. Let $\mathcal{F}(X, R)$ be the collection of functions from a set X to a ring R and define addition and multiplication of functions in the usual way, i.e. $(f + g)(x) = f(x) + g(x)$ and $(f \cdot g)(x) = f(x) \cdot g(x)$ using the operations in the ring R. Then $\mathcal{F}(X, R)$ is a ring which is commutative and with unity if R also has those properties.

7. Let $(G, +)$ be an abelian group and let $End(G)$ consist of all the endomorphisms of G. Define addition and multiplication as follows:

$$(\phi + \psi)(g) = \phi(g) + \psi(g) \qquad (\phi \cdot \psi)(g) = \phi(\psi(g)).$$

The $End(G)$ is a ring with unity (non commutative) where the unity is the identity map on G, i.e. 1_G, called the **ring of endomorphisms** of G.

8. Given rings $R_1, R_2, \ldots R_n$ we can define the **cartesian product** ring

$$R_1 \times R_2 \times \cdots \times R_n$$

similar to the construction for groups where operations are defined coordinate-wise.

9. *The ring* $R = \{0, 1\}$ *is called the* **trivial ring** *and* $R = \{0\}$ *is called the* **zero ring**.

Definition 7.2 *Let* R *be a ring with 1. An element* $r \in R$ *is a* **unit** *if it has a multiplicative inverse, i.e. there exists an* $s \in R$ *such that* $rs = 1 = sr$. *The collection of all units in* R *are denoted by* $U(R)$.

Remark 7.1 *When* r *is a unit one can show in the usual way that its inverse is unique, so from now on we will denote the inverse of* r *by the suggestive notation of reciprocal,* r^{-1}. *We leave it as an exercise to show that* $(U(R), \cdot)$ *forms a group.*

Example 7.2 *In each of the examples below it is easy to compute* $U(R)$.

1. $U(\mathbb{Z}) = \{-1, 1\}$.

2. $U(\mathbb{Q}) = \mathbb{Q}^*$, $U(\mathbb{R}) = \mathbb{R}^*$ *and* $U(\mathbb{C}) = \mathbb{C}^*$.

3. $U(M_n(\mathbb{R})) = GL_n(\mathbb{R})$.

4. $U(End(G)) = Aut(G)$.

5. $U(\mathbb{Z}_n) = \{m \in \mathbb{Z}_n \ : \ gcd(m, n) = 1\}$.

Using the multiplicative operation in a ring we can define exponentiation in the natural way. For $r \in R$ and n a positive integer, we define $r^n = r \cdot r \cdots \cdot r$ (n times). If R has unity, then $r^0 = 1$. If r is a unit, then $r^{-n} = r^{-1} \cdot r^{-1} \cdots \cdot r^{-1}$ (n times). We define an additive exponentiation using the additive operation in the ring. In other words for $r \in R$ and n a positive integer, we define $nr = r + r + \cdots + r$ (n times), $0r = 0$ and $(-n)r = (-r) + (-r) + \cdots + (-r)$ (n times). The standard properties of exponentiation hold here, namely, for $r \in R$ a ring a $m, n \in \mathbb{Z}$, we have

$$r^m r^n = r^{m+n} \qquad (r^m)^n = r^{mn} \qquad mr + nr = (m+n)r \qquad n(mr) = (nm)r.$$

Lemma 7.1 *Let* r *and* s *be elements of a ring* R. *Then*

1. $r0 = 0 = 0r$.

2. $r(-s) = (-r)s = -(rs)$.

3. $-(-r) = r$.

4. $(-r)(-s) = rs$.

5. $(-1)r = -r$.

Proof 7.1 *We leave the proof of these statements as a nice exercise.*

Definition 7.3 *Let* $(R, +, \cdot)$ *be a ring. A non-empty subset of* $S \subseteq R$ *is a* **subring** *of* R, *written* $S \leq R$, *if* $(S, +, \cdot)$ *is a ring (using the same two operations).*

Remark 7.2 *As with groups and subgroups, there is a short cut for determining when a non-empty subset is a subring. One simply shows that for all $r, s \in R$ we have $r - s \in R$ and $rs \in R$. The proof of this remark follows quickly from the one given for subgroups.*

Example 7.3 *Here, we list several examples of subrings.*

1. *For any integer n we have $n\mathbb{Z} \leq \mathbb{Z} \leq \mathbb{Q} \leq \mathbb{R} \leq \mathbb{C}$. For instance, to see that $n\mathbb{Z} \leq \mathbb{Z}$, using the shortcut, for $nk, nl \in n\mathbb{Z}$, we have*

   $$nk - nl = n(k - l) \in n\mathbb{Z} \quad and \quad nk \cdot nl = n(knl) \in n\mathbb{Z}.$$

   Since $(n\mathbb{Z}, +)$ are the only subgroups of $(\mathbb{Z}, +)$ it follows that $(n\mathbb{Z}, +, \cdot)$ are the only subrings of $(\mathbb{Z}, +, \cdot)$

2. *Differentiable real-valued functions are a subring of continuous real-valued functions which are a subring of real-valued functions. The notation we will use for these rings are*

   $$\mathcal{D}(\mathbb{R}) \leq \mathcal{C}(\mathbb{R}) \leq \mathcal{F}(\mathbb{R}).$$

3. *For any ring R, the **center** of R, written*

   $$Z(R) = \{r \in R \ : \ rs = sr \text{ for all } s \in R\}.$$

   For instance, one can show that $Z(M_n(R)) = \{rI_n \ : \ r \in R\}$, i.e. scalar matrices. We leave the verification that $Z(R) \leq R$ as an exercise.

Definition 7.4 *R is called a **division ring** if R is a ring with $1 \neq 0$ and $U(R) = R^*$. If in addition R is commutative, then R is called a **field** otherwise it is called a **skew-field**.*

*Analogous to subring, A non-empty subset S of a field $(R, +, \cdot)$ is a **subfield** if $(S, +, \cdot)$ is a field (with the same operations).*

Remark 7.3 *We list a couple of remarks about fields.*

1. *In other words, $(R, +, \cdot)$ is a field if both $(R, +)$ and $(R^*, \cdot)$ are abelian groups and R has the distributive property.*

2. *The shortcut for checking S is a subfield of R is as follows: Show that for all $r, s \in S$ we have $r - s \in S$ and $rs^{-1} \in S$.*

3. *It is not an easy matter to construct a skew-field. In fact, it will be shown that a finite division ring is always a field. In a later section, we will construct the **quaternions** which is an example of an infinite skew-field.*

4. *$\mathbb{Q}, \mathbb{R}, \mathbb{C}$ are examples of infinite fields and $\mathbb{Z}_p$ for any prime p are examples of finite fields. Later we will see that the only finite fields ones of prime power order.*

EXERCISES

1 Verify that each of the rings presented in Example 7.1 are indeed rings.

2 $R = \mathbb{Z}[\sqrt{2}]$. Prove that $m + n\sqrt{2} \in U(R)$ iff $m^2 - 2n^2 = \pm 1$

3 For any ring R, prove that $(U(R), \cdot)$ forms a group.

4 Let R be a ring with $0 \neq 1$ and fix an $r \in R$. Suppose there is a unique $s \in R$ such that $rs = 1$.

    a. Show that r cannot be a zero divisor.

    b. Show that r must be a unit.

5 Verify that the following subset of $M_2(\mathbb{R})$ is a subring:

$$\left\{ \begin{bmatrix} a & b \\ -b & a \end{bmatrix} : a, b \in \mathbb{R} \right\}.$$

6 R a commutative ring such that $a^2 = a$ for all $a \in R$. Prove $a + a = 0$ for all $a \in R$.

7 X a set and $R = \mathcal{P}(X)$ with $A + B = (A \cup B) - (A \cap B)$ and $A \cdot B = A \cap B$.

    a. Prove that R is a commutative ring.

    b. Write out the addition and multiplication table for R when $X = \{a, b\}$.

8 Prove all parts of Lemma 7.1.

9 Verify Remark 7.2.

10 For any ring R, prove that $Z(R) \leq R$.

11 Let R be a ring with unity. Prove that r is a unit iff $\exists s \in R$ such that $rsr = r$ and $sr^2s = 1$.

7.2 INTEGRAL DOMAINS

The generic example of a ring from which perhaps the general definition arose is the integers. An additional property which the integers have is that the only way for a product of two integers to equal zero is if one of the factors is zero. Now this is a very useful property, for instance when solving a factorable polynomial equation such as $x^2 - x - 2 = 0$ from basic algebra. One first factors it as $(x - 2)(x + 1) = 0$ and since the product is zero it must be the case that either $x - 2 = 0$ or $x + 1 = 0$. Thus $x = 2$ or $x = -1$ and we've solved the quadratic equation. In this section, we will generalize this property of the integers.

Definition 7.5 *An element r in a commutative ring R with $1 \neq 0$ is called a **zero divisor** if $r \neq 0$ and there exists an $s \neq 0$ in R such that $rs = 0$.*

Remark 7.4 *Equivalently, r in a commutative ring R with $1 \neq 0$ is* **not** *a zero divisor if whenever $rs = 0$ for some $s \in R$, then either $r = 0$ or $s = 0$. This restatement of the definition in the negation can come in handy for proofs involving zero divisors.*

Example 7.4 *In the ring $\mathbb{Z}_6$ the number 2 is a zero divisor, since $2 \cdot 3 = 0$. Likewise 3 is also a zero divisor.*

Definition 7.6 *A ring is called an* **integral domain** *(which we shall abbreviate as* **ID***) if*

1. *R is commutative with $1 \neq 0$, and*

2. *R has no zero divisors.*

Example 7.5 *Here, we list some examples on the topic of integral domains.*

1. *The ring $\mathbb{Z}_6$ is not an ID since it has zero divisors, such as 2 and 3.*

2. *$\mathbb{Z}$, $\mathbb{Q}$, $\mathbb{R}$, $\mathbb{C}$ and $\mathbb{Z}_p$ (p prime) are all ID's.*

Lemma 7.2 *Let R be a commutative ring with $1 \neq 0$.*

1. *No unit in R can be a zero divisor.*

2. *If R is a field, then R is an ID.*

Proof 7.2 *Let $r \in R$ be a unit and suppose that $rs = 0$. Since r^{-1} exists we can multiply both sides of the equation by r^{-1} to get $r^{-1}(rs) = r^{-1}0$ which simplifies to $s = 0$ and so r is not a zero divisor.*

If R is a field, then every non-zero element in R is a unit and hence R cannot have any zero divisors by the first part of this lemma.

The converse of the second part of Lemma 7.2 is false, i.e. not every ID is a field. Take, for instance, the integers which is an ID but is not a field. However, the result is true under the assumption that the field is finite.

Lemma 7.3 *Every finite ID is a field.*

Proof 7.3 *Let R be a finite ID. It's enough to show that every non-zero element in R is a unit. Take $r \in R^*$ and consider the map $f : R^* \to R^*$ by $f(s) = rs$. This function maps into R^*, since $s \neq 0$ implies that $f(s) = rs \neq 0$. This function is also one-to-one, for suppose that $f(s) = f(t)$ for some $s, t \in R^*$. Then $rs = rt$ and so $r(s - t) = 0$. Now $r \neq 0$ and since R has no zero divisors it must be the case that $s - t = 0$, i.e. $s = t$. Since R^* is finite, by Lemma 1.2.2, f also maps onto R^*. In particular, there exists an $s \in R^*$ such that $f(s) = 1$, i.e. $rs = 1$ and so r is a unit.*

Definition 7.7 *A non-empty subset S of an ID $(R, +, \cdot)$ is a* **subdomain** *if $(S, +, \cdot)$ is an ID.*

TABLE 7.1 ✓ means it has that property and x means it doesn't

Axiom	Ring	Commutative Ring	Ring w/1	Commutative Ring w/1	Division Ring	Skewfield	Field
closure +	✓	✓	✓	✓	✓	✓	✓
assoc +	✓	✓	✓	✓	✓	✓	✓
identity +	✓	✓	✓	✓	✓	✓	✓
inverse +	✓	✓	✓	✓	✓	✓	✓
commute +	✓	✓	✓	✓	✓	✓	✓
closure ·	✓	✓	✓	✓	✓	✓	✓
assoc ·	✓	✓	✓	✓	✓	✓	✓
identity ·			✓	✓	✓	✓	✓
inverse ·					✓	✓	✓
commute ·		✓		✓		x	✓
distribute ·	✓	✓	✓	✓	✓	✓	✓

Remark 7.5 *One can show that the shortcut for verifying subdomain is to show $1 \in S$ and for all $a, b \in S$ both $a - b \in S$ and $ab \in S$.*

Example 7.6 *Here are some examples that relate to subdomains.*

1. *The integers $\mathbb{Z}$ is a subdomain of the rationals $\mathbb{Q}$ since $1 \in \mathbb{Z}$.*

2. *For $n \neq 1$ the subring $n\mathbb{Z}$ of $\mathbb{Z}$ is **not** a subdomain of $\mathbb{Z}$ since $1 \notin n\mathbb{Z}$.*

We end this section with a summary of the structures we have seen in this chapter thus far and the properties they each have (Table 7.1).

EXERCISES

1 Let $R = \{m + 2ni \; : \; m, n \in \mathbb{Z}\}$. Verify that R is an ID by showing it's a subdomain of $\mathbb{Z}[i]$.

2 Prove that $\mathcal{F}(X, R)$ is not in general an integral domain.

3 Given an ID R, show that a non-empty subset $S \subseteq R$ is a subdomain iff $1 \in S$ and for all $a, b \in S$ both $a - b \in S$ and $ab \in S$.

4 Prove that given S is a subring of an integral domain R, if S has unity, then it must be the same unity that R has.

5 How many elements in an integral domain have the property that $a^2 = a$?

6 Let R be a ring with $|R| \geq 2$ and for all $a \in R*$ there exists a unique $b \in R$ such that $aba = a$. Prove that R must be a division ring.

7.3 THE QUATERNIONS

In this section, we carefully examine one example of a (necessarily infinite) skewfield called the **quaternions** discovered by the Irish mathematician William Rowan Hamilton. As the story goes he was so excited over his discover that he etched in a bridge he was crossing the basic relations which define this skewfield.

We will give two different presentations of this skewfield, since each of the two presentations has its benefits. The first representation is matrix based, while the second is a more formal representation.

Definition 7.8 *The following set of matrices with the usual matrix addition and matrix multiplication is called the* **quaternions***:*

$$Q = \left\{ \begin{bmatrix} z & w \\ -\overline{w} & \overline{z} \end{bmatrix} : z, w \in \mathbb{C} \right\}.$$

Theorem 7.1 *The quaternions form a skewfield.*

Proof 7.4 *Since $Q \subseteq M_2(\mathbb{C})$, we first check that Q is a subring of $M_2(\mathbb{C})$. Set*

$$A = \begin{bmatrix} z_1 & w_1 \\ -\overline{w}_1 & \overline{z}_1 \end{bmatrix} \quad and \quad B = \begin{bmatrix} z_2 & w_2 \\ -\overline{w}_2 & \overline{z}_2 \end{bmatrix}.$$

Then

$$A - B = \begin{bmatrix} z_1 - z_2 & w_1 - w_2 \\ -\overline{w}_1 + \overline{w}_2 & \overline{z}_1 - \overline{z}_2 \end{bmatrix} = \begin{bmatrix} z_1 - z_2 & w_1 - w_2 \\ -\overline{(w_1 - w_2)} & \overline{z_1 - z_2} \end{bmatrix} \in Q \quad and$$

$$AB = \begin{bmatrix} z_1 z_2 - w_1 \overline{w}_2 & z_1 w_2 + w_1 \overline{z}_2 \\ -z_2 \overline{w}_1 - \overline{z}_1 \overline{w}_2 & -\overline{w}_1 w_2 + \overline{z}_1 \overline{z}_2 \end{bmatrix} = \begin{bmatrix} z_1 z_2 - w_1 \overline{w}_2 & z_1 w_2 + w_1 \overline{z}_2 \\ -\overline{(z_1 w_2 + w_1 \overline{z}_2)} & \overline{z_1 z_2 - w_1 \overline{w}_2} \end{bmatrix} \in Q.$$

Q has unity, since

$$I_2 = \begin{bmatrix} 1 & 0 \\ -\overline{0} & \overline{1} \end{bmatrix} \in Q.$$

Every non-zero matrix in Q is a unit, since for any non-zero matrix

$$A = \begin{bmatrix} z & w \\ -\overline{w} & \overline{z} \end{bmatrix} = \begin{bmatrix} a + bi & c + di \\ -c + di & a - bi \end{bmatrix} \in Q,$$

the determinant is non-zero. Indeed, $|A| = z\overline{z} + \overline{w}w = a^2 + b^2 + c^2 + d^2 > 0$ when $A \neq 0$ and

$$A^{-1} = \left[\begin{array}{cc} \overline{z}/|A| & -w/|A| \\ \overline{w}/|A| & z/|A| \end{array} \right] = \left[\begin{array}{cc} \overline{z}/|A| & -w/|A| \\ -\overline{(-w)}/|A| & \overline{\overline{z}}/|A| \end{array} \right] \in Q.$$

Finally, Q is non-commutative, since for instance one can check the following two matrices in Q do not commute:

$$\left[\begin{array}{cc} i & 0 \\ 0 & -i \end{array} \right] \quad and \quad \left[\begin{array}{cc} 0 & 1 \\ -1 & 0 \end{array} \right].$$

Hence, Q forms a skewfield.

We now present an alternate more formal definition of the quaternions. Set i, j and k to be new symbols and define

$$Q = \{a + bi + cj + dk \;:\; a, b, c, d \in \mathbb{R}\}.$$

Addition will be defined component-wise, i.e.

$$(a_1+b_1i+c_1j+d_1k)+(a_2+b_2i+c_2j+d_2k) = (a_1+a_2)+(b_1+b_2)i+(c_1+c_2)j+(d_1+d_2)k.$$

In order to define multiplication, we first establish the following relations:

$$i^2 = j^2 = k^2 = -1, \quad ij = -ji = k, \quad jk = -kj = i \text{ and } ki = -ik = j.$$

Indeed, these are the very relations which helped define the groups of quaternions. A mnemonic for remembering some of the relations is to consider i, j and k as the unit vectors in 3-space and multiplication is cross product and use the right-hand rule. Using these relations, we can then define multiplication in Q by formal distribution and addition as defined above.

Example 7.7 *We will illustrate the two operations in Q by a couple of examples.*

1. *We can add two elements of Q,*

$$(2 + j - k) + (3i - k) = 2 + 3i + j - 2k$$

2. *We multiply two elements of Q,*

$$(2 + j - k) \cdot (3i - k) = 6i - 2k + 3ji - jk - 3ki + k^2$$

$$= 6i - 2k - 3k - i - 3j - 1 = -1 + 5i - 3j - 5k.$$

Now although this representation of Q is easier to manipulate, the verification that this representation of Q is a skewfield requires more work since it does not sit in any familiar setting having been formally defined (and we will omit this proof). Finally, the correspondence between the two representations is given as follows:

$$\begin{bmatrix} a + bi & c + di \\ -c + di & a - bi \end{bmatrix} = \begin{bmatrix} a & 0 \\ 0 & a \end{bmatrix} + \begin{bmatrix} bi & 0 \\ 0 & -bi \end{bmatrix} + \begin{bmatrix} 0 & c \\ -c & 0 \end{bmatrix} + \begin{bmatrix} 0 & di \\ di & 0 \end{bmatrix}$$

$$= a \begin{bmatrix} 1 & 0 \\ 0 & 1 \end{bmatrix} + b \begin{bmatrix} i & 0 \\ 0 & -i \end{bmatrix} + c \begin{bmatrix} 0 & 1 \\ -1 & 0 \end{bmatrix} + d \begin{bmatrix} 0 & i \\ i & 0 \end{bmatrix} \longleftrightarrow a + bi + cj + dk.$$

In other words,

$$i \longleftrightarrow \begin{bmatrix} i & 0 \\ 0 & -i \end{bmatrix}, \quad j \longleftrightarrow \begin{bmatrix} 0 & 1 \\ -1 & 0 \end{bmatrix} \text{ and } k \longleftrightarrow \begin{bmatrix} 0 & i \\ i & 0 \end{bmatrix}.$$

The reader may wish to check that the basic relations for i, j and k are satisfied for these three matrices.

Definition 7.9 *The following subset of the quaternions is called the* **Hamiltonian integers***:*

$$H = \{a + bi + cj + dk \ : \ a, b, c, d \in \mathbb{Z} \}.$$

Remark 7.6 *We list a few facts about the Hamiltonian integers.*

1. *One can easily see that H is a subring of Q.*

2. *The units of H, which we denote by $Q_8 = \{\pm 1, \pm i, \pm j, \pm k\}$ and is called the* **quaternion group***. To see that these eight elements of H are indeed the units of H, first note that certainly each of the eight elements is a unit. For instance, $i^{-1} = -i$. Now suppose that $A = a + bi + cj + dk \in U(H)$ (we switch now to the matrix representation of H). Then there is a $B \in H$ such that $AB = I_2$. Then*

$$1 = |I_2| = |AB| = |A||B| = (a^2 + b^2 + c^2 + d^2)|B|.$$

Hence, $a^2 + b^2 + c^2 + d^2 = 1$. Since $a, b, c, d \in \mathbb{Z}$, this implies that either $a = \pm 1$ and $b = c = d = 0$, $b = \pm 1$ and $a = c = d = 0$, $c = \pm 1$ and $a = b = d = 0$, or $d = \pm 1$ and $a = b = c = 0$. These eight cases correspond precisely the eight elements listed in Q_8.

We end this section with a diagram illustrating all the structures we've seen so far including examples for each type (Figure 7.2).

EXERCISES

1 Check the following two matrices in the quaternions do not commute:

$$\begin{bmatrix} i & 0 \\ 0 & -i \end{bmatrix} \text{ and } \begin{bmatrix} 0 & 1 \\ -1 & 0 \end{bmatrix}.$$

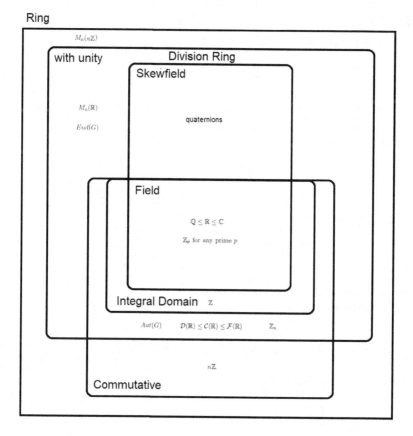

Figure 7.2 The world of rings.

2 Consider the quaternions $Q = \{a + bi + cj + dk \mid a, b, c, d \in \mathbb{R}\}$. Describe the elements of Q which commute with i (under multiplication).

3 Verify that the center of the quaternions consists of real scalar matrices, i.e. $Z(Q) = \{aI_2 \ : \ a \in \mathbb{R}\}$.

4 If we set

$$ i = \begin{bmatrix} i & 0 \\ 0 & -i \end{bmatrix}, \quad j = \begin{bmatrix} 0 & 1 \\ -1 & 0 \end{bmatrix} \text{ and } k = \begin{bmatrix} 0 & i \\ i & 0 \end{bmatrix}, $$

Verify the following relations:

$$ i^2 = j^2 = k^2 = -1, \quad ij = -ji = k, \quad jk = -kj = i \text{ and } ki = -ik = j. $$

5 Verify that the Hamiltonian integers form a subring of the quaternions.

7.4   RING HOMOMORPHISMS

Just as we had a notion of a group homomorphisms we now define a ring homomorphism. The parallels between the two algebraic structures groups and rings with regards to homomorphisms are extensive.

Definition 7.10 *Let $(R, +, \cdot)$ and $(R', +', \cdot')$ be two rings. A function $\phi : R \to R'$ is a **ring homomorphism** if for all $r, s \in R$, we have*

$$\phi(r + s) = \phi(r) +' \phi(s) \quad and \quad \phi(r \cdot s) = \phi(r) \cdot' \phi(s).$$

Example 7.8 *We list here several examples of ring homomorphisms.*

1. *Consider the ring $\mathbb{C}$ and the subring of $M_2(\mathbb{R})$ defined by*

$$R = \left\{ \begin{bmatrix} a & b \\ -b & a \end{bmatrix} : a, b \in \mathbb{R} \right\}.$$

The map $\phi : \mathbb{C} \to R$ by

$$\phi(a + bi) = \begin{bmatrix} a & b \\ -b & a \end{bmatrix},$$

is a ring homomorphism, since

$$\phi[(a + bi) + (c + di)] = \phi[(a + c) + (b + d)i] = \begin{bmatrix} a + c & b + d \\ -(b + d) & a + c \end{bmatrix}$$

$$= \begin{bmatrix} a & b \\ -b & a \end{bmatrix} + \begin{bmatrix} c & d \\ -d & c \end{bmatrix} = \phi(a + bi) + \phi(c + di).$$

$$\phi[(a + bi)(c + di)] = \phi[(ac - bd) + (ad + bc)i] = \begin{bmatrix} ac - bd & ad + bc \\ -(ad + bc) & ac - bd \end{bmatrix}$$

$$= \begin{bmatrix} a & b \\ -b & a \end{bmatrix} \begin{bmatrix} c & d \\ -d & c \end{bmatrix} = \phi(a + bi)\phi(c + di).$$

2. *The map $\phi : \mathbb{Z} \to \mathbb{Z}_n$ by $\phi(m) = r$, where $m = nq + r$ and $0 \leq r < n$, is a ring homomorphism.*

3. *Let R and R' be rings, and choose an $r \in R$. The map $\phi : \mathcal{F}(R, R') \to R'$ by $\phi(f) = f(r)$ is a ring homomorphism and is called the **evaluation** homomorphism.*

4. *We have the usual **identity** ring homomorphism, $1_R : R \to R$ by $1_R(r) = r$ for all $r \in R$ and the **trivial** (or **zero**) ring homomorphism $0_{RR'} : R \to R'$ defined by $\phi(r) = 0'$ for all $r \in R$, where $0'$ is the additive identity in the ring R'.*

Definition 7.11 *Let $\phi : R \to R'$ be a ring homomorphism. The **kernel** of ϕ, written*

$$\ker \phi = \{r \in R : \phi(r) = 0'\}.$$

Example 7.9 *Here, we discuss the kernel of specific examples of ring homomorphisms.*

1. *For the map given above in Example 7.8.1, one can check that the kernel is trivial, i.e.* $\ker \phi = \{0\}$.

2. *For the map given above* $\phi : \mathbb{Z} \to \mathbb{Z}_n$, *the kernel is* $n\mathbb{Z}$.

3. *The kernel of the evaluation homomorphism is the collection of all functions which take on the value 0 at* r.

4. $\ker 1_R = \{0\}$ *and* $\ker 0_{RR'} = R$.

The next result lists a few immediate consequences of the definitions given thus far in this section. Some can be verified by appealing to group theoretic results we have already proved.

Lemma 7.4 *Let* $\phi : R \to R'$ *be a ring homomorphism and* $r \in R$.

1. $\phi(0) = 0'$ *and* $\phi(-r) = -\phi(r)$.

2. $\ker \phi$ *is a subring of* R.

3. ϕ *is one-to-one iff* $\ker \phi = \{0\}$.

4. $\phi(R)$ *is a subring of* R'.

Definition 7.12 *Let* $\phi : R \to R'$ *be a ring homomorphism.*

1. *If* ϕ *is one-to-one, then it is called a* **monomorphism**.

2. *If* ϕ *maps onto* R', *then it is called a* **epimorphism**.

3. *If* ϕ *is one-to-one and maps onto* R', *then it is called a* **isomorphism** *and we say the rings* R *and* R' *are* **isomorphic** *and we write* $R \cong R'$.

Example 7.10 *Let's revisit Example 7.8.*

1. *For the map defined above* $\phi : \mathbb{C} \to R$ *we have already computed* $\ker \phi = \{0\}$ *and so by Lemma 7.4* ϕ *is one-to-one. Furthermore, it is clear from the definition of* ϕ *that it maps onto* R. *Hence,* ϕ *is an isomorphism and* $\mathbb{C} \cong R$.

2. *For the map* $\phi : \mathbb{Z} \to \mathbb{Z}_n$ *we have seen the kernel is non-trivial and so* ϕ *is not one-to-one. However, it is immediate from the definition of* ϕ *that it maps onto* $\mathbb{Z}_n$ *and so* ϕ *is an epimorphism.*

3. 1_R *is an isomorphism and* $0_{RR'}$ *is simply a homomorphism in general.*

EXERCISES

1 Verify that Example 7.8.2,.3,.4 are indeed ring homomorphisms.

2 Verify that the kernel of Example 7.8.1 is trivial.

3 Prove the statements listed in Lemma 7.4.

4 Prove the only ring automorphism of $\mathbb{Z}$ is the identity map.

5 If R is a commutative ring with unity and D is an integral domain and $\phi : R \to D$ is a ring homomorphism not the zero map, then $\phi(1) = 1$.

6 Let $\phi : R \to R'$ and $\psi : R' \to R''$ be two ring homomorphisms.

   a. If ϕ is an isomorphism, then so is ϕ^{-1}.

   b. $\psi \circ \phi$ is a homomorphism.

7.5   FACTOR RINGS AND IDEALS

Analogous to the concept of a factor group and a normal subgroup in group theory are the factor ring and *ideal* subring in ring theory. Let $(R, +, \cdot)$ be a ring and S a subring of R. The elements of the factor ring will be cosets of R modulo S using the additive structure of the ring $(R, +)$. In other words, $R/S = \{r + S \ : \ s \in R\}$. We remind the reader of some elementary facts about cosets that are used again and again.

1. $r_1 + S = r_2 + S$ iff $r_1 \in r_2 + S$ iff $r_1 - r_2 \in S$.

2. $r + S = S$ iff $r \in S$.

We would like to define the ring operations for this new structure *representative-wise*, that is,

$$(r_1 + S) + (r_2 + S) = (r_1 + r_2) + S \ \text{ and } \ (r_1 + S) \cdot (r_2 + S) = (r_1 \cdot r_2) + S.$$

As the reader will remember, the problem which can arise with these coset operations is that they may not be *well-defined* functions (from $R/S \times R/S$ to R/S). If we can get beyond this hurdle, then R/S can be easily seen to form a ring which we call the **factor** or **quotient** ring of R modulo S. The additive identity in this ring will be S and should R have unity, then R/S will have unity $1 + S$.

Recall that the cosets of a group modulo a subgroup had a well-defined operation iff the subgroup was normal. Now, since $(R, +)$ is an abelian group, we know $(S, +)$ is a normal subgroup and so coset addition in R/S is always well-defined. We now present the condition which guarantees that coset multiplication in R/S is well-defined.

Definition 7.13 *A subring I of R is called an* **ideal**, *written $I \lhd R$, if for all $r \in R$ and $a \in I$, we have $ra, ar \in I$.*

+	$3\mathbb{Z}$	$1+3\mathbb{Z}$	$2+3\mathbb{Z}$
$3\mathbb{Z}$	$3\mathbb{Z}$	$1+3\mathbb{Z}$	$2+3\mathbb{Z}$
$1+3\mathbb{Z}$	$1+3\mathbb{Z}$	$2+3\mathbb{Z}$	$3\mathbb{Z}$
$2+3\mathbb{Z}$	$2+3\mathbb{Z}$	$3\mathbb{Z}$	$1+3\mathbb{Z}$

$\cdot$	$3\mathbb{Z}$	$1+3\mathbb{Z}$	$2+3\mathbb{Z}$
$3\mathbb{Z}$	$3\mathbb{Z}$	$3\mathbb{Z}$	$3\mathbb{Z}$
$1+3\mathbb{Z}$	$3\mathbb{Z}$	$1+3\mathbb{Z}$	$2+3\mathbb{Z}$
$2+3\mathbb{Z}$	$3\mathbb{Z}$	$2+3\mathbb{Z}$	$1+3\mathbb{Z}$

Figure 7.3 The addition and multiplication table for $\mathbb{Z}/3\mathbb{Z}$.

Lemma 7.5 *Given a ring R and subring I of R. Coset multiplication for R/I is well-defined iff $I \lhd R$.*

Proof 7.5 *Assume first that $I \lhd R$. Suppose $r + I = r' + I$ and $s + I = s' + I$. Then, $r' \in r + I$ and $s' \in s + I$ and so $r' = r + a$ and $s' = s + b$ for some $a, b \in I$. We want to show $(r + I)(s + I) = (r' + I)(s' + I)$, i.e. $rs + I = r's' + I$ or $r's' \in rs + I$. Now $r's' = (r + a)(s + b) = rs + rb + as + ab$ and since $I \lhd R$ we know $rb, as, ab \in I$. Thus, $rb + as + ab \in I$ and so $rs + rb + as + ab \in rs + I$.*

Now we assume that coset multiplication is well-defined. For $a \in I$ and $r \in R$ note that

$$ra \in ra + I = (r + I)(a + I) = (r + I)I = (r + I)(0 + I) = r0 + I = 0 + I = I.$$

Similarly, one can show $ar \in I$ and so $I \lhd R$.

Corollary 7.1 *Let R be a ring with subring I. The cosets of R/I with coset addition and multiplication forms a ring iff $I \lhd R$.*

Proof 7.6 *Most of the work has already been done and we leave the details to the reader as an exercise*

Example 7.11 *Consider the ring $\mathbb{Z}$ and the subring $3\mathbb{Z}$. One can check that $3\mathbb{Z}$ is an ideal and so we may consider the factor ring $\mathbb{Z}/3\mathbb{Z}$. The addition and multiplication table for this ring are given in Figure 7.3.*

It is no coincidence that these tables look remarkably similar to the tables for the ring $\mathbb{Z}_3$. Later in the text we shall show they are isomorphic as rings.

Remark 7.7 *The shortcut for checking that $I \lhd R$ is to check for all $a, b \in I$ we have $a - b \in I$ and for all $r \in R$ we have $ar, ra \in I$ (or just $ar \in I$ if R is commutative). The reader should verify this as an exercise.*

Example 7.12 *We now give various examples of ideals.*

1. $n\mathbb{Z} \lhd \mathbb{Z}$ *(exercise). Recall from earlier that the only subrings of $\mathbb{Z}$ are subrings of the form $n\mathbb{Z}$, and thus they are also the only ideals in $\mathbb{Z}$.*

2. *Consider the ring of functions $\mathcal{F}(X, R)$ from any set X to a ring R. Fix an $x_0 \in X$ and set $I = \{f \in \mathcal{F}(X, R) : f(x_0) = 0\}$. Then $I \lhd \mathcal{F}(X, R)$, since if $f, g \in I$, then $(f - g)(x_0) = f(x_0) - g(x_0) = 0 - 0 = 0$ and if h is any function from X to R, then $(fh)(x_0) = f(x_0)h(x_0) = 0 \cdot h(x_0) = 0$ and in like manner $(hf)(x_0) = 0$.*

3. $Aut(G) \lhd End(G)$ *(exercise).*

4. *Let $R = M_2(\mathbb{R})$ and $I = \left\{ \begin{bmatrix} a & 0 \\ b & 0 \end{bmatrix} : a, b \in \mathbb{R} \right\}$. Now I is not an ideal of R, for although*

$$\begin{bmatrix} w & x \\ y & z \end{bmatrix} \begin{bmatrix} a & 0 \\ b & 0 \end{bmatrix} = \begin{bmatrix} aw + bx & 0 \\ ay + bz & 0 \end{bmatrix} \in I, \quad \text{however}$$

$$\begin{bmatrix} a & 0 \\ b & 0 \end{bmatrix} \begin{bmatrix} w & x \\ y & z \end{bmatrix} = \begin{bmatrix} aw & ax \\ bw & bx \end{bmatrix} \notin I \quad \text{when } x \neq 0.$$

5. *If $\phi : R \to R'$ be a ring homomorphism, then $\ker \phi \lhd R$ (exercise).*

6. *Let R be a commutative ring and fix an $a \in R$. Then $I = \{ra : r \in R\}$ is an ideal in R called the **principal** ideal **generated by** a. The notation employed for such as ideal is $I = (a)$ or $I = Ra$. Note that in $\mathbb{Z}$ all ideals are principal ideals, since $n\mathbb{Z} = (n)$ and as we pointed out these are the only ideals in $\mathbb{Z}$.*

7. *Let X be a subset of a commutative ring R. The ideal **generated by** X, written $I = (X)$ is the smallest ideal containing the set X. If $X = \{r_1, r_2, \ldots, r_n\}$ is finite then we write $I = (r_1, r_2, \ldots, r_n)$. Note that the ideal generated by a single element of R is a principal ideal.*

8. *For any ring R, $I = \{0\}$ is called the **trivial** ideal and R is called the **improper** ideal.*

Remark 7.8 *We make several remarks many of which are left as exercises to verify.*

1. *If R is a ring with 1 and I is an ideal of R containing 1, then $I = R$. Indeed, for any $r \in R$ we have $r = r1 \in I$.*

2. *If R is a ring with 1 and I is an ideal of R containing a unit, then $I = R$, for suppose $r \in I$ is a unit. Then r^{-1} exists and so $1 = r^{-1}r \in I$, and by above $I = R$.*

3. *A field has no proper non-trivial ideals.*

4. *Any ring homomorphism between two fields is either the zero map or a monomorphism.*

5. *For any ideal I in a ring R, the* **canonical** *map $\nu : R \to R/I$ by $\nu(r) = r + I$ is a ring homomorphism with kernel equal to I.*

6. *A subset X of a ring R is an ideal iff X is the kernel of some ring homomorphism with domain R.*

Theorem 7.2 (Fundamental Theorem of Ring Homomorphisms) *Let $\phi : R \to R'$ be a ring homomorphism and set $K = \ker \phi$. Then $R/K \cong \phi(R)$ and if ϕ is an epimorphism, then $R/K \cong R'$.*

Proof 7.7 *Analogous to the case for groups, one now shows that the map $\Psi : R/K \to \phi(R)$ by $\Psi(r + K) = \phi(r)$ is a ring isomorphism.*

Example 7.13 *Recall the epimorphism $\phi : \mathbb{Z} \to \mathbb{Z}_n$ in Example 7.8.2. We found the kernel to be $n\mathbb{Z}$ and so by FTH, $\mathbb{Z}_n \cong \mathbb{Z}/n\mathbb{Z}$ as rings just as they were as additive groups.*

EXERCISES

1 Prove Corollary 7.1.

2 Prove in the ring $(\mathbb{Z}, +, \cdot)$ that $n\mathbb{Z} \lhd \mathbb{Z}$ for any integer n.

3 Prove that $I \lhd R$ iff for all $a, b \in I$ we have $a - b \in I$ and for all $r \in R$ we have $ar, ra \in I$.

4 Prove that if $\phi : R \to R'$ be a ring homomorphism, then $\ker \phi \lhd R$.

5 Prove that $End(G) \lhd Aut(G)$.

6 Show for $r_1, r_2, \ldots, r_n \in R$ a commutative ring that $I = (r_1, r_2, \ldots, r_n)$ can be written as
$$\{s_1 r_1 + s_2 r_2 + \cdots + s_n r_n : s_1, s_2, \ldots, s_n \in R\}.$$

7 Fix a prime p and set $R = \{\frac{a}{b} : p \nmid b\}$ and $I = \{\frac{a}{b} \in R : p \mid a\}$.

   a. Prove that R is a ring.

   b. Verify that $I \lhd R$.

   c. Verify that $\phi : R \to \mathbb{Z}_p$ by $\phi(\frac{a}{b}) = ab^{-1}(mod\ p)$ is a ring epimorphism.

   d. Compute $ker\phi$.

   e. Apply the FTH to parts c and d.

8 Prove if $I, J \lhd R$ a commutative ring, then $I \cap J \lhd R$.

9 Let R be a commutative ring and $a \in R$. Define $I = \{r \in R \ : \ ra = 0\}$. Prove $I \lhd R$.

10 Consider the ideals $I, J \lhd R$ a ring and define

$$IJ = \{a_1b_1 + a_2b_2 + \cdots + a_nb_n \mid \text{each } a_i \in I, \ \text{each } b_i \in J, \ n \in \mathbb{Z}^{>0}\}.$$

Prove that $IJ \lhd R$.

11 Verify that a principal ideal is indeed an ideal.

12 Verify that the trivial ideal is indeed an ideal.

13 A field has no proper non-trivial ideals.

14 Any ring homomorphism between two fields is either the zero map or a monomorphism.

15 For any ideal I in a ring R, the **canonical** map $\nu : R \to R/I$ by $\nu(r) = r + I$ is a ring homomorphism with kernel equal to I.

16 A subset X of a ring R is an ideal iff X is the kernel of some ring homomorphism with domain R.

17 Prove Theorem 7.2.

18 **Correspondence Theorem:** Let $\phi : R \to R'$ be a ring epimorphism. There is a one-to-one and onto inclusion preserving map between the ideals of R' and the ideals of R containing $\ker \phi$.

   Hint: send each ideal I in R containing $\ker \phi$ to $\phi(I)$ and each ideal I' in R' to $\phi^{-1}(I')$, then verify that $\phi(\phi^{-1}(I')) = I'$, $\phi^{-1}(\phi(I)) = I$, $I \subseteq J$ implies $\phi(I) \subseteq \phi(J)$ and $I' \subseteq J'$ implies $\phi^{-1}(I') \subseteq \phi^{-1}(J')$.

19 **Second Isomorphism Theorem:** Let I be an ideal in a ring R and S a subring of R. Then

   a. $S + I \lhd R$

   b. $S \cap I \lhd R$

   c. $(S + I)/I \cong S/(S \cap I)$.

20 **Third Isomorphism Theorem:** If I and J are both ideals of a ring R with $I \subseteq J \subseteq R$, then $(R/I)/(J/I) \cong R/J$.

21 Let I and J both be ideals in a ring R. Define IJ to be the smallest ideal containing all products of the form ab where $a \in I$ and $b \in J$.

   a. Show that IJ consists of all possible sums $a_1b_1 + a_2b_2 + \cdots + a_nb_n$ where each $a_i \in I$, each $b_i \in J$ and n is any positive integer.

b. Give an example illustrating that it is not always the case that IJ is simply elements of the form ab where $a \in I$ and $b \in J$
   Hint: Look at the ring $\mathbb{Z}[x]$ with $I = (2, x)$ and $J = (3, x)$.

c. Show $IJ \subseteq I \cap J$.

d. $II \neq I$ in general.

e. $I + I = I$.

f. If K is another ideal of R, show that $I(JK) = (IJ)K$ and $I(J + K) = IJ + IK$.

7.6 QUOTIENT FIELD OF AN INTEGRAL DOMAIN

Our goal in this section is to generalize the construction of the rational numbers from the integers and pinpoint the exact connection between the two structures as it relates to ring theory. First off, one can think of the integers as the generic example of an integral domain and the rationals as the quotients of integers which form a field – we will call this the **quotient field**.

Let's begin the general construction: Let R be an integral domain and define the following relation on the set $R \times R^*$:

$$(a, b) \sim (c, d) \quad \text{iff} \quad ad = bc.$$

The reader should recognize this familiar relation, for it is the relation which identifies two fractions as being equal. This relation is, in fact, an equivalence relation which we prove now.

1. **Reflexive:** $(a, b) \sim (a, b)$ since $ab = ba$ and R is commutative.

2. **Symmetric:** If $(a, b) \sim (c, d)$, then $ad = bc$ and so $cb = da$ (since R is commutative) which implies $(c, d) \sim (a, b)$.

3. **Transitive:** If $(a, b) \sim (c, d)$ and $(c, d) \sim (e, f)$, then $ad = bc$ and $cf = de$. Notice $adf = bcf = bde$ and so using commutativity and cancellation in R we have $af = be$ which implies $(a, b) \sim (e, f)$.

We will define a fraction to be the class of an element in $R \times R^*$, i.e. $\frac{a}{b} = [(a, b)]$. Then the elements of the quotient field of R, which we will denote as $Q(R)$, will be the set of all fractions just defined. Now it should be of no surprise to the reader how we intend to define addition and multiplication in $Q(R)$.

$$\frac{a}{b} + \frac{c}{d} = \frac{ad + bc}{bd} \qquad \text{and} \qquad \frac{a}{b} \cdot \frac{c}{d} = \frac{ac}{bd}.$$

We have to be careful now for notice that we have defined two binary operations on equivalence classes and since an equivalence class has many difference representations we want to be sure that the operations are not dependent on the representations (i.e. the operations are *well-defined*). In other words,

$$\text{if} \quad \frac{a}{b} = \frac{a'}{b'} \quad \text{and} \quad \frac{c}{d} = \frac{c'}{d'}, \quad \text{then} \quad \frac{a}{b} + \frac{c}{d} = \frac{a'}{b'} + \frac{c'}{d'} \quad \text{and} \quad \frac{a}{b} \cdot \frac{c}{d} = \frac{a'}{b'} \cdot \frac{c'}{d'}.$$

Claim 7.1 *The addition and multiplication defined above for $Q(R)$ are well-defined.*

Proof 7.8 *If $\frac{a}{b} = \frac{a'}{b'}$ and $\frac{c}{d} = \frac{c'}{d'}$, then $ab' = ba'$ and $cd' = dc'$. Now to show addition is well-defined we need to verify that $\frac{ad+bc}{bd} = \frac{a'd'+b'c'}{b'd'}$ or equivalently that $(ad + bc)b'd' = bd(a'd' + b'c')$ or $adb'd' + bcb'd' = bda'd' + bdb'c'$ which follows, since $ab' = ba'$ and $cd' = dc'$. To show multiplication is well-defined we need to verify that $\frac{ac}{bd} = \frac{a'c'}{b'd'}$ or equivalently that $acb'd' = bda'c'$ which follows also from the fact that $ab' = ba'$ and $cd' = dc'$.*

To show $Q(R)$ with these two operations forms a field is identical to showing that the rational numbers $\mathbb{Q}$ with the usual addition and multiplication is a field and so we withhold the proof. We point out though that $\frac{0}{1}$ is the additive identity and $\frac{1}{1}$ is the multiplicative identity in $Q(R)$ (again, to no one surprise).

Now that we have constructed the quotient field of an integral domain we now wish that it has certain properties, namely that there is an isomorphic copy of R in $Q(R)$ and $Q(R)$ is the *smallest* field with this property. To say that $Q(R)$ contains an isomorphic copy of R means formally that there is a monomorphism from R into $Q(R)$ and we say R **embeds in** $Q(R)$. We are familiar with this property since the integers are contained isomorphically in the rational numbers, since any integer n can be viewed as the fraction $\frac{n}{1}$.

Claim 7.2 *R embeds in $Q(R)$.*

Proof 7.9 *Define the following map $\pi : R \to Q(R)$ by $\pi(a) = \frac{a}{1}$. First of all, π is a ring with unity homomorphism, since*

$$\pi(a + b) = \tfrac{a+b}{1} = \tfrac{a1+1b}{1 \cdot 1} = \tfrac{a}{1} + \tfrac{b}{1} = \pi(a) + \pi(b),$$

$$\pi(ab) = \tfrac{ab}{1} = \tfrac{ab}{1 \cdot 1} = \tfrac{a}{1}\tfrac{b}{1} = \pi(a)\pi(b) \text{ and}$$

$$\pi(1) = \tfrac{1}{1}.$$

Second, π is one-to-one, since $\pi(a) = \frac{0}{1}$ implies $\frac{a}{1} = \frac{0}{1}$ and so $a1 = 1 \cdot 0$ or $a = 0$. Hence, $\ker\pi = \{0\}$.

Now we show that $Q(R)$ is the *smallest* field into which R can embed. The next claim states formally what we mean by $Q(R)$ being *smallest*. Basically, it says that if R embeds in a field K, then, in fact, the entire quotient ring $Q(R)$ can embed in K.

Claim 7.3 *Suppose R is an integral domain and there exists a field K and a ring with unity monomorphism $\phi : R \to K$, then there exists a unique ring with unity monomorphism $\phi^* : Q(R) \to K$ such that $\phi^*\left(\frac{a}{1}\right) = \phi(a)$ for all $a \in R$, i.e. $\phi^* \circ \pi = \phi$ where π is defined as in Claim 7.2.*

Proof 7.10 *To show the existence of a map ϕ^* we define $\phi^* : Q(R) \to K$ by $\phi^*\left(\frac{a}{b}\right) = \phi(a)\phi(b)^{-1}$. First note that we have the desired property that $\phi^*(\frac{a}{b}) = \phi(a)\phi(1)^{-1} =$*

$\phi(a)$ and so ϕ^* does indeed extend ϕ. Now ϕ^* is a well-defined (this needs to be checked, since the domain of this map is equivalence classes) ring with unity monomorphism, since

Well-defined and One-to-one:

$\frac{a}{b} = \frac{c}{d}$ iff $ad = bc$ iff $\phi(ad) = \phi(bc)$ iff $\phi(a)\phi(d) = \phi(b)\phi(c)$

iff $\phi(a)\phi(b)^{-1} = \phi(c)\phi(d)^{-1}$ iff $\phi^*\left(\frac{a}{b}\right) = \phi^*\left(\frac{c}{d}\right)$

Ring with Unity Homomorphism:

(a.) $\phi^*\left(\frac{a}{b} + \frac{c}{d}\right) = \phi^*\left(\frac{ad+bc}{bd}\right) = \phi(ad + bc)\phi(bd)^{-1} = [\phi(a)\phi(d)$

$+\phi(b)\phi(c)]\phi(b)^{-1}\phi(d)^{-1} = \phi(a)\phi(b)^{-1} + \phi(c)\phi(d)^{-1} = \phi^*\left(\frac{a}{b}\right) + \phi^*\left(\frac{c}{d}\right)$

(b.) $\phi^*\left(\frac{a}{b}\frac{c}{d}\right) = \phi^*\left(\frac{ac}{bd}\right) = \phi(ac)\phi(bd)^{-1} = [\phi(a)\phi(c)]\phi(b)^{-1}\phi(d)^{-1}$

$= \phi(a)\phi(b)^{-1}\phi(c)\phi(d)^{-1} = \phi^*\left(\frac{a}{b}\right)\phi^*\left(\frac{c}{d}\right)$

(c.) $\phi^*\left(\frac{1}{1}\right) = \phi(1)\phi(1)^{-1} = 1 \cdot 1^{-1} = 1$

To show the uniqueness of ϕ^*, suppose ψ were another such monomorphism extending ϕ, then for any element of $Q(R)$

$$\psi\left(\frac{a}{b}\right) = \psi\left(\frac{a}{1} \cdot \frac{1}{b}\right) = \psi\left(\frac{a}{1}\left(\frac{b}{1}\right)^{-1}\right) = \psi\left(\frac{a}{1}\right)\psi\left(\left(\frac{b}{1}\right)^{-1}\right) = \psi\left(\frac{a}{1}\right)\psi\left(\frac{b}{1}\right)^{-1}$$

$$= \phi(a)\phi(b)^{-1} = \phi^*\left(\frac{a}{b}\right).$$

In summary, the quotient field of an integral domain is the *smallest* field which contains an isomorphic image of the integral domain. One last observation we make is that a ring with $1 \neq 0$ embedding in a field is a characterization of being an integral domain.

Theorem 7.3 Let R be a ring with $1 \neq 0$. Then R is an integral domain iff R embeds in a field.

Proof 7.11 *One direction of the proof follows from our work of embedding R in its quotient field. For the reverse direction, assume there is a field K and a ring with unity monomorphism $\phi : R \to K$. To show that R is an integral domain it is enough to verify that R is commutative with no zero divisors.*

To show that R is commutative, take any $a, b \in R$. In the field K we know that $\phi(a)\phi(b) = \phi(b)\phi(a)$ and so $\phi(ab) = \phi(ba)$. Now since ϕ is one-to-one we get $ab = ba$.

To show that R has no zero divisors, suppose that there were an $a, b \in R$ with $ab = 0_R$. Then

$$\phi(a)\phi(b) = \phi(ab) = \phi(0_R) = 0_K.$$

Now K being a field we know that it has no zero divisors and so either $\phi(a) = 0_K$ or $\phi(b) = 0_K$. But then either $a \in \ker\phi$ or $b \in \ker\phi$ which is trivial, since ϕ is one-to-one. Therefore, either $a = 0$ or $b = 0$.

EXERCISES

1 For each of the following integral domain R, compute $Q(R)$:

  a. $R = \mathbb{Z}[i]$.

  b. $R = \mathbb{Z}[\sqrt{2}]$.

2 Show that if R is a field, then $Q(R) = R$.

7.7 CHARACTERISTIC OF A RING

Although the notion of ring characteristic is especially relevant for fields we introduce it now in more generality.

Definition 7.14 *The **characteristic** of a ring R, written $char(R)$, is defined as follows:*

  1. *If there exists a positive integer n such that for all $r \in R$ we have $nr = \underbrace{r + r + \cdots + r}_{n} = 0$, then $char(R)$ is the smallest such positive integer.*

  2. *If no such integer exists such that for all $r \in R$ we have $nr = \underbrace{r + r + \cdots + r}_{n} = 0$, then $char(R) = 0$.*

Example 7.14 *One can easily verify the characteristic for each of the following rings:*

  1. *$char(\mathbb{Z}_n) = n$*

  2. *$char(\mathbb{Z}) = char(\mathbb{Q}) = char(\mathbb{R}) = char(\mathbb{C}) = 0$.*

  3. *$char(\mathbb{Z}_6 \times \mathbb{Z}_{15}) = 30$.*

The rest of the section deals with the notion of a prime subfield in a field. We will start with a broader context and narrow our way down to this notion. Let R be a ring with 1. We start by considering the following map $\phi : \mathbb{Z} \to R$ by $\phi(n) = n1 = \underbrace{1 + 1 + \cdots + 1}_{n}$. It is readily seen, using a proof by cases, that this map is a ring homomorphism. We show the case when $n, m > 0$:

Proof 7.12

$$\phi(n + m) = (n + m)1 = \underbrace{1 + 1 + \cdots + 1}_{n+m} = \underbrace{1 + 1 + \cdots + 1}_{m} + \underbrace{1 + 1 + \cdots + 1}_{n}$$

$$= m1 + n1 = \phi(m) + \phi(n) \quad and$$

$$\phi(nm) = (nm)1 = \underbrace{1 + 1 + \cdots + 1}_{nm} = (\underbrace{1 + 1 + \cdots + 1}_{m})(\underbrace{1 + 1 + \cdots + 1}_{n})$$

$$= (m1)(n1) = \phi(m)\phi(n).$$

Lemma 7.6 *Let R be a ring with 1.*

1. *If $char(R) = n > 0$, then R contains an isomorphic copy of $\mathbb{Z}_n$, namely the cyclic subgroup of $(R, +)$ generated by 1.*

2. *If $char(R) = 0$, then R contains an isomorphic copy of $\mathbb{Z}$, namely the cyclic subgroup of $(R, +)$ generated by 1.*

Proof 7.13 *When $char(R) = n > 0$ then the map ϕ defined above has kernel $n\mathbb{Z}$ and image $\langle 1 \rangle$ and so by FTH,*

$$\mathbb{Z}_n \cong \mathbb{Z}/n\mathbb{Z} \cong \langle 1 \rangle \subseteq R.$$

When $char(R) = 0$, the map ϕ is a monomorphism and by FTH,

$$\mathbb{Z} \cong \mathbb{Z}/\{0\} \cong \langle 1 \rangle \subseteq R.$$

Corollary 7.2 *Let F be a field.*

1. *If $char(F) = p$ prime, then F contains an isomorphic copy of $\mathbb{Z}_p$.*

2. *If $char(F) = 0$, then R contains an isomorphic copy of $\mathbb{Q}$.*

Proof 7.14 *The case when $char(F) = p$ prime follows immediately from Lemma 7.6.1. For the case when $char(F) = 0$, by Lemma 7.6.2, F contains an isomorphic copy of $\mathbb{Z}$ via the monomorphism $\phi : \mathbb{Z} \to F$ by $\phi(n) = n1$. Now, by our work on quotient fields, we can extend ϕ to a monomorphism $\phi^* : \mathbb{Q} \to F$. Thus, $\mathbb{Q}$ embeds in F.*

Definition 7.15 *The subfield of F which is either $\mathbb{Z}_p$ or $\mathbb{Q}$ in Corollary 7.2 above is called the **prime subfield** of F.*

Remark 7.9 *One can show that the prime subfield of a field F is the smallest subfield in F, i.e. any other subfield of F must contain the prime subfield.*

EXERCISES

1 If R is a ring with 1, then

   a. *$char(R) = n > 0$ iff n is the smallest positive integer such that $n1 = \underbrace{1 + 1 + \cdots + 1}_{n} = 0$, i.e. the order of 1 in $(R, +)$ equals n.*

   b. *$char(R) = 0$ iff the order of 1 in $(R, +)$ is infinite.*

2 If R is a finite ring with 1, then $char(R)$ divides $|R|$.

3 If R is an ID, then the non-zero elements of R are either all have finite order or all have infinite order in $(R, +)$.

4 If R is an ID, then either $char(R) = 0$ or a prime number

note: be sure to rule out $char(R) = 1$.

5 Complete the proof by cases that map $\phi : \mathbb{Z} \to R$ by $\phi(n) = n1 = \underbrace{1 + 1 + \cdots + 1}_{n}$ is a ring homomorphism.

6 Prove that the prime subfield of a field F is the smallest subfield in F, i.e. any other subfield of F must contain the prime subfield.

7.8   THE RING OF POLYNOMIALS

Our focus in this section is the ring of polynomials in an indeterminant, $R[x]$, first introduced in Section 7.1. The following properties are readily verified for a ring of polynomials which we leave as exercises:

1. If R is a ring, then $R[x]$ is a ring.

2. If R is commutative a ring, then $R[x]$ is a commutative ring.

3. If R is a ring with 1, then $R[x]$ is a ring with 1.

Later on in the text, we will show that other properties of R are carried over to $R[x]$. In this section, we will show among other things that ID is carried over from R to $R[x]$.

Definition 7.16 *Consider the ring of polynomials $R[x]$ in the indeterminant x and let $p(x) = a_n x^n + \cdots + a_1 x + a_0 \in R[x]$ with $a_n \neq 0$.*

1. *The* **degree** *of $p(x)$, written $deg(p) = n$. Note that the degree of the zero polynomial will be assigned the value $-\infty$ and we will extend addition to include $-\infty$ as follows:*

$$(-\infty) + (-\infty) = -\infty \qquad\qquad (-\infty) + n = -\infty = n + (-\infty).$$

2. *The* **leading coefficient** *of $p(x)$, written $L(p) = a_n$.*

3. *A polynomial is called* **monic** *if $L(p) = 1$.*

Lemma 7.7 *Let R be a commutative ring with 1 with $p(x)$ and $q(x)$ two polynomials in $R[x]$.*

1. *$deg(pq) \leq deg(p) + deg(q)$*

2. *If $L(p)$ is not a zero divisor, then $L(pq) = L(p)L(q)$ and $deg(pq) = deg(p) + deg(q)$.*

Proof 7.15 *First note that if $p(x)$ or $q(x)$ is the zero polynomial, then $p(x)q(x)$ is as well with $L(pq) = 0 = L(p)L(q)$ and $deg(pq) = -\infty = deg(p) + deg(q)$. Assume now that $p(x)$ and $q(x)$ are not the zero polynomial. Set $p(x) = a_n x^n + \cdots + a_1 x + a_0$ and $q(x) = b_m x^m + \cdots + b_1 x + b_0$. Then $p(x)q(x) = (a_n b_m)x^{m+n} + \cdots + (a_1 b_0 + a_0 b_1)x + (a_0 b_0)$. It's clear then that the degree of $p(x)q(x)$ can be at most $m + n$ and when $L(p)$ is not a zero divisor, then $L(pq) = a_n b_m = L(p)L(q)$ and $deg(pq) = n + m = deg(p) + deg(q)$.*

Theorem 7.4 *Let R be an ID with $p(x)$ and $q(x)$ two polynomials in $R[x]$.*

1. *$deg(pq) = deg(p) + deg(q)$.*

2. *$R[x]$ is an ID.*

3. *$U(R[x]) = U(R)$.*

Proof 7.16 *Since R has no zero-divisors, the first part follows immediately from Lemma 7.7.2.*

For the second statement, suppose that $p(x)$ and $q(x)$ are not the zero polynomial. Thus, $deg(p), deg(q) \geq 0$ and so $deg(pq) = deg(p) + deg(q) \geq 0$. Hence, neither is $p(x)q(x)$ the zero polynomial and $R[x]$ has no zero divisors making it an ID.

For the third statement, it suffices to show that every unit in $R[x]$ must be a constant polynomial. Suppose that $p(x) \in U(R[x])$ and $q(x) \in R[x]$ is such that $p(x)q(x) = 1$. Then

$$0 = deg(1) = deg(pq) = deg(p) + deg(q),$$

and so it is necessarily the case that $deg(p) = deg(q) = 0$ making $p(x)$ a constant polynomial.

Example 7.15 *These examples discuss the units in certain polynomial rings.*

1. *$U(\mathbb{Z}[x]) = U(\mathbb{Z}) = \{-1, 1\}$.*

2. *For any field F the units in $F[x]$ consist of the non zero constant polynomials.*

3. *An element $r \in R$ a ring is **nilpotent** if there is a positive integer n such that $r^n = 0$. For example, $1, 3, 5, 7$ are units in $\mathbb{Z}_8$ while $0, 2, 4, 6$ are nilpotent, since $0^1 = 0$, $2^3 = 0$, $4^2 = 0$ and $6^3 = 0$.*

4. *If R is a commutative ring with 1 and $p(x) = a_n x^n + \cdots + a_1 x + a_0 \in R[x]$, then $p(x)$ is a unit in $R[x]$ iff a_0 is a unit in R and $a_1, \ldots, a_n$ are nilpotent in R. For instance in $\mathbb{Z}_8[x]$, an example of a unit in $\mathbb{Z}_8[x]$ is $2x^5 + 6x^2 + 4x + 3$.*

Definition 7.17 *Let F be a field. A non-constant polynomial in $F[x]$ is called **irreducible** if it cannot be factored into two polynomials of lesser degree. Otherwise the polynomial is said to be **reducible** (or **factorable**).*

Example 7.16 *In these examples we discuss reducibility in certain polynomial rings.*

1. *Consider the polynomial ring $\mathbb{Z}_2[x]$ and polynomials of degree two. The reducible polynomials are*

$$x^2 = x \cdot x, \quad x^2 + x = x(x+1), \quad x^2 + 1 = (x+1)(x+1).$$

   The only irreducible quadratic is x^2+x+1, for suppose $x^2+x+1 = (x+a)(x+b)$. Then $x^2 + x + 1 = x^2 + (a+b)x + ab$ which implies $a+b = 1$ and $ab = 1$, but this system of equations has no solution in $\mathbb{Z}_2$.

2. *Every polynomial of degree one in $F[x]$ is irreducible, for suppose $p(x) \in F[x]$ of degree one and $p(x) = r(x)s(x)$ with $deg(r), deg(s) < deg(p)$. But then it must be that $deg(r) = deg(s) = 0$, however $1 = deg(p) = deg(rs) = deg(r)+deg(s) = 0$, a contradiction.*

The next result will be useful in the upcoming chapter on integral domains. It show that polynomials over a field have a division algorithm just the way integers do.

Theorem 7.5 *If F is a field and $f(x), g(x)$ are two polynomials in $F[x]$ with $g(x)$ non-zero, then there exist unique polynomials $q(x), r(x)$ in $F[x]$ such that*

$$f(x) = g(x)q(x) + r(x) \quad where \quad deg(r) < deg(g).$$

Proof 7.17 *This proof divides logically into two parts.*
Existence: *Let's first dispense with the trivial case of $f = 0$ in which case set $q = r = 0$. Now we prove the rest by induction on $m = deg(f)$. If $m = 0$, then $f(x) = a$ a constant polynomial. If $deg(g) > 0$, then take $q(x) = 0$ and $r(x) = f(x)$ to satisfy the theorem, otherwise $g(x) = b$ a non-zero constant and since F is a field, we can assign $q(x) = b^{-1}a$ and $r(x) = 0$ to satisfy the theorem. Now assume $m > 0$. Again, should it be the case that $deg(f) < deg(g)$, then $q(x) = 0$ and $r(x) = f(x)$ satisfies the theorem. For the case that $deg(f) \geq deg(g)$, first set $a = L(f), b = L(g)$ and $d = deg(g)$. Notice that $f_1(x) = f(x) - ab^{-1}x^{m-d}g(x)$ has degree less than m, since we have effectively eliminated the leading term of $f(x)$. By induction, there exist polynomials $q_1(x), r_1(x)$ such that $f_1(x) = g(x)q_1(x) + r_1(x)$ with $deg(r_1) < deg(g)$. But then*

$$f(x) = f_1(x) + ab^{-1}x^{m-d}g(x) = g(x)q_1(x) + r_1(x) + ab^{-1}x^{m-d}g(x)$$

$$= g(x)[q_1(x) + ab^{-1}x^{m-d}] + r_1(x).$$

Therefore, $q(x) = q_1(x) + ab^{-1}x^{m-d}$ and $r(x) = r_1(x)$ satisfy the theorem.
Uniqueness: *Suppose we also have that $f(x) = g(x)\hat{q}(x) + \hat{r}(x)$ where $deg(\hat{r}) < deg(\hat{g})$. Equating yields $[q(x)-\hat{q}(x)]g(x) = r(x)-\hat{r}(x)$. Therefore, $deg(r-\hat{r})+deg(g) = deg(r - \hat{r}) < deg(g)$. The only way this is possible is for $deg(q - \hat{q}) = -\infty$, i.e. $q(x) - \hat{q}(x)$ is the zero polynomial and so $q(x) = \hat{q}(x)$. But then $r(x) - \hat{r}(x)$ equals the zero polynomial as well making $r(x) = \hat{r}(x)$.*

We point out that the above result can be generalized to $R[x]$ where R is a commutative ring with 1 and the divisor $g(x) \in R[x]$ has the property that $L(g) \in U(R)$. We leave it as an exercise for the reader to verify this fact (following the same proof as above).

Definition 7.18 *Let* $a \in E \supseteq F$ *where* E *and* F *are fields and* $f(x)$ *a non-zero polynomial in* $F[x]$. *If* $f(a) = 0$, *then* a *is called a* **root** *or* **zero** *of* $f(x)$.

Corollary 7.3 *Let* F *be a field.*

1. *The remainder when dividing* $f(x) \in F[x]$ *by* $x - a \in F[x]$ *is* $f(a)$.

2. $a \in F$ *is a zero of a non-zero polynomial* $f(x) \in F[x]$ *iff* $f(x) = (x-a)q(x)$ *for some polynomial* $q(x) \in F[x]$.

3. *If* $a_1, \ldots, a_k \in F$ *are zeros of* $f(x) \in F[x]$, *then* $f(x) = (x-a_1) \cdots (x-a_k)q(x)$ *for some polynomial* $q(x) \in F[x]$.

4. *The number of distinct zeros of a polynomial* $f(x) \in F[x]$ *does not exceed the degree of* $f(x)$.

Proof 7.18 *For the first statement, divide* $f(x)$ *by* $x - a$ *and conclude* $r(x) = f(a)$. *The second statement follows immediately from the first statement. The third statement is by induction using the second statement. The fourth statement follows from the third statement.*

Remark 7.10 *Here, we make some remarks regarding Corollary 7.3.*

1. *We can illustrate Corollary 7.3.2 with the polynomial* $f(x) = x^2 - x - 2 \in \mathbb{R}[x]$. *One can check* -1 *and* 2 *are roots of* $f(x)$ *and* $f(x) = (x-2)(x+1)$.

2. *We can illustrate Corollary 7.3.1 with the polynomial* $f(x) = x^2 + 1 \in \mathbb{R}[x]$. *If we divide* $f(x)$ *by* $x - 1$, *one can check that the division algorithm yields* $f(x) = (x-1)(x+1) + 2$, *so the remainder is* $2 = f(1)$.

3. *We can illustrate Corollary 7.3.4 with the polynomial* $f(x) = x^2 - 2x + 1 \in \mathbb{R}[x]$ *which has only one distinct root not exceeding its degree of* 2.

4. *The final part of the Corollary relies heavily on the fact that* F *is a field. For instance,* $f(x) = x^2 + x \in \mathbb{Z}_6$ *has four distinct zeros in* $\mathbb{Z}_6$, *namely* $0, 2, 3, 5$. *In fact, for the quaternions* Q *one can show the polynomial* $f(x) = x^2 + 1 \in Q[x]$ *has an infinite number of zeros.*

5. *We point out that the last part of the corollary can be proved in* $R[x]$ *where* R *is an integral domain. We outline the proof and leave the details as an exercise.*

   (a) *For any positive integer* n, *the linear polynomial* $x - a$ *is a factor of* $x^n - a^n$.

   (b) *The linear polynomial* $x - a$ *is a factor of* $f(x) - f(a)$.

   (c) *Conclude that* $f(x) = (x-a)q(x) + f(a)$.

EXERCISES

1 Prove that if R is a ring, then $R[x]$ is a ring.

2 Prove that if R is a commutative a ring, then $R[x]$ is a commutative ring.

3 Prove that if R is a ring with 1, then $R[x]$ is a ring with 1.

4 Prove that if R is a commutative ring with 1 and $p(x) = a_n x^n + \cdots + a_1 x + a_0 \in R[x]$, then $p(x)$ is a unit in $R[x]$ iff a_0 is a unit in R and $a_1, \ldots, a_n$ are nilpotent in R.

5 $R[x]$ where R is a commutative ring with 1. Prove that if $f(x), g(x)$ are two polynomials in $F[x]$ with $g(x)$ having the property that $L(g) \in U(R)$, then there exist unique polynomials $q(x), r(x)$ in $F[x]$ such that

$$f(x) = g(x)q(x) + r(x) \text{ where } deg(r) < deg(g).$$

6 Fill in the details of the proof of Corollary 7.3.

7 Use the following outline to generalize Corollary 7.3 to $R[x]$ where R is an integral domain:

   a. For any positive integer n, the linear polynomial $x - a$ is a factor of $x^n - a^n$.

   b. The linear polynomial $x - a$ is a factor of $f(x) - f(a)$.

   c. Conclude that $f(x) = (x - a)q(x) + f(a)$.

8 Let $\phi : F[x] \to F[x]$ be a ring automorphism such that $\phi(a) = a$ for all $a \in F$ a field.

   a. For a given $f(x) \in F[x]$, Prove $deg f = deg\ \phi(f)$
       (**hint:** You will need to show that $deg\ \phi(x) = 1$)

   b. For a given $f(x) \in F[x]$, Prove that f is irreducible iff $\phi(f)$ is irreducible.
       (**hint:** Use contrapositive)

7.9 SPECIAL IDEALS

We remind the reader that much of the study of rings in this text is based on generalizations of properties of the integers. Continuing in this vein, we consider certain properties of ideals in the integers and generalize them to arbitrary commutative rings.

Definition 7.19 *Let R be a commutative ring and $I \lhd R$ with $I \neq R$.*

   1. I is called **prime** *ideal if for all $r, s \in R$, whenever $rs \in I$ either $r \in I$ or $s \in I$.*

2. I is called a **maximal** ideal if it is not properly contained in another proper ideal of R, i.e. If $J \lhd R$ and $I \subseteq J \subseteq R$, then either $J = I$ or $J = R$.

Example 7.17 Here, we present several examples in order to illustrate the definitions just given.

1. If $I = (p) = p\mathbb{Z}$ for p prime in $\mathbb{Z}$, then I is both maximal and prime We first show I is a prime ideal. If $mn \in I$, then $mn = pk$ and so p divides mn. Since p is prime either p divides m or n and so either $m \in (p) = I$ or $n \in (p) = I$. Second we show I is maximal. Suppose $J \lhd \mathbb{Z}$ and $I \subseteq J \subseteq \mathbb{Z}$. Now $J = (n)$ for some integer n. Since $p = p \cdot 1 \in (p) \subseteq (n)$, this implies $p = mn$ for some integer m. Since p is prime, either $n = \pm p$ and so $J = (p) = I$ or $n = \pm 1$ and so $J = (1) = R$. Later we will show these are the only prime (and maximal) ideals in $\mathbb{Z}$ and hence in $\mathbb{Z}$ the notions of prime and maximal ideal coincide.

2. Let $R = \mathbb{Z}[i] = \{m + ni : m, n \in \mathbb{Z}\}$ and $i = \sqrt{-1}$, called the **Gaussian integers**, and set $I = \{m + ni \in R : 3|m \ \& \ 3|n\}$. One can easily show that $I \lhd R$, but we wish to show that I is, in fact, a maximal ideal in R.

   Proof 7.19 Suppose there was an ideal $J \lhd R$ such that $I \subset J \subseteq R$. Then there exists an $m + ni \in J$ but not in I, so that either $3 \nmid m$ or $3 \nmid n$. In other words, $m = 3k + r$ and $n = 3l + s$ where $0 \leq r, s < 3$ but not both r and s equal 0. But then, since $3k + 3li \in I \subset J$ we have $(m + ni) - (3k + 3li) = r + si \in J \lhd R$. Again, since J is an ideal in R, we have $r^2 + s^2 = (r + si)(r - si) \in J$. Now because of the constraint on r and s we know that $r^2 + s^2$ is either 1, 2, 4, 5 or 8. In either of these five cases we can always subtract an appropriate multiple of 3 (which is in I and so in J) to get a difference of 1. Therefore, $1 \in J$ and so $J = R$.

The next result connects these special ideals with their corresponding factor rings.

Theorem 7.6 Let R be a commutative ring with 1 and $I \lhd R$.

1. I is prime iff R/I is an ID.

2. I is maximal iff R/I is a field.

Proof 7.20 First assume that I is prime and we show R/I has no zero divisors. If $(r + I)(s + I) = I$, then $rs + I = I$ and so $rs \in I$. Since I is prime, either $r \in I$ or $s \in I$ and so either $r + I = I$ or $s + I = I$. Now assume that R/I is an integral domain. If $rs \in I$, then $rs + I = I$ and so $(r + I)(s + I) = I$. Since R/I has no zero divisors, either $r + I = I$ or $s + I = I$ which implies either $r \in I$ or $s \in I$.

For the second statement we first assume that I is maximal and show every nonzero element in R/I is a unit. If $r + I \neq I$, then $r \notin I$. Define the following set:

$$J = \{sr + a : s \in R \text{ and } a \in I\}.$$

First note that $J \lhd R$, since if $sr + a, s'r + b \in J$, then $(sr + a) - (s'r + b) = (s - s')r + (a - b) \in J$ and for $t \in R$ we have $t(sr + a) = (ts)r + (ta) \in J$. Second, certainly J contains I since we may set $s = 0$ to obtain all the elements of I in J. Finally, $I \neq J$, since $r = 1 \cdot r + 0 \in J$ while $r \notin I$. Thus, since I is maximal, it must be that $J = R$ and so in particular $1 \in J$. Then we can express $1 = sr + a$ for $s \in R$ and $a \in I$ and so $1 \in sr + I$; this implies $1 + I = sr + I = (s + I)(r + I)$ and so $r + I$ has an inverse (note that we cannot simply take $r^{-1} + I$ as the inverse, since r may not be a unit in R). Now assume that R/I is a field and suppose we have $J \lhd R$ with $I \subseteq J \subseteq R$. First note that $J/I \lhd R/I$ since if $r + I, s + I \in J/I$, then $(r+I)-(s+I) = (r-s)+I \in J/I$ and if $t+I \in R/I$, then $(t+I)(r+I) = (tr)+I \in J/I$. Since R/I is a field, and we've seen that a field has no proper non-trivial ideals, it follows that either $J/I = \{I\}$ or $J/I = R/I$ and so either $J = I$ or $J = R$.

Example 7.18 *We can now give an alternate proof to the one presented in Example 7.17.2.*

Proof 7.21 *We will show that R/I is a field and so by Theorem 7.6.2, it follows that I is maximal. Since R is a commutative ring with 1 it is enough to show that evey non-zero element in R/I is a unit. Therefore, take an $(m + ni) + I \in R/I$ with $m+ni \notin I$ (and so either 3 does not divide m or 3 does not divide n). We wish to find an $(x + yi) + I \in R/I$ such that $[(m+ni) + I][(x+yi) + I] = 1+I$. This is equivalent to saying that $(m + ni)(x + yi) - 1 \in I$, i.e. $(mx - ny - 1) + (nx + my)i \in I$. In other words, 3 must divide $mx - ny - 1$ and must divide $nx + my$. This is equivalent to saying that the following linear system of equations has a solution in $\mathbb{Z}_3$:*

$$\begin{aligned} mx - ny &= 1 \\ nx + my &= 0 \end{aligned}.$$

Now the coefficient matrix for this system is

$$A = \begin{bmatrix} m & -n \\ n & m \end{bmatrix}.$$

Therefore, the system has a solution in $\mathbb{Z}_3$ iff A is invertible in $M_2(\mathbb{Z}_3)$ iff $|A|$ is invertible in $\mathbb{Z}_3$. Notice that $|A| = m^2 + n^2 \neq 0$ in $\mathbb{Z}_3$, since 3 does not divide both m and n. Therefore $|A|$ is indeed invertible in $\mathbb{Z}_3$ and so backtracking through the argument we see that $(m + ni) + I$ is indeed a unit in R/I.

Corollary 7.4 *Let R be a commutative ring with 1.*

1. *Every maximal ideal is prime.*

2. *R is a field iff R has no proper non-trivial ideals.*

Proof 7.22 *For the first statement, I maximal implies R/I a field implies R/I an ID implies I prime.*

For the second statement, we have already shown one direction, so assume that R has no proper non-trivial ideals. Therefore, the trivial ideal $\{0\}$ is maximal and so $R \cong R/\{0\}$ is a field.

Remark 7.11 *The converse of Corollary 7.4.1 is false in general. For instance, consider the ring with 1, $R = \mathbb{Z} \times \mathbb{Z}$ and the ideal $I = \mathbb{Z} \times \{0\}$. One can easily check that I is prime, however it is not maximal, since for instance $J = \mathbb{Z} \times 2\mathbb{Z}$ is an ideal properly between I and R.*

We shall show below, however, that maximal and prime are always equivalent notions in the ring of integers. We remind the reader that every ideal in the ring of integers is principal.

Corollary 7.5 *In the ring of integers, any ideal $I = (n)$ is maximal iff n is prime.*

Proof 7.23 *I maximal implies $\mathbb{Z}_n \cong \mathbb{Z}/(n)$ is a field and so n must be prime. On the other hand, n prime implies $\mathbb{Z}/(n) \cong \mathbb{Z}_n$ is a field, and so (n) must be maximal.*

Proposition 7.1 *In the ring of integers the following are equivalent for any ideal $I = (n)$ in $\mathbb{Z}$:*

1. *I is prime.*

2. *I is maximal.*

3. *n is prime.*

Proof 7.24 *What is left to show after all the work thus far is to show that I prime implies I maximal. Assume that I is prime and suppose there is an ideal $J = (m)$ with $I \subseteq J \subseteq \mathbb{Z}$. Without loss of generality we may assume that $m, n > 0$. Since $n \in I \subseteq J$ this implies $n = mk$. Thus $mk \in I$ and so either $m \in I$ or $k \in I$. If $m \in I$ we have $m = nr$ and so $n = nrk$ and so $rk = 1$ which implies $k = 1$. Therefore, $n = m$ and $I = J$. If $k \in I$ we have $k = ns$ and so $n = mns$ and so $ms = 1$ which implies $m = 1$. Therefore, $J = R$.*

EXERCISES

1 Let $R = \mathcal{F}(\mathbb{R})$ and $I = \{f \in R \; : \; f(1) = 0\}$. Prove that I is a maximal ideal in R.

2 Prove that for the ring with 1, $R = \mathbb{Z} \times \mathbb{Z}$, the ideal $I = \mathbb{Z} \times \{0\}$ is prime.

3 Let R be a ring with unity and $I, J, M \lhd R$ with M maximal. Prove that if $I \cap J \subseteq M$, then either $I \subseteq M$ or $J \subseteq M$.

   (hint: Consider $I + M, J + M \lhd R$)

4 Let P be a prime ideal of a commutative ring R. Show that for any ideals I and J with $I \cap J \subseteq P$ we have that either $I \subseteq P$ or $J \subseteq P$.

5 Prove that a commutative ring in which every ideal is principal must have unity.

6 Let R be a ring with unity and $r \in R$ such that $r \neq 0, 1$ and $r^2 = r$. Show that $R = (r) + (1 - r)$ and $(r) \cap (1 - r) = \{0\}$.

CHAPTER 8

Integral Domain Theory

IN THIS CHAPTER, we further explore properties of the integers and by doing so we define some special cases of integral domains. In Section 8.1, we introduce two of these special integral domains called Euclidean domain (ED) and principal ideal domain (PID). In Section 8.2, we introduce the third special integral domain called a unique factorization domain (UFD). In Section 8.3, we look at one particular integral domain which fails to be any of these three special integral domains. One area of study already discussed was to determine which property of a ring R carries over to the corresponding polynomial ring $R[x]$. In Section 8.4, we prove that the UFD property indeed carries over to the polynomial ring.

8.1   EUCLIDEAN AND PRINCIPAL IDEAL DOMAINS

In this section, we explore two properties of the ring of integers and look at other rings which share these properties.

Definition 8.1 *An integral domain R is called a* **Euclidean domain** *(**ED***) if there exists a function $\delta : R^* \to \mathbb{N}$ having the property that for all $a, b \in R$ with $b \neq 0$, there exists $q, r \in R$ such that $a = bq + r$ where either $\delta(r) < \delta(b)$ or $r = 0$.*

Example 8.1 *Here, we list some examples of integral domains.*

1. *We have seen that the natural numbers has this property described in the definition above, but alas it is not a ring. However, the ring of integers forms a ED with $\delta(n) = |n|$. This is a slightly different formalization of the Division Algorithm and is easy to verify as was done in Theorem 1.2.*

2. *We have seen that $F[x]$ where F is a field is an ED with $\delta(f(x)) = deg(f(x))$. This was verified in Theorem 7.5.*

3. *The Gaussian integers, $\mathbb{Z}[i]$, forms an ED with $\delta(m + ni) = m^2 + n^2$. We will verify this in detail at the end of the section.*

Definition 8.2 *An integral domain R is called a* **principal ideal domain** *(**PID***) if every ideal in R is a principal ideal. In other words, for every $I \lhd R$ there exists an $a \in R$ such that $I = (a)$.*

DOI: 10.1201/9781003335283-8

Example 8.2 *We have already seen that the ring of integers is a PID, since any ideal in $\mathbb{Z}$ has the form $n\mathbb{Z} = (n)$. But this argument will become moot, since we now show that every ED is a PID and hence the three examples given above must all be PIDs. We remark that it is no easy matter to come up with a PID which is not an ED. One can show that $\mathbb{Z}[z_0]$ where z_0 is the complex number $(1 + \sqrt{-19})/2$ is an example of a PID which is not an ED (we omit the proof).*

Theorem 8.1 *Every ED is also a PID.*

Proof 8.1 *Let R be an ED. If I is the trivial ideal, then $I = (0)$ is principal, so we may assume that I is a non-trivial ideal in R. Since δ maps into the natural numbers, there must be an element $b \in I$ of minimal δ-value. We show that $I = (b)$ and thus complete the proof. Certainly, $(b) \subseteq I$, since $b \in I$. For the reverse inclusion, take any $a \in I$. Since R is an ED, there exist $q, r \in R$ such that $a = bq + r$ where either $\delta(r) < \delta(b)$ or $r = 0$. Notice that $r = a - bq \in I$ and since b has minimal δ-value in I it must be that $r = 0$ and so $a = bq \in (b)$.*

Example 8.3 *We need to point out that not every integral domain is principal. One good example is $\mathbb{Z}[x]$. It's enough to show there is an ideal in $\mathbb{Z}[x]$ which is not principal. Select a prime number p and set $I = (p, x) = \{p \cdot f(x) + x \cdot g(x) : f, g \in \mathbb{Z}[x]\}$. Note that $I \lhd \mathbb{Z}[x]$ since it is the ideal generated by p and x (see Exercise 6 in Section 7.5).*

We need to point out that I is a proper ideal in $\mathbb{Z}[x]$, for suppose that $I = \mathbb{Z}[x]$, then in particular $1 \in I$ and so $1 = pf(x) + xg(x)$ for some $f, g \in \mathbb{Z}[x]$. Set $f(x) = a_k x^k + \cdots + a_1 x + a_0$ and equate constant coefficients in the equation $1 = pf(x) + xg(x)$ to get $1 = pa_0$ and so $p = \pm 1$ which contradicts that p is a prime.

Now we show I is not principal. Suppose it were, i.e. $I = (h(x))$ for some $h \in \mathbb{Z}[x]$. Now since $p = p \cdot 1 + x \cdot 0$, this implies $p \in I$ and so $p = h(x)r(x)$ for some $r \in \mathbb{Z}[x]$. Notice then that $0 = deg(p) = deg(hr) = deg(h) + deg(r)$ and so $deg(h) = 0$ and $deg(r) = 0$. Set $h(x) = m$ and $r(x) = n$ and so we have the integer equation $p = mn$. Now p is prime so either $m = \pm 1$ or $m = \pm p$. If $m = \pm 1$, then $I = (1) = \mathbb{Z}[x]$ which we have shown is not possible. Hence, $m = \pm p$ and $I = (p)$ an ideal generated by the prime p. Now $x = p \cdot 0 + x \cdot 1$ and so $x \in I$. Hence, $x = p \cdot s(x)$ for some $s \in \mathbb{Z}[x]$. Notice that $1 = deg(x) = deg(ps) = deg(p) + deg(s) = deg(s)$. Hence, $s(x) = kx + l$ a linear polynomial. Therefore, $x = p(kx + l) = (pk)x + (pl)$. Equating x coefficients yields the equation $1 = pk$ again contradicting that p is a prime.

For the remainder of this section we generalize some notions found in $\mathbb{Z}$ to rings, namely *divides, greatest common divisor* and *prime*.

Definition 8.3 *Let R be a commutative ring with 1 and $r, s \in R$ with $r \neq 0$. We say r **divides** s and write $r|s$, if there exists an $a \in R$ such that $s = ra$.*

Remark 8.1 *We make several remarks which we leave to the reader to verify.*

1. We note that $r|s$ is equivalent to $s \in (r)$ which is equivalent to $(s) \subseteq (r)$.

2. In a field F every non-zero element divides every element in the field, since if $a, b \in F$ with $a \neq 0$, then $b = a(a^{-1}b)$ and so $a|b$.

Example 8.4 *Here, we list several examples in different rings illustrating the notion of dividing.*

1. *In $\mathbb{Z}$, the integer -6 divides 30, since $30 = (-6)(-5)$*

2. *In $F[x]$, for some field F, the polynomial $x^2 + 1$ divides $x^4 - 1$, since $x^4 - 1 = (x^2 + 1)(x^2 - 1)$.*

3. *In $R = \mathbb{Z}[\sqrt{-5}]$ we have that $2 + \sqrt{-5}$ divides 9, since $9 = (2 + \sqrt{-5})(2 - \sqrt{-5})$.*

Here are some additional properties of divides which we leave to the reader to verify. The proofs are nearly identical to the ones presented for $\mathbb{Z}$. For R be a commutative ring with 1,

1. For all $r \in R$ we have $r|r$.

2. If for $r, s, t \in R$ we have $r|s$ and $s|t$, then $r|t$.

3. If for $r, s, t \in R$ we have $r|s$ and $r|t$, then $r|(sx + ty)$ for any $x, y \in R$.

Lemma 8.1 *Let $r, s \in R$ be a commutative ring with 1.*

1. *If $s = ru$ for some $u \in U(R)$, then $(s) = (r)$.*

2. *When R is an ID, then $s = ru$ for some $u \in U(R)$ iff $(s) = (r)$.*

Proof 8.2 *For the first statement, given $s = ru$, be definition $r|s$ and so $(s) \subseteq (r)$ (see Remark 8.1.1). Since $u \in U(R)$ we know u^{-1} exists so that we can rewrite equation $s = ru$ as $r = su^{-1}$ so that $s|r$ which implies $(r) \subseteq (s)$ as well.*

For the second statement, we already have one direction from the first statement, so assume that $(s) = (r)$. Since $r \in (r) = (s)$ we know $r = sa$ for some $a \in R$, and since $s \in (s) = (r)$ we also know $s = rb$ for some $b \in R$. Hence, $r = sa = (rb)a$ and, by cancellation in an ID, $1 = ba$ which implies $b \in U(R)$ with $s = ba$.

Definition 8.4 *Let R be a commutative ring with 1. We say $r, s \in R$ are **associates** if both $r|s$ and $s|r$.*

Example 8.5 *We illustrate the notion of associate in several different settings.*

1. *In $\mathbb{Z}$ any number and its opposite are associates, such as 3 and -3. These are the only kind of associates in $\mathbb{Z}$ (see Lemma 8.1.1).*

2. *In $F[x]$ any polynomial and a constant multiple of itself are associates, such as $2x + 1$ and $-6x - 3$.*

3. *In a field all non-zero elements are associates, since every non-zero element divides any element in the field.*

Remark 8.2 *Here are some additional remarks about associates some of which we leave to the reader to verify.*

1. *r and s are associates imples that $r = su$ for some $u \in U(R)$.*

2. *r and s are associates iff $(r) = (s)$.*

3. *The notion of associates defines an equivalence relation on R.*

4. *When R is an ID, then Lemma 8.1 says r and s are associates iff they differ by a unit. In the examples above both rings were IDs and notice that the associates did indeed differ by a unit.*

Definition 8.5 *Let R be an ID and $r, s \in R^*$. An element $d \in R$ is a **greatest common divisor** (**gcd**) of r and s, if*

1. *$d \neq 0$*

2. *$d|r$ and $d|s$ (common divisor)*

3. *If $e \in R$ with $e|r$ and $e|s$, then $e|d$ (greatest)*

Remark 8.3 *We make several remarks which we leave to the reader to verify.*

1. *The gcd of two ring elements is never unique. In fact, from the definition of gcd one can quickly show that any two elements of a ring are gcd's of the same pair of ring elements iff they are associates.*

2. *If F is a field, then the gcd of any pair of elements is any non-zero element in F.*

Example 8.6 *Here, we illustrate the notion of gcd in several settings.*

1. *Take the case of the ring $\mathbb{Z}$. Both 6 and -6 are gcd's of 30 and 12. Notice that they are associates differing by the unit -1. We typically say the $\gcd(30, 12) = 6$, the positive one. Thus, if we require the gcd to be positive, then in this ring it becomes unique.*

2. *Take the case of the ring $F[x]$ for some field F. Both $x^2 + 1$ and $-2x^2 - 2$ are gcd's of $x^4 - 1$ and $x^3 + 2x^2 + x + 2$. Notice that they are associates differing by the unit -2. We typically say the $\gcd(30, 12) = x^2 + 1$, the monic one. Thus, if we require the gcd to be monic, then in this ring it becomes unique.*

3. *It is not true that every ID has a well-defined gcd. Take the case of $\mathbb{Z}[\sqrt{-5}]$. The elements 9 and $6 + 3\sqrt{-5}$ do not have a gcd. we leave the details for a later section, but basically ± 3 and $\pm(2 + \sqrt{-5})$ are the only common divisors of 9 and $6 + 3\sqrt{-5}$ and yet 3 does not divide $2 + \sqrt{-5}$ nor does $2 + \sqrt{-5}$ divide 3.*

Theorem 8.2 *In a PID the gcd exists.*

Proof 8.3 *Let $r, s \in R^*$ where R is a PID. Consider the ideal $(r) + (s)$. Since R is a PID, there exists $d \in R$ such that $(d) = (r) + (s)$. We show that d is a gcd of r and s. First of all $d \neq 0$, for otherwise $(r) + (s) = \{0\}$, but then $r = r \cdot 1 + 0 \in (r) + (s) = \{0\}$ which implies $r = 0$, a contradiction. Second, since $r \in (r) + (s) = (d)$ this implies that $d|r$. Similarly $d|s$. Finally, suppose that $e|r$ and $e|s$. Then $r = ea$ and $s = eb$ for some $a, b \in R$. Now, $d \in (d) = (r) + (s)$ and so $d = rx + sy$ for some $x, y \in R$. Therefore, $d = eax + eby = e(ax + by)$ and so $e|d$.*

Corollary 8.1 *Let R be a PID and $r, s \in R^*$.*

1. *If d is a gcd of r and s, then there $x, y \in R$ such that $d = rx + sy$.*

2. *1 is a gcd of r and s iff there exist $x, y \in R$ such that $1 = rx + sy$.*

Proof 8.4 *As in the proof above, $(d) = (r) + (s)$ with d a gcd of r and s (note, since gcd's are associates, they generate the same principal ideal). Since $d \in (d) = (r) + (s)$ the first statement follows.*

For the second statement, we already have one direction, so suppose there $x, y \in R$ such that $1 = rx + sy$. If d is a gcd of r and s, then $d|r$ and $d|s$ and so $d|(rx + sy)$, i.e. $d|1$ and there exists $t \in R$ such that $1 = dt$. In other words, d is a unit, an associate of 1, and so 1 is a gcd of r and s.

Definition 8.6 *Let R be a commutative ring with 1 and $p \in R$ with $p \neq 0$ and p not a unit.*

1. *p is **prime** if for all $r, s \in R$, whenever $p|rs$ either $p|r$ or $p|s$.*

2. *p is **irreducible** if whenever $r|p$ for some $r \in R$, then either r is a unit or r is an associate of p.*

Remark 8.4 *We make several remarks which we leave to the reader to verify.*

1. *Let R be a commutative ring with 1 and $p \in R$ with $p = ab$. Then p is irreducible if exactly one of a and b is a unit and the other is an associate of p.*

2. *If R is an ID, then p is irreducible iff the only way to factor p is as a unit times an associate.*

3. *If p is prime, by induction one can easily show that $p|(r_1 r_2 \cdots r_n)$ implies $p|r_i$ for some $i \in \{1, 2, \ldots, n\}$.*

4. *There are no primes nor irreducibles in a field.*

Example 8.7 *In these examples we illustrate prime and irreducible, and we leave several remarks as exercises.*

1. *When $R = \mathbb{Z}$, then prime and irreducible are equivalent – for instance 3 is both prime and irreducible*

2. When $R = F[x]$, then prime and irreducible are equivalent – for instance all linear polynomials are both prime and irreducible. When $F = \mathbb{R}$, the polynomial $x^2 + 1$ is both prime and irreducible.

3. When $R = \mathbb{Z}[\sqrt{-5}]$, the notion of prime and irreducible are not equivalent – for instance, 3 is irreducible, but not prime. The details of this fact are given in a separate section, but basically since $3 \cdot 3 = (2 + \sqrt{-5})(2 - \sqrt{-5}) = 9$, we have that 3 divides $(2 + \sqrt{-5})(2 - \sqrt{-5})$ yet 3 does not divide $(2 + \sqrt{-5})$ nor $(2 - \sqrt{-5})$.

4. When $R = \mathbb{Z}[x]$ linear polynomials are not all irreducible – for instance, $2x + 2$ is not irreducible, since $2x + 2 = 2(x + 1)$ where 2 is neither a unit nor associate of $2x + 2$. Neither is $2x + 2$ prime, since $2x + 2$ divides $2(x + 1)$, but $2x + 2$ does not divide 2 nor $x + 1$ in R.

The next result gives some insight into the general relationship between prime and irreducible.

Lemma 8.2 In an ID, prime implies irreducible and in a PID, prime and irreducible are equivalent.

Proof 8.5 Assume that $p \in R$ an ID and p is a prime, and suppose that $r|p$. Then there exists s such that $p = rs$. Now certainly p divides rs so that either $p|r$ or $p|s$. Since both r and s divide p, then either r or s is an associate of p (and the other is a unit).

Since R a PID is an ID, we already have one direction, so assume $p \in R$ is irreducible. We show that p is prime. If for some $r, s \in R$ we have $p|rs$, then either $p|s$ (and we're done) or $p \nmid s$. In the latter case we now show that $p|r$. Since R a PID we know the gcd, say d, of p and s exists. So then $d|p$ and $d|s$ which implies $(p) \subseteq (d)$ and $s \in (d)$. Since $p \nmid s$ we know $s \notin (p)$ and so (p) is properly contained in (d). But $p \in (d)$ and so $p = da$ where a is not a unit (for otherwise $(p) = (d)$). But then d must be a unit, and every unit is an associate of 1, which means 1 is a gcd of p and s. Therefore, by Corollary 8.1, there exist $x, y \in R$ such that $1 = px + sy$ and so $r = prx + rsy$. Since p divides prx and p divides rs it follows that p divides $prx + rsy = r$.

Here now is the promised proof that the Gaussian integers are an ED.

Proposition 8.1 The integral domain $\mathbb{Z}[i]$ is an ED.

Proof 8.6 Define the map $\delta : \mathbb{Z}[i] \to \mathbb{N}$ by $\delta(m + ni) = m^2 + n^2$. We need to first show that δ is multiplicative, i.e. for all $m + ni, x + yi \in \mathbb{Z}[i]$ we have $\delta(m + ni)\delta(x + yi) = \delta[(m + ni)(x + yi)]$. We verify this directly:

$$\delta[(m + ni)(x + yi)] = \delta[(mx - ny) + (my + nx)i] = (mx - ny)^2 + (my + nx)^2$$

$$= m^2x^2 - 2mnxy + n^2y^2 + m^2y^2 + 2mnxy + n^2x^2 = m^2x^2 + n^2y^2 + m^2y^2 + n^2x^2$$

$$= (m^2 + n^2)(x^2 + y^2) = \delta(m + ni)\delta(x + yi).$$

Take any $z_1, z_2 \in \mathbb{Z}[i]$ with $z_2 \neq 0$. Our goal is to divide z_1 by z_2 and obtain a quotient q and remainder r such that $z_1 = z_2 q + r$ with $\delta(r) < \delta(z_2)$. To do this, first note that $z_1/z_2 \in \mathbb{Q}[i]$ (multiply top and bottom by the conjugate of z_2). Set $z_1/z_2 = r + si$ for some $r, s \in \mathbb{Q}$. Imagine the elements of $\mathbb{Z}[i]$ as a grid of points in the complex plane with $r + si$ falling within some one-by-one square of that grid. Viewed this way it is clear that there is an $m + ni \in \mathbb{Z}$ with $|m - r| < 1/2$ and $|n - s| < 1/2$. Set $u = r - m$ and $v = s - n$. Then

$$z_1 = z_2(r + si) = z_2[(m + u) + (n + v)i] = z_2(m + ni) + z_2(u + vi).$$

We show that $q = m + ni$ and $r = z_2(u + vi)$ are the q and the r we seek. Certainly $q \in \mathbb{Z}[i]$ and so $r = z_1 - z_2 q \in \mathbb{Z}[i]$ by closure in the ring $\mathbb{Z}[i]$. There is the possibility that $z_1/z_2 \in \mathbb{Z}[i]$ and so $r = 0$. Otherwise,

$$\delta(r) = \delta[z_2(u + vi)] = \delta(z_2)\delta(u + vi) = \delta(z_2)(u^2 + v^2)$$

$$\leq \delta(z_2)\left(\frac{1}{4} + \frac{1}{4}\right) = \frac{1}{2}\delta(z_2) < \delta(z_2).$$

EXERCISES

1 Verify the three statements in Remark 8.1.

2 Prove the following: Let R be a commutative ring with 1.

   a. For all $r \in R$ we have $r|r$.

   b. If for $r, s, t \in R$ we have $r|s$ and $s|t$, then $r|t$.

   c. If for $r, s, t \in R$ we have $r|s$ and $r|t$, then $r|(sx + ty)$ for any $x, y \in R$.

3 Let $S \leq R$ a commutative ring with $a, b \in S^*$. Suppose d is a gcd of a and b in S and $d = ra + sb$ for some $r, s \in S$. Prove that d is a gcd of a and b in R.

4 Prove that the gcd of two integers in the ring $\mathbb{Z}$ is the same in $\mathbb{Z}[i]$.

5 Let $f(x) \in R[x]$, where R is a commutative ring. Prove that for $a \in R$, if $x - a$ is a common divisor of $f(x)$ and $f'(x)$, then $(x - a)^2$ divides $f(x)$.

   (hint: all derivative rules from Calculus 1 apply here)

6 Verify the first two statements in Remark 8.2.

7 Verify the two statements in Remark 8.3.

8 Verify the four statements in Remark 8.4.

9 Verify the statements given in first two examples of Example 8.7.

10 Let F be a field and R be the subset of $F[x]$ consisting of polynomials with no x-term.

a. Carefully verify that R is an ID by showing it's a subdomain of $F[x]$.

b. Verify that $I = \{x^2 f(x) + x^3 g(x) \mid f, g \in R\} \lhd R$.

c. Prove that R is **not** a PID by showing that I in part b is not principal.

11 Prove that in a PID every non-trivial proper ideal is contained in a maximal ideal.

12 Prove that every non-trivial prime ideal in a PID is maximal.

13 Let R be a commutative ring with 1. Prove the following are equivalent for p a non-zero, non-unit in R:

a. p is a prime.

b. (p) is prime.

c. $R/(p)$ is an ID.

14 Let R be an ID. Prove that if $R[x]$ is a PID, then R is a field.

15 Let $a, b \in R$ an ED.

a. Prove if a and b are associates, then $\delta(a) = \delta(b)$.

b. Prove that $\delta(a) < \delta(ab)$ iff b is not a unit in R.

c. Prove $\delta(1) \le \delta(a)$ for all non-zero $a \in R$.

d. Prove a non-zero $a \in R$ is a unit iff $\delta(a) = \delta(1)$.

e. Prove that for any non-zero, non-unit $a \in R$ that $\delta(a^n) < \delta(a^{n+1})$ for any non-negative integer n.

16 Let R be an ED and suppose $a, b \in R$ with $b \mid a$ but $a \nmid b$. Prove that $\delta(b) < \delta(a)$.

17 Let R be an ID and $p \in R$ non-zero, non-unit. p is irreducible iff (p) is maximal among principal ideals, i.e. if $(p) \subseteq (r) \subseteq R$, then either $(r) = (p)$ or $(r) = R$ (you'll need to consider two cases: when r is a unit and when r is not).

8.2 UNIQUE FACTORIZATION DOMAINS

In this section, we explore one final property of the ring of integers and define a special family of integral domains which is a larger family of IDs than the two already investigated: ED and PID.

Definition 8.7 *Let R be an ID and $r \in R^* - U(R)$.*

1. *We say r has* **factorization** *if there exist irreducibles $p_1, p_2, \ldots, p_n \in R$ such that $r = p_1 p_2 \cdots p_n$ for $n \ge 1$.*

2. *We say R has **factorization** if every non-zero, non-unit in R has factorization.*

3. *We say r has **unique factorization** if it has factorization and whenever $r = p_1 p_2 \cdots p_n$ and $r = q_1 q_2 \cdots q_m$ for some irreducibles $p_1, \ldots, p_n, q_1, \ldots, q_m \in R$, it must be that $n = m$ and there exists a permutation $\sigma \in S_n$ such that p_i and $q_{\sigma(i)}$ are associates for all $i = 1, 2, \ldots, n$.*

4. *R is a **unique factorization domain** (**UFD**), if every non-zero, non-unit in R has unique factorization.*

Example 8.8 *Our immediate goal is to show that PID implies UFD and so ED implies PID implies UFD. Thus, all PID examples we have already seen are examples of UFDs, such as $\mathbb{Z}$, $F[x]$, and $\mathbb{Z}[i]$.*

1. *For instance, in $\mathbb{Z}$, two factorizations of -90 are $(5)(-2)(3)(3)$ and $(-2)(-3)(3)(-5)$, but $\sigma = (1\ 4\ 2) \in S_4$ matches up the first set of irreducibles with their associates in the second.*

2. *$\mathbb{Z}[\sqrt{-5}]$ is an ID which is not a UFD. For instance, 9 can be factored as $(3)(3)$ and $(2+\sqrt{-5})(2-\sqrt{5})$, but 3 is not an associate of either $2+\sqrt{-5}$ or $2-\sqrt{-5}$. We will cover the details of this in the next section.*

3. *Consider the polynomials in $F[x]$ for a field F which have no x-term. One can show that R is an ID (show its a subdomain of $F[x]$) and that x^2 and x^3 are irreducibles in R which are certainly not associates (since $U(F[x]) = F^*$). Notice then that $x^6 = (x^2)(x^2)(x^2) = (x^3)(x^3)$ are two different factorizations of x^6 into irreducibles.*

Definition 8.8 *A ring has the **ascending chain condition** (or **ACC**) on principal ideals if there are no infinite chains of the form*

$$(r_1) \subsetneq (r_2) \subsetneq (r_3) \subsetneq \cdots .$$

Example 8.9 *We will illustrate the failure of ACC in a couple of examples.*

1. *Consider the ring $R = \mathbb{Z} + x\mathbb{Q}[x]$, polynomials with rationals coefficients except for the constant coefficient which must be an integer. Then R fails to have ACC on principal ideals, since*

$$(x) \subsetneq \left(\frac{1}{2}x\right) \subsetneq \left(\frac{1}{4}x\right) \subsetneq \left(\frac{1}{8}x\right) \cdots .$$

2. *For any polynomial ring $R[x]$ with R a ring with 1, ACC fails for ideals (not assuming principal). Take the ideal I_n to be polynomials of degree $\leq n$, then*

$$I_0 \subsetneq I_1 \subsetneq I_2 \subsetneq \cdots .$$

Lemma 8.3 *Let R be an ID. If R has ACC on principal ideals, then R has factorization.*

Proof 8.7 *We prove the contrapositive statement. If R does not have factorization, there exists an $r_1 \in R^* - U(R)$ which does not have factorization. Certainly r_1 cannot be an irreducible so that $r_1 = ab$ for some non-units $a, b \in R$. Since r_1 does not have factorization, at least one of a and b does not as well. Let's say its b and set $r_2 = b$.*

Claim 8.1 *$(r_1) \subsetneq (r_2)$.*

First, since r_2 divides r_1 we have $(r_1) \subseteq (r_2)$. Second, $(r_1) \neq (r_2)$ for otherwise we've seen that $r_1 = ur_2$ for some $u \in U(R)$. Since $r_1 = ar_2$ by equating and cancellation we would have $a = u \in U(R)$, a contradiction.

The method above can be repeated on r_2 to produce an r_3 with $(r_1) \subsetneq (r_2) \subsetneq (r_3)$. And this method can be repeated to construct an infinite chain of principal ideals with each properly contained in the next. Hence, R does not have ACC on principal ideals.

Theorem 8.3 *If R is a PID, then R has factorization.*

Proof 8.8 *We show that R has ACC on principal ideals and then appeal to Lemma 8.3. Suppose to the contrary that R did not have ACC on principal ideals. Then there would exist an infinite chain of the form*

$$(r_1) \subsetneq (r_2) \subsetneq (r_3) \subsetneq \cdots .$$

Set $I = \bigcup_{n=1}^{\infty}(r_n)$ which one can show is an ideal in R (Exercise 2). Since R is a PID, there exists an $r \in R$ such that $I = (r)$. Since $r \in I$ this implies $r \in (r_k)$ for some k and so $(r) \subseteq (r_k)$. But then

$$(r) \subsetneq (r_k) \subsetneq (r_{k+1}) \subsetneq \cdots \subsetneq (r),$$

which is an obvious contradiction.

Lemma 8.4 *If R is an ID which has factorization, then R has unique factorization iff every irreducible in R is a prime.*

Proof 8.9 *First assume that R have unique factorization. Suppose $p \in R$ is irreducible and $p | rs$ for some $r, s \in R$. So then $rs = pt$ for some $t \in R$. We may assume that neither r nor s are units, for if r or s is a unit, the result easily follows. For instance, if r is a unit, then $s = ptr^{-1}$ and $p | s$. Now t is not a unit, for otherwise $p = r(st^{-1})$ which would contradict that p is irreducible (since neither r nor st^{-1} are units). By assumption we can express each of r, s and t as a product of irreducibles, i.e.*

$$r = p_1 p_2 \cdots p_k, \qquad s = p_{k+1} p_{k+2} \cdots p_n, \qquad t = q_1 q_2 \cdots q_m.$$

Then $p_1 p_2 \cdots p_n = p q_1 q_2 \cdots q_m$, two factorizations of a ring element into irreducibles. Since R has unique factorization, in particular, p must be an associate of some p_i and so $p | p_i$. If $1 \leq i \leq k$, then $p | r$ and if $k + 1 \leq i \leq n$, then $p | s$.

Second, assume that every irreducible in R is also prime. Suppose that $r \in R^* - U(R)$ with $r = p_1 p_2 \cdots p_n$ and $r = q_1 q_2 \cdots q_m$ where the p_i and q_i are irreducibles.

We will prove this direction of the proof by induction on $n + m$. When $n + m = 2$, then $r = p_1 = q_1$ and the result follows trivially. When $n + m > 2$, note that since $p_1 p_2 \cdots p_n = q_1 q_2 \cdots q_m$ this implies p_1 divides $q_1 q_2 \cdots q_m$. Since p_1 is also prime, this implies that $p_1 | q_k$ for some k and so $(q_k) \subseteq (p_1)$. Since (q_k) and (p_1) are maximal among principal ideals (see Exercise 17 in Section 8.1), it must be that $(q_k) = (p_1)$ which makes q_k and p_1 associates. By reordering the q_i in the factorization of r we may assume that $q_k = q_1$ (since unique factorization is up to a permutation). Write $q_1 = p_1 u$ for some $u \in U(R)$. By cancellation, $p_2 \cdots p_n = (u q_2) \cdots q_m$. We can rename $u q_2$ as q_2 since we are simply looking for associate pairs. Now by induction, $n = m$ and the irreducibles $p_2, \cdots, p_n$ can be put into one-to-one correspondence with $q_2, \ldots, q_m$ so that each pair are associates.

Corollary 8.2 *Every PID is a UFD.*

Proof 8.10 *By Theorem 8.3, R has factorization. By Lemma 8.2, the notion of prime and irreducible are equivalent in a PID and so, by Theorem 8.4, R is a UFD.*

Corollary 8.3 *In a UFD the notion of prime and irreducible are equivalent.*

Proof 8.11 *This is an immediate consequence of Theorem 8.4.*

One final goal concerning UFDs is to show that the property of UFD is carried over from a ring R to the ring of polynomials $R[x]$. For instance, it will follow that $\mathbb{Z}[x]$ is a UFD, since $\mathbb{Z}$ is a UFD. This will be achieved in Section 8.4.

Figure 8.1 is a picture of the world of integral domains.

EXERCISES

1 Verify that in Example 8.9.1 that R is a ring.

2 Prove that if R is a ring and $I_1 \subseteq I_2 \subseteq I_3 \subseteq \cdots$ is an infinite chain of ideals in R, then the union

$$\bigcup_{i=1}^{\infty} I_n \vartriangleleft R.$$

3 Show that in a UFD the gcd of two non-zero elements exists. (hint: if $r, s \in R^*$ factor each into irreducibles, $r = p_1^{e_1} p_2^{e_2} \cdots p_n^{e_n}$ and $s = p_1^{f_1} p_2^{f_2} \cdots p_n^{f_n}$. For each i set $m_i = \min \{e_i, f_i\}$ and show $p_1^{m_1} p_2^{m_2} \cdots p_n^{m_n}$ is a gcd of r and s).

8.3 ONE PARTICULAR INTEGRAL DOMAIN

We now look carefully at $\mathbb{Z}[\sqrt{-5}]$. It gives us a good example of when things go wrong for an ID, for this ID is not a UFD and the notion of prime and irreducible are not equivalent in this ID. The proofs use a notion of a *norm*, which has not been formally introduced, but easy to understand. First off, it is an integral domain. To see this we point out that $\mathbb{Z}[\sqrt{-5}]$ is a subset of the integral domain $\mathbb{C}$ and so it is enough to show it is a subdomain of $\mathbb{C}$. There are three things to check:

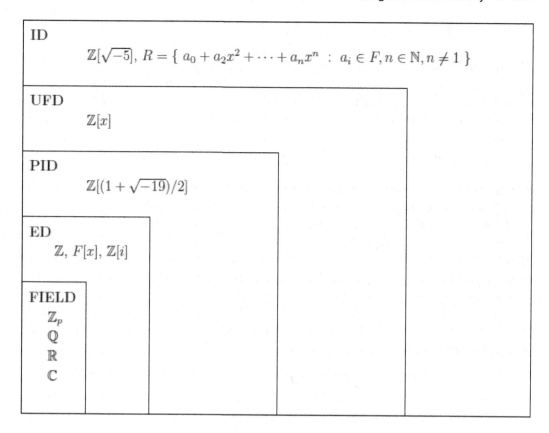

Figure 8.1   A snapshot of the landscape of integral domains.

1. $1 \in R$, since $1 = 1 + 0\sqrt{-5}$.

2. $(a + b\sqrt{-5}) - (c + d\sqrt{-5}) = (a - c) + (b - d)\sqrt{-5} \in R$.

3. $(a + b\sqrt{-5})(c + d\sqrt{-5}) = (ac - 5bd) + (ad + bc)\sqrt{-5} \in R$.

We now introduce an important function which is very useful in analyzing subrings of $\mathbb{C}$.

Definition 8.9 *The* **norm** *of a complex number, written* $\delta(z) = z\bar{z}$ *for* $z \in \mathbb{C}$.

In other words, the norm of a complex number is the square of its magnitude. For instance, for $a + b\sqrt{-5} \in \mathbb{Z}[\sqrt{-5}]$,

$$\delta(a + b\sqrt{-5}) = (a + b\sqrt{-5})(a - b\sqrt{-5}) = a^2 + 5b^2.$$

The norm function is multiplicative, i.e. $\delta(z_1 z_2) = \delta(z_1)\delta(z_2)$, since

$$\delta(z_1 z_2) = z_1 z_2 \overline{z_1 z_2} = z_1 z_2 \bar{z}_1 \bar{z}_2 = z_1 \bar{z}_1 z_2 \bar{z}_2 = \delta(z_1)\delta(z_2).$$

Let's put the norm to work right away.

Claim 8.2 *The only units in $\mathbb{Z}[\sqrt{-5}]$ are ± 1.*

Proof 8.12 *Suppose $a + b\sqrt{-5} \in U(\mathbb{Z}[\sqrt{-5}])$. Then there exists a $c + d\sqrt{-5} \in \mathbb{Z}[\sqrt{-5}]$ such that $(a + b\sqrt{-5})(c + d\sqrt{-5}) = 1$. Taking the norm of both sides yields $(a^2 + 5b^2)(c^2 + 5d^2) = 1$, but then as a positive integer $a^2 + 5b^2 = 1$ and the only way this is possible for integers a and b is $a = \pm 1$ and $b = 0$ and so $a + b\sqrt{-5} = \pm 1$.*

Claim 8.3 *The gcd cannot be defined on $\mathbb{Z}[\sqrt{-5}]$.*

Proof 8.13 *To see this we give an example of two numbers in $\mathbb{Z}[\sqrt{-5}]$ which have no gcd, namely 9 and $6 + 3\sqrt{-5}$. We shall see this in several steps.*

First of all, we need to show that the only proper divisors of $6 + 3\sqrt{-5}$ are ± 3 and $\pm(2 + \sqrt{-5})$. For suppose $6 + 3\sqrt{-5} = (a + b\sqrt{-5})(c + d\sqrt{-5})$. Taking the norm of both sides yields $81 = (a^2 + 5b^2)(c^2 + 5d^2)$ and so $a^2 + 5b^2 = 1, 3, 9, 27$ or 81. By symmetry of the factors it is enough to consider the values $1, 3$ and 9. The value 1 has only solution $a = \pm 1$ and $b = 0$ and so $a + b\sqrt{-5} = \pm 1$. The value 3 has no solution for integers a and b. The value 9 has two possible solutions: $a = \pm 3$, $b = 0$ and so $a + b\sqrt{-5} = \pm 3$ or $a = \pm 2$, $b = \pm 1$ and so $a + b\sqrt{-5} = \pm(2 + \sqrt{-5})$ or $\pm(2 - \sqrt{-5})$. But in the latter case we would have

$$6 + 3\sqrt{-5} = (2 - \sqrt{-5})(c + d\sqrt{-5}) = (2c + 5d) + (2d - c)\sqrt{-5}.$$

This gives the following linear system of equations:

$$2c + 5d = 6$$
$$2d - c = 3$$

But this linear system does not have an integer solution and so $2 - \sqrt{-5}$ cannot be a factor of $6 + 3\sqrt{-5}$ (similarly, neither can $-(2 - \sqrt{-5})$ be a factor). Second, notice the following factorizations:

$$9 = 3 \cdot 3 = (2 + \sqrt{-5})(2 - \sqrt{-5}) \qquad 6 + 3\sqrt{-5} = 3(2 + \sqrt{-5}).$$

Thus we see that ± 3 and $\pm(2 + \sqrt{-5})$ are common divisors of 9 and $6 + 3\sqrt{-5}$ and by what we have just shown above they are the only common divisors and thus are the only candidates for the greatest common divisor.

Third, we now show that 3 does not divide $2 + \sqrt{-5}$ nor does $2 + \sqrt{-5}$ divide 3. We first suppose that 3 divides $2 + \sqrt{-5}$. It follows that $2 + \sqrt{-5} = 3(a + b\sqrt{-5})$ and taking the norm of both sides yields $9 = 9(a^2 + 5b^2)$ and so $a^2 + 5b^2 = 1$. Therefore, $a + b\sqrt{-5} = \pm 1$ and so $2 + \sqrt{-5} = \pm 3$, a contradiction. Similarly, if $3 = (2 + \sqrt{-5})(a + b\sqrt{-5})$, then taking the norm yields once more that $9 = 9(a^2 + 5b^2)$.

Now we see the $\gcd(3, 6 + 3\sqrt{-5})$ cannot exist, since if we choose one of the candidates to be the greatest, then the other candidate should divide it, but it doesn't.

Claim 8.4 *In $\mathbb{Z}[\sqrt{-5}]$ the notions of prime and irreducible are not equivalent.*

Proof 8.14 *We shall show that the number 3 is irreducible, but not prime. First, 3 is irreducible, since if $3 = (a + b\sqrt{-5})(c + d\sqrt{-5})$, then taking norms of both sides yields $9 = (a^2 + 5b^2)(c^2 + 5d^2)$ and so $a^2 + 5b^2 = 1, 3$ or 9. It's enough to consider the values 1 or 3. In the case of value 1 as we have seen $a + b\sqrt{-5} = \pm 1$ a unit. In the case of value 3 we've seen there is no integer solution. Second, 3 is* **not** *prime, for notice 3 divides $9 = (2 + \sqrt{-5})(2 - \sqrt{-5})$ yet we've seen that 3 does not divide $2 + \sqrt{-5}$ and with a similar argument one can show 3 does not divide $2 - \sqrt{-5}$.*

Claim 8.5 $\mathbb{Z}[\sqrt{-5}]$ *is not a UFD*

Proof 8.15 *To see this we shall produce a number in $\mathbb{Z}[\sqrt{-5}]$ which has more than one factorization and these factorization are truly distinct. Consider*

$$9 = 3 \cdot 3 \qquad\qquad 9 = (2 + \sqrt{-5})(2 - \sqrt{-5}).$$

We've seen already that 3 is irreducible. We show that $2 + \sqrt{-5}$ is irreducible (the argument for $2 - \sqrt{-5}$ being irreducible is similar). Suppose $2 + \sqrt{-5} = (a + b\sqrt{-5})(c + d\sqrt{-5})$. Taking norm of both sides yields $9 = (a^2 + 5b^2)(c^2 + 5d^2)$ and so as in Claim 8.4, the only possibilities for $a + b\sqrt{-5}$ are ± 1, a unit. Finally, 3 is not an associate of $2 \pm \sqrt{-5}$, since the only units are ± 1 and they certainly do not differ by ± 1.

EXERCISES

1 Consider the set $R = \mathbb{Z}[\sqrt{-p}]$ for some prime number $p \in \mathbb{Z}$.

 a. Prove that R is an integral domain, by showing it is a subdomain of $\mathbb{C}$ (what does a typical element in R look like?).

 b. Define a norm N on R as follows: $N(a + ib\sqrt{p}) = a^2 + b^2 p$. Prove that N is multiplicative, i.e. $N(z_1 z_2) = N(z_1)N(z_2)$.

 c. Prove for $z \in R$ that z is a unit iff $N(z) = 1$.

 d. Show that every non-unit in R has a factorization into irreducibles.

2 Explore the Gaussian integers $R = \mathbb{Z}[i]$.

 a. Define a norm on R and show it's multiplicative.

 b. Define a norm N on R as follows: $N(a + ib) = a^2 + b^2$. Prove that N is multiplicative.

 c. What are the primes (also irreducibles) in R?

8.4 POLYNOMIALS OVER A UFD

Our goal in this section is two-fold. First, we wish to show that polynomials whose coefficients come from a UFD is again a UFD, hence another property preserved when we go from a ring R to $R[x]$. We point out that this does not hold for ED nor PID. Simply consider the ring $\mathbb{Z}$ which is both an ED and PID, yet $\mathbb{Z}[x]$ is neither an ED nor PID.

Second, we explore when a polynomial over a UFD is irreducible. Both goals stem from a result attributed to Gauss which we will first prove. We will often refer to the quotient field of a UFD, so if R is a UFD, then Q will represent the quotient field of R.

Definition 8.10 *Let R be a UFD and $f(x) \in R[x]$.*

1. *The **content** of f, written $C(f)$ is the gcd of all the coefficients of $f(x)$.*

2. *f is **primitive** if its content is 1.*

Example 8.10 *Let $f(x) = 30x^2 - 12x + 60 \in \mathbb{Z}[x]$. The content of f in this case is 6 (or -6). Notice that $f(x) = 6\tilde{f}(x)$ where $\tilde{f}(x) = 5x^2 + 2x + 10$. It's easy to see that for any $f(x) \in R[x]$ we can express $f(x) = C(f)\tilde{f}(x)$ where $\tilde{f}(x) \in R[x]$ is primitive and this representation is unique up to a unit, since content is unique up to a unit (since this is true of gcd). This fact remains true in $Q[x]$ as we prove now.*

Lemma 8.5 *If R is a UFD and $Q = Q(R)$ with $f(x) \in Q[x]$, then $f(x) = c\hat{f}(x)$ where $c \in Q$ and $\hat{f}(x) \in R[x]$ is primitive. Furthermore, this representation is unique up to a unit.*

Proof 8.16 *Write $f(x) = \frac{a_n}{b_n}x^n + \cdots + \frac{a_1}{b_1}x + \frac{a_0}{b_0} \in Q[x]$. Set $b = b_0 b_1 \cdots b_n$. Then certainly $bf(x) \in R[x]$. Set $g(x) = bf(x)$. As in the example we can express $g(x) = C(g)\tilde{g}(x)$ where $\tilde{g}(x) \in R[x]$ is primitive. Then $f(x) = \frac{C(g)}{b}\tilde{g}(x)$ and we have the required representation.*

Now suppose $f(x) = c\hat{f}(x)$ and $f(x) = dh(x)$ where $c, d \in Q$ and $\hat{f}(x), h(x) \in R[x]$ primitive. Write $c = \frac{a}{b}$ and $d = \frac{r}{s}$ and equate to get $\frac{a}{b}\hat{f}(x) = \frac{r}{s}h(x)$. Then as $\hat{f}(x) = brh(x)$. Set $p(x) = as\hat{f}(x) = brh(x)$. Since $\hat{f}(x)$ and $h(x)$ are primitive, we have $C(p) = as$ and $C(p) = br$. Since content is unique up to a unit, $as = ubr$ for some unit $u \in U(R)$. Then $c = ud$ and so $ud\hat{f}(x) = dh(x)$ which implies $u\hat{f}(x) = h(x)$.

Example 8.11 *We will illustrate the proof with an example. Let $f(x) = \frac{35}{3}x^2 + \frac{28}{5}x + \frac{21}{2} \in \mathbb{Q}[x]$. Then $30f(x) = 350x^2 + 168x + 315$. Now $C(350x^2 + 168x + 315) = 7$ so that $30f(x) = 7(50x^2 + 27x + 45)$ and hence $f(x) = \frac{7}{30}(50x^2 + 27x + 45)$, where $\hat{f}(x) = 50x^2 + 27x + 45 \in \mathbb{Z}[x]$ is primitive.*

Definition 8.11 *For $f(x) \in Q[x]$, the **content** of f, written again as $C(f)$ will be any associate of the $c \in Q$ of Lemma 8.5.*

Example 8.12 *In Example 8.11, $C(f) = \frac{7}{30}$.*

Note that this definition of content agree with the former definition in the case that $f(x) \in R[x]$. We now prove the fundamental result of this section.

Lemma 8.6 (Gauss) *In a UFD R, the product of two primitives in $R[x]$ is again primitive.*

Proof 8.17 *Let $f(x), g(x) \in R[x]$ be primitive and set $h(x) = f(x)g(x)$. Express $f(x) = a_n x^n + \cdots + a_1 x + a_0$, $g(x) = b_m x^m + \cdots + b_1 x + b_0$ and $h(x) = c_{m+n} x^{m+n} + \cdots + c_1 x + c_0$. Now suppose to the contrary that $h(x)$ were not primitive. Then $C(h)$ is a non-unit and so in R a UFD we may factor $C(h)$ into irreducibles which are also primes. Hence, there exists a prime $p \in R$ dividing $C(h)$. Let's fix this prime and consider the factor ring $R/(p)$. For $a \in R$, we will use the shorthand notation $\bar{a}$ to signify $a + (p) \in R/(p)$. Since p is prime, this implies the ideal (p) is prime and so $R/(p)$ is an ID. Now consider the polynomial ring $(R/(p))[x]$ which is thus also an ID. Let's establish the notation $\bar{f}(x), \bar{g}(x), \bar{h}(x) \in (R/(p))[x]$, where, for instance, $\bar{f}(x) = \overline{a_n} x^n + \cdots \overline{a_1} x + \overline{a_0}$. Just as $h(x) = f(x)g(x)$, so does $\bar{h}(x) = \bar{f}(x)\bar{g}(x)$, since*

$$\overline{c_i} = \overline{\sum_{i=k+l} a_k b_l} = \sum_{i=k+l} \overline{a_k b_l} = \sum_{i=k+l} \overline{a_k}\, \overline{b_l}.$$

Notice that since p divides $C(h)$ this implies $\bar{c} = \bar{0}$ for all i and so $\bar{h}(x)$ is the zero polynomial in $(R/(p))[x]$, however neither $\bar{f}(x)$ nor $\bar{g}(x)$ are the zero polynomial. Indeed, since $f(x)$ is primitive, its content is 1 and so no prime can divide $C(f)$. In particular p does not divide $C(f)$ and so there exists an a_i such that p does not divide a_i. Thus, $\overline{a_i} \neq \bar{0}$ and therefore, $\bar{f}(x)$ has non-zero coefficients (the same argument works for $\bar{g}(x)$). Therefore, $\bar{f}(x)$ and $\bar{g}(x)$ are zero divisors in $(R/(p))[x]$, which contradicts that $(R/(p))[x]$ is an ID.

Corollary 8.4 *Let R be a UFD and $Q = Q(R)$ with $f(x), g(x) \in Q[x]$. Then $C(fg) = C(f)C(g)$.*

Proof 8.18 *By Lemma 8.5, we can write $f(x) = c\hat{f}(x)$ and $g(x) = d\hat{g}(x)$ for some $c, d \in Q$ and $\hat{f}(x), \hat{g}(x) \in R[x]$ primitive. So then $f(x)g(x) = cd\hat{f}(x)\hat{g}(x)$ and $\hat{f}(x)\hat{g}(x)$ is primitive by Gauss' Lemma. Therefore, $C(fg) = cd = C(f)C(g)$.*

Example 8.13 *Consider the polynomial $f(x) = x^2 - 5x + 6 \in \mathbb{Z}[x]$. The factorization $x^2 - 5x + 6 = (\frac{3}{5}x - \frac{6}{5})(\frac{5}{3}x - 5)$ attests to the fact that $f(x)$ is not irreducible in $\mathbb{Q}[x]$. Now $f(x)$ is, in fact, not irreducible in $\mathbb{Z}[x]$, since $x^2 - 5x + 6 = (x-2)(x-3)$. Using Gauss' Lemma, we will show that it is always the case that if a polynomial factors over the quotient field, then it factors over the UFD. We can illustrate the proof with this example:*

$$x^2 - 5x + 6 = \left(\frac{3}{5}x - \frac{6}{5}\right)\left(\frac{5}{3}x - 5\right) = \frac{3}{5}(x-2)\frac{5}{3}(x-3) = (x-2)(x-3).$$

Corollary 8.5 *Let R be a UFD and $Q = Q(R)$ with $f(x) \in R[x]$ of degree at least one. If $f(x)$ is irreducible in $R[x]$, then $f(x)$ is irreducible in $Q[x]$. In particular, if $f(x)$ is irreducible in $\mathbb{Z}[x]$, then $f(x)$ is irreducible in $\mathbb{Q}[x]$.*

Proof 8.19 *We prove the contrapositive statement (as illustrated in the example above). Suppose that $f(x) = g(x)h(x)$ with $g(x), h(x) \in Q[x]$ non-units. In $Q[x]$ this means the degree of $g(x)$ and $h(x)$ are at least one. By Lemma 8.5, we can write $g(x) = c\hat{g}(x)$ and $h(x) = d\hat{h}$ for some $c, d \in Q$ and $\hat{g}(x), \hat{h}(x) \in R[x]$ primitive. Note that by a simple degree argument it follows that $\deg(\hat{g}) = \deg(g)$ and $\deg(\hat{h}) = \deg(h)$. Furthermore, since $f(x) \in R[x]$ we know on the one hand $C(f) \in R$, and on the other hand $C(f) = C(g)C(h) = cd$. Thus, $cd \in R$ and $f(x) = [cd\hat{g}(x)]\hat{h}(x)$ where $cd\hat{g}(x), \hat{h}(x) \in R[x]$ are non-units.*

Example 8.14 *The converse of Corollary 8.5 is nearly true, i.e. if $f(x)$ is irreducible in $Q[x]$, then $f(x)$ is irreducible in $R[x]$. The problem is that sometimes the content of $f(x)$ can act as a non-unit factor. Consider $f(x) = 2x - 4 \in \mathbb{Z}[x]$. This is certainly irreducible in $\mathbb{Q}[x]$, since all degree one polnomials over a field are irreducible. However, $2x - 4 = 2(x - 2)$ where neither 2 nor $x - 2$ are units in $\mathbb{Z}[x]$ and so $f(x)$ is not irreducible in $\mathbb{Z}[x]$. However, if we assume $f(x)$ is primitive, then we can show the converse holds.*

Corollary 8.6 *Let R be a UFD and $Q = Q(R)$ with $f(x) \in R[x]$ primitive of degree at least one. If $f(x)$ is irreducible in $Q[x]$, then $f(x)$ is irreducible in $R[x]$.*

Proof 8.20 *We prove this again using the contrapositive statement. Suppose that $f(x) = g(x)h(x)$ where $g(x), h(x) \in R[x]$ non-units. We show that $g(x)$ (and similarly $h(x)$) has degree at least one. If $\deg(g) = 0$, then $g(x) = a \notin R^* - U(R)$. Then, by Gauss' Lemma, $C(f) = C(gh) = C(g)C(h) = aC(h)$. But then $C(f)$ is not a unit which contradicts that $f(x)$ is primitive. Thus, since $g(x), h(x)$ have degree at least one, this implies they are non-units in $Q[x]$ and so the same factorization illustrates that $f(x)$ is not irreducible in $Q[x]$.*

Remark 8.5 *If R is a commutative ring with 1 and $p \in R$ is prime. Then p is prime in $R[x]$ as well. Indeed, p prime in R implies $R/(p)$ is an ID. Now $R[x]/(p) \cong (R/(p))[x]$ via the map $(a_n x^n + \cdots + a_1 x + a_0) + (p) \mapsto (a_n + (p))x^n + \cdots + (a_1 + (p))x + (a_0 + (p))$. Therefore, $R[x]/(p)$ is an ID which in turn implies that p is prime in $R[x]$.*

Theorem 8.4 *If R is a UFD, then $R[x]$ is a UFD.*

Proof 8.21 *First of all, since R is a ID, then so is $R[x]$. Second, we show that $R[x]$ has factorization. Let $f(x)$ be a non-zero non-unit in $R[x]$. Should the degree of $f(x)$ be zero, then $f(x) \in R$ which is a UFD and so factors into irreducibles. If the degree of $f(x)$ is greater than zero, then $f(x)$ is a non-unit in $Q[x]$. Since $Q[x]$ is an ED, it is also a UFD and so $f(x)$ factors into irreducibles in $Q[x]$, i.e. $f(x) = p_1(x)p_2(x)\cdots p_n(x)$ where each $p_i(x) \in Q[x]$ is irreducible. By Lemma 8.5, write each $p_i(x) = c_i\hat{p}_i(x)$ where each $c_i \in Q$ and $\hat{p}_i(x) \in R[x]$ primitive. By Corollary 8.6, each $\hat{p}_i(x)$ is irreducible in $R[x]$. By Gauss' Lemma, $\hat{p}_1(x)\hat{p}_2(x)\cdots\hat{p}_n(x)$ is primitive, and so since $f(x) = c_1c_2\cdots c_n\hat{p}_1(x)\hat{p}_2(x)\cdots\hat{p}_n(x)$, it follows that $C(f) = c_1c_2\cdots c_n$. Now $f(x) \in R[x]$ so then $c_1c_2\cdots c_n = C(f) \in R$*

a UFD. Thus, $c_1 c_2 \cdots c_n = q_1 q_2 \cdots q_m$ a product of irreducibles in R (and so in $R[x]$ as well). Hence, $f(x) = q_1 q_2 \cdots q_m \hat{p}_1(x) \hat{p}_2(x) \cdots \hat{p}_n(x)$ is a factorization of $f(x)$ into irreducibles in $R[x]$.

Finally, we show that $R[x]$ has unique factorization. By Lemma 8.4, it's enough to show that every irreducible in $R[x]$ is also prime. Let $f(x)$ be irreducible in $R[x]$. Should $f(x)$ have degree zero, then $f(x) = p \in R$ and so, since R is a UFD, p is prime in R. By Remark 8.5, $f(x) = p$ is also prime in $R[x]$. If $\deg(f) > 0$, then by Corollary 8.5, $f(x)$ is irreducible in $Q[x]$. Since $Q[x]$ is a UFD, it follows that $f(x)$ is prime in $Q[x]$. Now suppose $f(x)$ divides $g(x)h(x)$ in $R[x]$, where $g(x), h(x) \in R[x]$ Then $f(x)$ certainly divides $g(x)h(x)$ in $Q[x]$. Now $f(x)$ is prime in $Q[x]$ so that either $f(x)$ divides $g(x)$ or divides $h(x)$ in $Q[x]$. Without loss of generality, assume $f(x)$ divides $g(x)$ so that $g(x) = f(x)p(x)$ for some $p(x) \in Q[x]$. By Lemma 8.5, write $p(x) = c\hat{p}(x)$ where $c \in Q$ and $\hat{p}(x) \in R[x]$ primitive. On the one hand, $C(g) = C(fg) = C(f)C(p) = c$ (note that $f(x)$ irreducible in $R[x]$ implies $f(x)$ is primitive, for otherwise its content would be a non-unit factor). On the other hand, since $g(x) \in R[x]$, its content is in R. Hence, $c \in R$ which implies $p(x) \in R[x]$ and so $f(x)$, in fact, divides $g(x)$ in $R[x]$. Thus, we have shown that $f(x)$ is prime in $R[x]$.

To finish this section we present two results which can be useful in determining irreducibility called the Rational Root Theorem and Eisenstein's Criterion.

Theorem 8.5 (Rational Root Theorem) *Let R be a UFD and $Q = Q(R)$ with $f(x) = a_n x^n + \cdots + a_1 x + a_0 \in R[x]$ of degree at least one ($a_n \neq 0$). If $\frac{s}{t} \in Q$ is a root of $f(x)$ with $\gcd(s, t) = 1$, then $s | a_0$ and $t | a_n$.*

Proof 8.22 *Since $f(s/t) = 0$, this implies that $a_n (s/t)^n + \cdots + a_1(s/t) + a_0 = 0$ and multiplying through by t^n yields $a_n s^n + a_{n-1} s^{n-1} t + \cdots + a_1 s t^{n-1} + a_0 t^n = 0$.*

On the one hand $a_n s^n = -t(a_{n-1} s^{n-1} + \cdots + a_1 s t^{n-2} + a_0 t^{n-1})$ which implies t divides $a_n s^n$, but since $\gcd(s, t) = 1$ it follows that t must divide a_n.

On the other hand $(a_n s^{n-1} + a_{n-1} s^{n-2} t + \cdots + a_1 t^{n-1})s = a_0 t^n$ which implies s divides $a_0 t^n$, but since $\gcd(s, t) = 1$ it follows that s must divide a_0.

Example 8.15 *Consider the polynomial $f(x) = 4x^3 + 2x^2 - x - 5 \in \mathbb{Z}[x]$. The factors of $a_0 = 5$ are $\pm 1, \pm 5$ and the factors of $a_n = 4$ are $\pm 1, \pm 2, \pm 4$. Hence, should $f(x)$ have a root in Q it would have to be among the candidates*

$$\pm \frac{1}{1}, \ \pm \frac{1}{2}, \ \pm \frac{1}{4}, \ \pm \frac{5}{1}, \ \pm \frac{5}{2}, \ \pm \frac{5}{4}.$$

The only candidate that works is 1. Should no candidate have worked, then we could conclude that $f(x)$ is irreducible in $Q[x]$. Indeed, should $f(x)$ factor it would have to have a linear factor (since $\deg(f) = 3$) and hence a root in Q.

Theorem 8.6 (Eisenstein's Criterion) *Let R be a UFD and $Q = Q(R)$ with $f(x) = a_n x^n + \cdots + a_1 x + a_0 \in R[x]$ of degree at least one. Suppose there is a prime $p \in R$ such that*

1. $p \nmid a_n$.

2. $p \mid a_i$ for $i = 0, 1, \cdots, n-1$.

3. $p^2 \nmid a_0$.

Then $f(x)$ is irreducible in $Q[x]$.

Proof 8.23 *Suppose, to the contrary, that $f(x)$ were not irreducible in $Q[x]$. Then by Corollary 8.5, $f(x)$ would not be irreducible in $R[x]$. Thus, $f(x) = g(x)h(x)$ where $g(x) = b_k x^k + \cdots + b_1 x + b_0 \in R[x]$ with $b_k \neq 0$ and $h(x) = c_m x^m + \cdots + c_1 x + c_0 \in R[x]$ with $c_m \neq 0$. Since $p \mid a_0$ and $a_0 = b_0 c_0$ and p is prime, it follows that $p \mid b_0$ or $p \mid c_0$. But p^2 does not divide a_0 so we may conclude that $p \mid b_0$ or $p \mid c_0$, but not both. Without loss of generality, assume that p divides c_0 but not b_0. Since p does not divides $a_n = b_k c_m$, it follows that p does not divide c_m. Let r be smallest such that p does not divide c_r (note that $1 \leq r \leq m < n$). Consider the coefficient $a_r = b_0 c_r + b_1 c_{r-1} + \cdots + b_r c_0$ and rewrite as $a_r - b_1 c_{r-1} - \cdots - b_r c_0 = b_0 c_r$. Since $p \mid a_r$, $p \mid b_1 c_{r-1}, \ldots, p \mid b_r c_0$, it must be that $p \mid b_0 c_r$ and so $p \mid b_0$ or $p \mid c_r$ neither or which is a true statement. Hence, $f(x)$ must be irreducible in $Q[x]$.*

Example 8.16 *We illustrate Eisenstein's Criterion with several examples.*

1. *Consider the polynomial $f(x) = 10x^2 + 15x + 6 \in Z[x]$ and the prime $p = 3$. Since $3 \nmid 10$, $3 \mid 15$, $3 \mid 6$ and $9 \nmid 6$, by Eisenstein's Criterion, $f(x)$ must be irreducible in $Q[x]$ (and since $f(x)$ is primitive it is also irreducible in $Z[x]$).*

2. *If no prime satisfies Eisenstein's Criterion for a given polynomial, we cannot conclude that the polynomial is not irreducible. Take, for instance, $x^2 + 1 \in Z[x]$ which is certainly irreducible (for otherwise $i \in Z$).*

3. *We will apply the following statement which is left as an exercise: Suppose $f(x) \in R[x]$ and there exists an $r \in R$ such that $f(x+r)$ is irreducible in $Q[x]$. Then $f(x)$ is irreducible in $Q[x]$.*

   Consider again $f(x) = x^2 + 1 \in Z[x]$. Notice that $f(x+1) = (x+1)^2 + 1 = x^2 + 2x + 2$ which is irreducible in $Q[x]$ (Eisenstein with $p = 2$). Therefore, $f(x)$ is irreducible in $Q[x]$ (and since $f(x)$ is primitive it is also irreducible in $Z[x]$).

EXERCISES

1 Decide whether or not each of the following polynomials in $Z[x]$ is irreducible over $Q[x]$ and irreducible over $Z[x]$.

   a. $10x^3 + 6x^2 - 18x + 12$

   b. $x^3 + x^2 - x + 2$

   c. $x^4 - 10x^2 + 1$

   d. $x^4 + 1$

   e. $3x^5 - 6x^3 + 28x - 14$

f. $2x^4 - 3x^3 - x^2 - x + 1$

g. $x^4 + 2x^2 - 1$

2 Let $R = \{m + 2ni \; : \; m, n \in \mathbb{Z}\}$.

  a. Prove that $Q(R) = Q(i)$.

  b. Show that $x^2 + 1$ is reducible over $Q(R)$ but not over R

  c. Use the previous parts to conclude that R is not a UFD.

3 Prove if $a_n x^n + \cdots + a_1 x + a_0 \in F[x]$ is irreducible, then so is $a_0 x^n + \cdots + a_{n-1} x + a_n$.

4 Suppose $f(x) \in R[x]$ and there exists an $r \in R$ such that $f(x+r)$ is irreducible in $Q[x]$. Prove that $f(x)$ is irreducible in $Q[x]$.

CHAPTER 9

Field Theory

I N THIS CHAPTER, we set the stage for the next chapter on Galois theory. In Section 9.1, we remind the reader of some definitions, concepts and results as well as introduce the definition of *algebraic*. In Section 9.2, we investigate field extensions which allows us to look at chains of fields, an important idea in Galois theory. This also allows us to investigate some famous geometric impossibilities in Section 9.3. In Section 9.4, we look at some particular and important field extensions, and in Section 9.5, we prove the existence of some of these structures we introduced. Finally, in Section 9.6, we completely classify finite fields.

9.1 REVIEW AND ALGEBRAICITY

We need to review some ideas from general ring theory, but present them from a different perspective. First, there is the notion of a **polynomial extension** of a ring.

Definition 9.1 *Let S be any subring of a commutative ring R and take $r \in R$. The* **ring of polynomials in r over** S, *written*

$$S[r] = \{a_d r^d + \cdots + a_1 r + a_0 \ : \ a_0, a_1, \ldots, a_d \in S \text{ and } d > 0\}.$$

We leave it to the reader to show that $S[r]$ (see Figure 9.1) is indeed a ring (show it's a subring of R). In fact, one can show that $S[r]$ is the smallest ring containing both S and r in the sense that if T is a ring and $S \subseteq T$ and $r \in T$, then $S[r] \subseteq T$.

Example 9.1 *We have run across several examples of this structure:* $\mathbb{Z}[i]$, $\mathbb{Z}[\sqrt{2}]$, $\mathbb{Z}[\sqrt{-5}]$, $\mathbb{Q}[i]$.

Second, we have $S[x]$, the **ring of polynomials in x over** S, where x is an indeterminate and S is a commutative ring.

Third, we have a map which connects these two structures above. Recall the **evaluation homomorphism**

$$\Phi_r : S[x] \to S[r] \quad \text{by} \quad \Phi_r(a_d x^d + \cdots + a_1 x + a_0) = a_d r^d + \cdots + a_1 r + a_0.$$

DOI: 10.1201/9781003335283-9

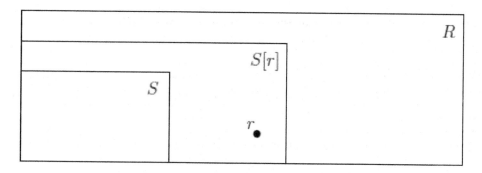

Figure 9.1 A visual representation of $S[r]$.

The reader should note that the evaluation map is indeed a homomorphism which fixes S and maps onto $S[r]$. For brevity, we may rephrase this epimorphism using functional notation as follows: If we set $f(x) = a_d x^d + \cdots + a_1 x + a_0$, then $f(r) = a_d r^d + \cdots + a_1 r + a_0$ and so $\Phi_r(f(x)) = f(r)$.

An important concept in field theory is the notions of algebraic versus transcendent, which we now define.

Definition 9.2 *Let $r \in R \supseteq S$ rings.*

1. *We say r is **algebraic over** S if there exists a non-zero polynomial $f(x) \in S[x]$ such that $f(r) = 0$, i.e. r is a **zero** of some non-zero polynomial in $S[x]$*

2. *We say r is **transcendental over** S if no such polynomial exists.*

Example 9.2 *Here are several examples which illustrate the definitions.*

1. *$i \in \mathbb{C}$ is algebraic over $\mathbb{Z}$, since i is a zero of $f(x) = x^2 + 1 \in \mathbb{Z}[x]$.*

2. *$\sqrt{2} \in \mathbb{R}$ is algebraic over $\mathbb{Z}$, since $\sqrt{2}$ is a zero of $f(x) = x^2 - 2 \in \mathbb{Z}[x]$.*

3. *$i\sqrt[4]{2}$ is algebraic over $\mathbb{Z}$, since $i\sqrt[4]{2}$ is a zero of $x^4 - 2 \in \mathbb{Z}[x]$.*

4. *$i\sqrt[4]{2}$ is algebraic over $\mathbb{Z}[\sqrt{2}]$, since $i\sqrt[4]{2}$ is a zero of $x^2 + \sqrt{2} \in \mathbb{Z}[\sqrt{2}][x]$.*

5. *It can be shown that both π and e in $\mathbb{R}$ are transcendental over $\mathbb{Z}$.*

Note that, r being algebraic over S is equivalent to there being an element $a_d r^d + \cdots + a_1 r + a_0 \in S[r]$ with $a_d \neq 0$ and $a_d r^d + \cdots + a_1 r + a_0 = 0$, i.e. there is a non-trivial element in $S[r]$ which equals zero. This also implies that elements in $S[r]$ are not uniquely represented as elements of $S[r]$ (just consider zero which can be represented as 0 and also as $a_d r^d + \cdots + a_1 r + a_0$ for some $a_d r^d + \cdots + a_1 r + a_0 \in S[r]$ with $a_d \neq 0$ – see Theorem 9.1.3 for what is meant by unique representation in $S[r]$). This is yet another equivalent way to define r being algebraic over S. We now summarize these statements in a theorem and leave the details of the proof to the reader.

Theorem 9.1 *Let S be any subring of a commutative ring R and take $r \in R$. The following statements are equivalent:*

1. *r is transcendental over S.*

2. *There is no element $a_d r^d + \cdots + a_1 r + a_0 \in S[r]$ with $a_d \neq 0$ and $a_d x^d + \cdots + a_1 x + a_0 = 0$.*

3. *If $(a_d r^d + \cdots + a_1 r + a_0) = (b_e r^e + \cdots + b_1 r + b_0)$ in $S[r]$, then $d = e$ and $a_i = b_i$ for $i = 1, 2, \ldots, d$.*

4. *$ker \Phi_r = \{0\}$.*

5. *$S[x] \cong S[r]$.*

One can interpret the last statement in the theorem as saying if r is transcendental over S, then in a sense (with respect to S) r is just like an indeterminate.

Example 9.3 *Here are some examples which further illustrate unique representation and Theorem 9.1.*

1. *$i \in \mathbb{C}$ is algebraic over $\mathbb{Z}$ since $i^2 + 1$ is a non-trivial element of $\mathbb{Z}[i]$ which equals zero. Hence, we do not have unique representation in $\mathbb{Z}[i]$ (for instance, observe how $i^3 + 1 = -i + 1$). Notice also that $i^2 + 1 = 0$ implies that $i^2 = -1$, $i^3 = -i$, $i^4 = 1$, etc., so that*

$$\mathbb{Z}[i] = \{m + ni \; : \; m, n \in \mathbb{Z}\},$$

and thus no higher powers of i are needed to represent elements in $\mathbb{Z}[i]$. Furthermore, this description of $\mathbb{Z}[i]$ has unique representation.

2. *$\sqrt{2} \in \mathbb{R}$ is algebraic over $\mathbb{Z}$ since $\left(\sqrt{2}\right)^2 - 2$ is a non-trivial element of $\mathbb{Z}[\sqrt{2}]$ which equals zero. In this case,*

$$\mathbb{Z}[\sqrt{2}] = \{m + n\sqrt{2} \; : \; m, n \in \mathbb{Z}\},$$

and this description has unique representation.

3. *One can show there is no non-trivial element of $\mathbb{Z}[\pi]$ which equals zero. Therefore, there is no simpler way to describe $\mathbb{Z}[\pi]$ besides with the original definition,*

$$\mathbb{Z}[\pi] = \{a_d \pi^d + \cdots + a_1 \pi + a_0 \; : \; a_0, a_1, \ldots, a_d \in \mathbb{Z} \text{ and } d > 0\},$$

and this description has unique representation.

We now focus our attention on fields and address the same material. Our choice of variable names will change a bit. Instead of $r \in R \supseteq S$, we will have $a \in E \supseteq F$ with E and F fields. Let's look again at the evaluation epimorphism in this context, namely

$$\Phi_a : F[x] \to F[a] \quad \text{by} \quad \Phi_a(f(x)) = f(a).$$

In this context we know a bit more. First, since F is a field, we know that $F[x]$ is a PID. Therefore, the ideal, $\ker\Phi_a = (p(x))$, for some $p(x) \in F[x]$. Now since Φ_a is an epimorphism, by the Fundamental Theorem of Homomorphisms, we have

$$F[x]/(p(x)) = F[x]/\ker\Phi_a \cong F[a].$$

We now simply restate Theorem 9.1 in the context of fields with no further proof required.

Theorem 9.2 *Let F be any subfield of a field E and take $a \in E$. The following statements are equivalent:*

1. *a is transcendental over F.*

2. *There is no element $a_d a^d + \cdots + a_1 a + a_0 \in F[a]$ with $a_d \neq 0$ and $a_d x^d + \cdots + a_1 x + a_0 = 0$.*

3. *If $(a_d a^d + \cdots + a_1 a + a_0) = (b_e a^e + \cdots + b_1 a + b_0)$ in $F[a]$, then $d = e$ and $a_i = b_i$ for $i = 1, 2, \ldots, d$.*

4. *$\ker\Phi_a = \{0\}$.*

5. *$F[x] \cong F[a]$.*

EXERCISES

1 For each $r \in R \supseteq S$ a commutative ring, verify that r is algebraic over S:

   a. $\sqrt{-5} \in \mathbb{C} \supseteq \mathbb{Z}$.

   b. $\sqrt[6]{3} \in \mathbb{R} \supseteq \mathbb{Z}[\sqrt[3]{3}]$.

   c. $\sqrt[3]{2} \in \mathbb{R} \supseteq \mathbb{Z}$.

   d. $i\sqrt[3]{2} \in \mathbb{C} \supseteq \mathbb{Z}$.

   e. $\sqrt[6]{3} \in \mathbb{R} \supseteq \mathbb{Z}[\sqrt{3}]$.

2 For each problem in Exercise 1, express elements of $S[r]$ with unique representation.

3 Let S be any subring of a commutative ring R and take $r \in R$.

   a. Prove that $S[r]$ is a ring.

   b. Prove that $S[r]$ is the smallest ring containing S and r.

   c. Verify that the evaluation map

   $$\Phi_r : S[x] \to S[r] \quad \text{by} \quad \Phi_r(a_d x^d + \cdots + a_1 x + a_0) = a_d r^d + \cdots + a_1 r + a_0,$$

   is a homomorphism which fixes S and maps onto $S[r]$.

4 Prove Theorem 9.1.

9.2   VECTOR SPACES & EXTENSION FIELDS

The goal of this section is to make a connection between fields and vector spaces and develop some results based on this connection. The reader may wish to review some basic concepts in Linear Algebra such as *vector space* over a field, *basis* and *dimension.*

Definition 9.3 *If F is a subfield of a field E we call E an **extension field** of F and F is called the **base field**. The notation we shall use is $F \subseteq E$.*

In this context one can view E as a vector space. The vectors will be the elements of E, scalars the elements of F, vector addition will be the field addition in E, and scalar multiplication will be the field multiplication in E. The dimension of the vector space E over the field F will be denoted by $[E : F]$ and is called the **degree** of E over F. If $[E : F] < \infty$ we say that E is a **finite extension** of F. There is a very good reason why this notation is employed and is identical to $[G : H]$, the index of a group G over a subgroup H, but this will take some time to develop.

Example 9.4 $[\mathbb{C} : \mathbb{R}] = 2$ *with basis $1, i$. Indeed, $1, i$ span $\mathbb{C}$ since every element of $\mathbb{C}$ can be written as a real linear combination of 1 and i. They are linearly independent over $\mathbb{R}$, since if for $a, b \in \mathbb{R}$ we have $a \cdot 1 + b \cdot i = 0$, then $a + bi = 0$ and so $a = b = 0$ (otherwise i would be an element of the reals or $\mathbb{C}$ would have zero divisors, a contradiction).*

We leave it to the reader as exercises to show that

1. $[E : F] = 1$ iff $E = F$.

2. If $F \subseteq K \subseteq E$ are fields and $[E : F] < \infty$, then $[E : K] \leq [E : F]$ and $[K : F] \leq [E : F]$.

This next result hints at a connection between $[E : F]$ and $[G : H]$.

Theorem 9.3 *If $F \subseteq E \subseteq K$ are fields with $[K : E], [E : F] < \infty$, then $[K : F] < \infty$ and $[K : F] = [K : E][E : F]$.*

Proof 9.1 *Set $[K : E] = m$ and $[E : F] = n$ and suppose that E has basis $a_1, a_2, \ldots, a_n$ over F and K has basis $b_1, b_2, \ldots, b_m$ over E. We will show that the set of products $\{a_i b_j : 1 \leq i \leq n, 1 \leq j \leq m\}$ forms a basis for K over F, and so the result will be proved. We first note that these products $a_i b_j$ are distinct, becauseof the linear independence of $b_1, b_2, \ldots, b_m$ over E.*

To show they are linearly independent over F, suppose that $\sum_{j=1}^{m} \sum_{i=1}^{n} c_{ij}(a_i b_j) = 0$ for some $c_{ij} \in F$. Then $\sum_{j=1}^{m} \left(\sum_{i=1}^{n} c_{ij} a_i \right) b_j = 0$ where $\sum_{i=1}^{n} c_{ij} a_i \in E$. Since the b_j's are linearly independent over E, we have that $\sum_{i=1}^{n} c_{ij} a_i = 0$, for all j. But since the a_i's are linearly independent over F, we have that $c_{ij} = 0$ for all i and for all j. To show the set of products span K over F, take any element $c \in K$. Since the b_j's span K over E, we can write $c = \sum_{j=1}^{m} c_j b_j$, for some $c_j \in E$. Since the a_i's span E

over F, for each j we can write $c_j = \sum_{i=1}^{n} d_{ij} a_i$ for some $d_{ij} \in F$. Putting all this together we have

$$c = \sum_{j=1}^{m} c_j b_j = \sum_{j=1}^{m} \sum_{i=1}^{n} d_{ij} a_i b_j = \sum_{i=1}^{n} \sum_{j=1}^{m} d_{ij} (a_i b_j),$$

and so we have written c as a linear combination of the products with scalars from F.

Notice the parallel to the group theory result $[G : H] = [G : K][K : H]$ for $H \le K \le G$, which was proved in a very different way appealing to an equivalence relation.

We need a bit of notation to continue our discussion. Let $F(x)$ represent the field of quotients of the integral domain $F[x]$, i.e.

$$F(x) = \mathcal{Q}\left(F[x]\right) = \left\{ \frac{f(x)}{g(x)} : f(x), g(x) \in F[x] \text{ and } g(x) \text{ is not the zero polynomial} \right\}.$$

For $a \in E \supseteq F$, let $F(a)$ represent the field of quotients of the integral domain $F[a]$, i.e.

$$F(a) = \mathcal{Q}\left(F[a]\right) = \left\{ \frac{f(a)}{g(a)} : f(a), g(a) \in F[a] \text{ and } g(a) \ne 0 \right\}.$$

Lemma 9.1 *For $a \in E \supseteq F$ fields, the following statements are equivalent:*

1. *a is algebraic over F.*

2. *$F[a] \cong F(a)$.*

3. *$[F(a) : F] < \infty$.*

Proof 9.2 *We first show the first statement implies the second. As we pointed out earlier, given that a is algebraic over F it follows that $F[a] \cong F[x]/(p(x))$ for some non-zero polynomial $p(x) \in F[x]$. Since $F[a] \subseteq E$, a field, the ring $F[a]$ is, in fact, an integral domain, and via the isomorphism above, $F[x]/(p(x))$ is also an integral domain. Therefore, $p(x)$ is prime (see Exercise 13 in Section 8.1). Since $F[x]$ is an integral domain, $p(x)$ is irreducible in $F[x]$ as well. This in turn makes $(p(x))$ maximal among principle ideals (see Exercise 17 in Section 8.1). Since $F[x]$ is a PID, $(p(x))$ is, in fact, a maximal ideal and so $F[a] \cong F[x]/(p(x))$ is a field. Now if $F[a]$ is a field, then its quotient field is no larger than itself; more precisely, $F(a) = \mathcal{Q}(F[a]) \cong F[a]$.*

*Now we show the second statement implies the third. We are given that $F(a) \cong F[a] \cong F[x]/(p(x))$. Now $p(x)$ cannot be the zero polynomial, for otherwise the field $F(a)$ would be isomorphic to $F[x]$ which we know is **not** a field. Furthermore, $p(x)$ is not of degree zero for then it could not have any zeros. Set $d = deg(p) \ge 1$. For brevity, let's establish the coset notation $\overline{f(x)} = f(x) + (p(x)) \in F[x]/(p(x))$. We can view $F[x]/(p(x))$ as a vector space over F as follows: Vector addition will be the usual coset addition and scalar multiplication will be defined as*

$$c\,\overline{f(x)} = \overline{cf(x)} \quad \text{which equals} \quad \overline{c}\,\overline{f(x)}.$$

We show that the vectors $\overline{1}, \overline{x}, \ldots, \overline{x^{d-1}}$ form a basis for $F[x]/(p(x))$ over F. Once we show this, the third statement is then true, since the dimension of $F[x]/(p(x))$ over F equals d and $F[x]/(p(x)) \cong F(a)$ as vector spaces over F, so that the dimension of $F(a)$ over F also equals d, i.e. $[F(a) : F] = d < \infty$. First, they are linearly independent since if

$$a_0 \overline{1} + a_1 \overline{x} + \cdots + a_{d-1} \overline{x^{d-1}} = \overline{0}, \text{ for some } a_i \in F, \text{ then}$$

$$\overline{a_0 + a_1 x + \cdots + a_{d-1} x^{d-1}} = \overline{0}, \text{ which implies } a_0 + a_1 x + \cdots + a_{d-1} x^{d-1} \in (p(x))$$

Set $f(x) = a_0 + a_1 x + \cdots + a_{d-1} x^{d-1}$ and so we have $f(x) = p(x)g(x)$ for some $g(x) \in F[x]$. Notice that

$$d > d - 1 \geq \deg(f) = \deg(p) + \deg(g) = d + \deg(g),$$

which implies that $\deg(g) < 0$, i.e. g is the zero polynomial. But then so is $f = pg$ the zero polynomial, i.e. $a_0 = a_1 = \cdots = a_{d-1} = 0$, and so linear independence is shown.

These vectors $\overline{1}, \overline{x}, \ldots, \overline{x^{d-1}}$ also span $F[x]/(p(x))$. Take any $\overline{f(x)} \in F[x]/(p(x))$. Since $F[x]$ is an ED we have $f(x) = p(x)q(x) + r(x)$ for some $q, r \in F[x]$ and $\deg(r) < \deg(p)$. Set $r(x) = a_0 + a_1 x + \cdots + a_{d-1} x^{d-1}$. Then

$$\overline{f(x)} = \overline{p(x)} \cdot \overline{q(x)} + \overline{r(x)} = \overline{r(x)} = a_0 \overline{1} + a_1 \overline{x} + \cdots + a_{d-1} \overline{x^{d-1}}.$$

Hence, we see that $\overline{f(x)}$ is a linear combination of $\overline{1}, \overline{x}, \ldots, \overline{x^{d-1}}$.

Now we show third statement implies the first by proving the contrapositive statement. If a is transcendental over F, we've shown that $F[a] \cong F[x]$. Since $1, x, x^2, \ldots$ forms an infinite basis for $F[x]$ we have $[F[x] : F] = \infty$ and so $[F[a] : F] = \infty$. Now since $F(a)$ contains an isomorphic copy of $F[a]$, then it's also the case that $[F(a) : F] = \infty$ as well.

The polynomial $p(x)$ mentioned in the proof of the Lemma is called the **irreducible** (or **minimal**) **polynomial of** a **over** F. We need one last result before we look at some specific examples in detail. This result will make it easier for us to identify this $p(x)$ in concrete settings.

Lemma 9.2 *For $a \in E \supseteq F$ fields and $p(x) \in F[x]^*$, the following are statements equivalent:*

1. *$\ker \Phi_a = (p(x))$*

2. *p is irreducible with $p(a) = 0$*

3. *For all $f \in F[x]$, $f(a) = 0$ iff $p|f$.*

Proof 9.3 *The third statement being equivalent to the first is straight forward. First assume the third statement is true and notice that*

$$f \in \ker \Phi_a \text{ iff } f(a) = 0 \text{ iff } p|f \text{ iff } f \in (p(x)).$$

Now assume the first statement is true and observe that

$$f(a) = 0 \text{ iff } f \in ker\Phi_a \text{ iff } f \in (p(x)) \text{ iff } p|f.$$

Now we show that the first statement is equivalent to the second. First assume the first statement is true. The second statement follows from our earlier work, for we proved in Lemma 9.1 that $F(a) \cong F[a] \cong F[x]/(p(x))$ where p is irreducible (note that a is algebraic over F, since p is not the zero polynomial, so that $ker\Phi_a$ is non-trivial). Furthermore, $p(a) = 0$ since $p \in ker\Phi_a$. Now assume that the second statement is true. Set $ker\Phi_a = (f(x))$ and we will show that f and p are associates. Since $p(a) = 0$ this implies that $p \in ker\Phi_a = (f(x))$ and so $f|p$. Since p is irreducible, f is either a unit or an associate of p. However, f cannot be a unit, for this would imply that $ker\Phi_a = (f(x)) = F[x]$ and this would mean that every polynomial in $F[x]$ has the element a as a zero, an obvious contradiction. Hence, f is an associate of p, and so $(p(x)) = (f(x)) = ker\Phi_a$.

Observe that the third statement in the Lemma is what gives p the alternate name of **minimal** polynomial of a over F.

Lemma 9.2.2 gives us a practical way of finding the irreducible polynomial of a over F. We simply need to find an irreducible polynomial which has a as a zero. This in turn allows us to compute the value of $[F(a) : F]$, since as in the proof of Lemma 9.1, $[F(a) : F]$ is equal to the degree of the irreducible polynomial of a over F. The reader should verify that such an irreducible polynomial is unique up to associates.

Example 9.5 *We compute the degree of some extension fields.*

1. *Now $[\mathbb{Q}(i) : \mathbb{Q}] = 2$, since $x^2 + 1 \in \mathbb{Q}[x]$ is irreducible over $\mathbb{Q}$ with i as a zero (see Example 8.16.3).*

2. *Now $[\mathbb{Q}(\sqrt[4]{2}) : \mathbb{Q}] = 4$, since $x^4 + 2 \in \mathbb{Q}[x]$ is irreducible over $\mathbb{Q}$ (Eisenstein with $p = 2$).*

3. *We compute $[\mathbb{Q}(\sqrt{1 + \sqrt{3}}) : \mathbb{Q}] = 4$ as follows: Set $a = \sqrt{1 + \sqrt{3}}$. Then $a^2 = 1 + \sqrt{3}$ and $(a^2 - 1)^2 = 3$ and so $a^4 - 2a^2 - 2 = 0$. Therefore, a is a zero of $x^4 - 2x^2 - 2 \in \mathbb{Q}[x]$ which is irreducible over $\mathbb{Q}$ (Eisenstein with $p = 2$).*

4. *We compute $[\mathbb{Q}(\sqrt{1 + \sqrt{2}}) : \mathbb{Q}] = 4$ as follows: Set $a = \sqrt{1 + \sqrt{2}}$. Then $a^2 = 1 + \sqrt{2}$ and $(a^2 - 1)^2 = 2$ and so $a^4 - 2a^2 - 1 = 0$. Therefore, a is a zero of $x^4 - 2x^2 - 1 \in \mathbb{Q}[x]$ which is irreducible over $\mathbb{Q}$. To see this evaluate $p(x + 1) = x^4 + 4x^3 + 4x^2 - 2$ which is irreducible over $\mathbb{Q}$ (Eisenstein with $p = 2$).*

Remark 9.1 *It's easy to prove that $a \in E \supseteq F$ fields and a is a zero of $f(x) \in F[x]$ implies $[F(a) : F] \leq deg(f)$ (exercise).*

Example 9.6 *Let's illustrate the use of Remark 9.1. First,* $[\mathbb{Q}(\sqrt[4]{2}) : \mathbb{Q}(\sqrt{2})] \leq 4$, *since* $\sqrt[4]{2}$ *is a root of* $x^4 - 2 \in \mathbb{Q}[x]$ *which has degree 4. In fact, we can do better,* $[\mathbb{Q}(\sqrt[4]{2}) : \mathbb{Q}(\sqrt{2})] \leq 2$, *since* $\sqrt[4]{2}$ *is a root of* $x^2 - \sqrt{2} \in \mathbb{Q}(\sqrt{2})[x]$ *which has degree 2.*

Before we look at some examples in depth, we need to define the some additional concepts.

Definition 9.4 *Let* $a_1, a_2, \ldots, a_n \in E \supseteq F$ *fields. We define* $F(a_1, a_2, \ldots, a_n)$ *to be the smallest subfield of* E *containing* $a_1, a_2, \ldots, a_n$ *and* F. *We say that* E *is* **finitely generated over** F *if* $E = F(a_1, a_2, \ldots, a_n)$ *for some* $a_1, a_2, \ldots, a_n \in E$, *and the* $a_1, a_2, \ldots, a_n$ *are called the* **generators of** E **over** F.

Remark 9.2 *Observe that this definition agrees with and generalizes the definition of* $F(a)$. *Furthermore, the reader should check that for any* k, $1 \leq k \leq n$ *that*

$$(F(a_1, a_2, \ldots, a_k))(a_{k+1}, \ldots, a_n) = F(a_1, a_2, \ldots, a_n).$$

Example 9.7 *We illustrate how one can reduce the number of generators for an extension field. These facts will be useful in Example 9.8.*

1. *Set* $E = \mathbb{Q}(\sqrt{2}, i\sqrt[4]{2})$. *We show that* $E = \mathbb{Q}(i\sqrt[4]{2})$. *Indeed, since* $-(i\sqrt[4]{2})^2 = \sqrt{2}$, *the field* $\mathbb{Q}(i\sqrt[4]{2})$ *contains* $\sqrt{2}, i\sqrt[4]{2}$ *and* $\mathbb{Q}$. *Therefore, since* E *is the smallest field containing* $\sqrt{2}, i\sqrt[4]{2}$ *and* $\mathbb{Q}$, *we have that* $E \subseteq \mathbb{Q}(i\sqrt[4]{2})$. *The reverse inclusion is evident, since* E *contains* $i\sqrt[4]{2}$ *and* $\mathbb{Q}$ *and* $\mathbb{Q}(i\sqrt[4]{2})$ *is the smallest field containing* $i\sqrt[4]{2}$ *and* $\mathbb{Q}$.

2. *Set* $E = \mathbb{Q}(\sqrt{2}, \sqrt[3]{2})$. *We show that* $E = \mathbb{Q}(\sqrt[6]{2})$. *Indeed, notice that*

$$\sqrt[6]{2} = 2^{1/6} = 2^{1/2}(2^{-1/3}) = \sqrt{2}(\sqrt[3]{2})^{-1} \in E,$$

so that $\mathbb{Q}(\sqrt[6]{2}) \subseteq E$. *Likewise* $\sqrt{2} = (\sqrt[6]{2})^3$ *and* $\sqrt[3]{2} = (\sqrt[6]{2})^2$ *so that* $E \subseteq \mathbb{Q}(\sqrt[6]{2})$.

Example 9.8 *We illustrate how one can compute the degree of a finite extension.*

1. *Set* $E = \mathbb{Q}(\sqrt{2}, i\sqrt[4]{2})$. *We wish to compute* $[E : \mathbb{Q}]$. *We will compute this in two different ways (for sometimes only one of these ways may be available in certain situations).*

   For the first approach, we've seen in Example 9.7.1 that $E = \mathbb{Q}(i\sqrt[4]{2})$. *Hence, to compute* $[E : \mathbb{Q}]$ *we need only find the irreducible polynomial of* $i\sqrt[4]{2}$ *over* $\mathbb{Q}$. *Now this is easy, since* $x^4 + 2$ *is irreducible over* $\mathbb{Q}$ *(use Eisenstein's Criterion with* $p = 2$*) with* $i\sqrt[4]{2}$ *as a zero. Therefore,*

$$[E : \mathbb{Q}] = [\mathbb{Q}(i\sqrt[4]{2}) : \mathbb{Q}] = deg(x^4 + 2) = 4.$$

   A second approach to computing $[E : \mathbb{Q}]$ *is to make use of Theorem 9.3. Set* $E_1 = \mathbb{Q}(\sqrt{2})$ *and note that* $\mathbb{Q} \subseteq E_1 \subseteq E$ *with* $E = E_1(i\sqrt[4]{2})$. *Now* $[E_1 : \mathbb{Q}] = 2$ *since* $x^2 - 2$ *is irreducible over* $\mathbb{Q}$ *(Eisenstein with* $p = 2$*) and has* $\sqrt{2}$ *as a zero.*

The index $[E : E_1] = 2$ as well. To see this, first note that $i\sqrt[4]{2}$ is a root of $x^2 + \sqrt{2} \in E_1[x]$ so that

$$[E : e_1] \leq deg(x^2 + \sqrt{2}) = 2.$$

Now, $[E : E_1]$ cannot equal 1, for if it did, then $E = E_1$. But then $i\sqrt[4]{2} \in E_1 = \mathbb{Q}(\sqrt{2})$, a contradiction (Since $\mathbb{R}$ is a field containing $\mathbb{Q}$ and $\sqrt{2}$, then $\mathbb{R}$ contains $\mathbb{Q}(\sqrt{2})$ the smallest such field. Hence, if $i\sqrt[4]{2} \in \mathbb{Q}(\sqrt{2})$, then $i\sqrt[4]{2} \in \mathbb{R}$ and so $i \in R$, a contradiction). Therefore,

$$[E : \mathbb{Q}] = [E : E_1][E_1 : \mathbb{Q}] = (2)(2) = 4.$$

2. *Set $E = \mathbb{Q}(\sqrt{2}, \sqrt[3]{2})$. We wish to compute $[E : \mathbb{Q}]$. This example illustrates yet another style of approach for computing indexes. Once again we consider the chain of fields $\mathbb{Q} \subseteq \mathbb{Q}(\sqrt{2}) \subseteq E$. We just saw in Example 9.8.1 that $[\mathbb{Q}(\sqrt{2}) : \mathbb{Q}] = 2$. Thus,*

$$[E : \mathbb{Q}] = [E : \mathbb{Q}(\sqrt{2})][\mathbb{Q}(\sqrt{2}) : \mathbb{Q}] = 2 \cdot [E : \mathbb{Q}(\sqrt{2})],$$

and so 2 divides $[E : \mathbb{Q}]$. In a similar manner, consider the chain of fields $\mathbb{Q} \subseteq \mathbb{Q}(\sqrt[3]{2}) \subseteq E$. Observe that $[\mathbb{Q}(\sqrt[3]{2}) : \mathbb{Q}] = 3$ since $x^3 - 2 \in \mathbb{Q}[x]$ is the irreducible polynomial of $\sqrt[3]{2}$ over $\mathbb{Q}$ (Eisenstein with $p = 2$). Therefore, as above, we get that 3 divides $[E : \mathbb{Q}]$. Since 2 and 3 are relatively prime, we have that 6 divides $[E : \mathbb{Q}]$ and so $[E : \mathbb{Q}] \geq 6$. We now show the reverse inequality (and hence, $[E : \mathbb{Q}] = 6$). Let's set $E_1 = \mathbb{Q}(\sqrt{2})$ so that $\mathbb{Q} \subseteq E_1 \subseteq E$ with $E = E_1(\sqrt[3]{2})$. Then

$$[E : E_1] = [E_1(\sqrt[3]{2}) : E_1] \leq 3,$$

since $x^3 - 2 \in E_1[x]$ of degree 3 with $\sqrt[3]{2}$ as a zero (not necessarily irreducible and Eisenstein doesn't apply here). Therefore,

$$[E : \mathbb{Q}] = [E : \mathbb{Q}(\sqrt{2})][\mathbb{Q}(\sqrt{2}) : \mathbb{Q}] = 2 \cdot [E : \mathbb{Q}(\sqrt{2})] \leq (2)(3) = 6.$$

There is yet another (easiest) way to compute this degree. We've seen in Example 9.7.1 that $E = \mathbb{Q}(\sqrt[6]{2})$. Now $x^6 - 2 \in \mathbb{Q}[x]$ with $x^6 - 2$ irreducible (Eisenstein with $p = 2$) and having $\sqrt[6]{2}$ as a root. Therefore,

$$[E : \mathbb{Q}] = [\mathbb{Q}(\sqrt[6]{2}) : \mathbb{Q}] = 6.$$

The next result besides its theoretical importance has practical implications, for it allows us to construct finite fields.

Theorem 9.4 *Let $a \in E \supseteq F$ fields with a algebraic over F. Let $p(x) \in F[x]$ be the irreducible polynomial for a over F with $d = deg(p)$. Then $1, a, a^2, \ldots, a^{d-1}$ forms a basis for $F(a)$ over F.*

Proof 9.4 *As was mentioned at the beginning of this section, we know that* $[F(a) : F] = d$, *so it's enough to show that* $1, a, a^2, \ldots, a^{d-1}$ *are linearly independent over* F. *Suppose, to the contrary, that* $1, a, a^2, \ldots, a^{d-1}$ *are linearly dependent. Then there would be scalars* $a_0, a_1, \ldots, a_{d-1} \in F$ *not all zero with* $a_0 1 + a_1 a + \cdots + a_{d-1} a^{d-1} = 0$. *Set* $f(x) = a_0 + a_1 x + \cdots + a_{d-1} x^{d-1} \in F[x]$ *so that* $f(a) = 0$. *By Lemma 9.2 we know* $[F(a) : F] \leq \deg(f) \leq d - 1$, *a contradiction.*

Example 9.9 *We use Theorem 9.4 to given explicit descriptions of* $F(a)$ *when a is algebraic over* F.

1. *Now* $i \in \mathbb{C}$ *is algebraic over* $\mathbb{R}$ *with irreducible polynomial* $p(x) = x^2 + 1$. *By Theorem 1, the elements* $1, i$ *form a basis for* $\mathbb{R}(i)$ *over* $\mathbb{R}$. *Therefore,* $\mathbb{R}(i) = \{a1 + bi \ : \ a, b \in \mathbb{R}\}$. *In other words,* $\mathbb{R}(i) = \mathbb{C}$, *i.e* $\mathbb{C}$ *is the smallest field containing both* i *and* $\mathbb{R}$, *which should come as no surprise to the reader. Note that in general, the number of generators of a field will be different than the number of elements in a basis. For instance, in this example there is one generator* i *and two basis elements* $1, i$.

2. *We've seen that* $\sqrt[3]{2}$ *is algebraic over* $\mathbb{Q}$ *with* $[\mathbb{Q}(\sqrt[3]{2}) : \mathbb{Q}] = 3$ *and so by Theorem 9.4, a basis for* $\mathbb{Q}(\sqrt[3]{2})$ *over* $\mathbb{Q}$ *is* $1, 2^{1/3}, 2^{2/3}$. *Therefore,*

$$\mathbb{Q}(\sqrt[3]{2}) = \{a + b2^{1/3} + c2^{2/3} \ : \ a, b, c \in \mathbb{Q}\}.$$

3. *Theorem 9.4 allows us to construct finite fields of any prime power order (which we will see later are the only sizes allowed for finite fields). To illustrate this we will construct a field of order* $4 = 2^2$. *Set* $F = \mathbb{Z}_2$ *and* $p(x) = x^2 + x + 1 \in \mathbb{Z}_2[x]$. *Notice that* $p(x)$ *is irreducible over* $\mathbb{Z}_2$, *for otherwise* $p(x)$ *would have a linear factor, say* $x - c$ *for some* $c \in \mathbb{Z}_2$ *and so* c *would be a zero of* $p(x)$. *However* $p(0) = 1 \neq 0$ *and* $p(1) = 1 \neq 0$, *so this cannot be true. Therefore, let's formally define a new symbol a to be a zero of* $p(x)$ *so that* $p(x)$ *is the irreducible polynomial of a over* $\mathbb{Z}_2$. *By the theorem we know that* $1, a$ *forms a basis for* $\mathbb{Z}_2(a)$ *over* $\mathbb{Z}_2$ *so that*

$$\mathbb{Z}_2(a) = \{c \cdot 1 + d \cdot a \ : \ c, d \in \mathbb{Z}_2\} = \{c + da \ : \ c, d \in \mathbb{Z}_2\} = \{0, 1, a, 1 + a\}.$$

Note that since there are two choices for c *and for* d, *we see that* $\mathbb{Z}_2(a)$ *is a field with 4 elements. Up to this point, we have never run across a field of such a size. Recall that the only finite fields we were aware of were* $\mathbb{Z}_p$ *for some prime* p. *Let's write the addition and multiplication tables for this field. Let's see a few examples of adding and multiplying before we exhibit the whole table (which the reader should be able to reproduce). First note that since a is a zero of* $p(x) = x^2 + x + 1$, *this implies that* $a^2 + a + 1 = 0$ *or* $a^2 = a + 1$ *in* $\mathbb{Z}_2$. *For instance, we can add* $(1 + a) + a = 1 + 2a = 1$ *and multiply* $(1 + a) \cdot a = a + a^2 = a + (a + 1) = 1$. *Figure 9.2 exhibits the complete addition and multiplication table.*

+	0	1	a	$1+a$
0	0	1	a	$1+a$
1	1	0	$1+a$	a
a	a	$1+a$	0	1
$1+a$	$1+a$	a	1	0

·	0	1	a	$1+a$
0	0	0	0	0
1	0	1	a	$1+a$
a	0	a	$1+a$	1
$1+a$	0	$1+a$	1	a

Figure 9.2 The addition and multiplication table for a field with four elements.

EXERCISES

1 Compute each of the following:

  a. $[\mathbb{Q}(\sqrt{1+i}) : \mathbb{Q}]$

  b. $[\mathbb{Q}(\sqrt{3}, \sqrt[3]{3}) : \mathbb{Q}]$

  c. $[\mathbb{Q}(\sqrt{2}, i\sqrt{3}) : \mathbb{Q}]$

  d. $[\mathbb{Q}(\sqrt{2}, \sqrt[3]{3}) : \mathbb{Q}]$

2 Let $F = \mathbb{Z}_5$ and $f(x) = x^3 + x^2 + 1 \in F[x]$.

  a. Carefully explain why $f(x)$ is irreducible in $F[x]$.

  b. Let a be a root of $f(x)$ in some extension of F. Give a description of $F(a)$ and compute its size.

  c. Compute $(3a^2 + 2a + 1) + (3a^2 + 3a + 2)$ and $(a^2 + 2)(a^2 + 3a)$ in $F(a)$.

3 If F is a subfield of a field E, prove that

  a. $[E : F] = 1$ iff $E = F$.

  b. If $F \subseteq K \subseteq E$ are fields and $[E : F] < \infty$, then $[E : K] \leq [E : F]$ and $[K : F] \leq [E : F]$.

4 Suppose $F \subseteq K \subseteq E$ are fields with $[E : F] < \infty$. Prove that if K is properly between F and E, then $[E : K]$ and $[K : F]$ are strictly less than $[E : F]$.

5 Verify that the irreducible polynomial of $a \in E \supseteq F$ is unique up to associates.

6 Verify Remark 9.1.

7 Verify the statement made in Remark 9.2.

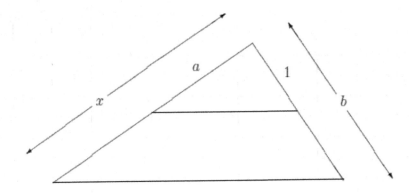

Figure 9.3 Diagram used for constructing ab.

8 Let $a \in E \supseteq F$ be algebraic over F with $[F(a) : F] = n$ an odd number. Prove that a^2 is also algebraic over F with $[F(a^2) : F] = n$ an odd number. Furthermore, show that $F(a^2) = F(a)$.

9.3 GEOMETRIC CONSTRUCTIONS

Imagine you have at your disposal a straightedge with two marks on it so that the length between the two marks is one unit length of measure (pick your favorite units of measure). You also have an unmarked compass and a writing instrument (like a pencil) and paper.

Recall from high school geometry that certain constructions are possible with these tools:

1. two line at right angles to each other.

2. two parallel lines.

3. the midpoint of a line.

Definition 9.5 *A real number a will be considered* **constructible** *if using these tools you can construct a line segment of length $|a|$. A subset X of the real numbers is* **constructible** *if every $a \in X$ is constructible.*

Theorem 9.5 *If a, b are two positive constructible real numbers, then so are $a + b$, $a - b$, ab and a/b (when $b \neq 0$).*

Proof 9.5 *To construct $a + b$ simply mark off the length of a and of b side by side. To construct $a - b$ mark off the length of a and then at the end of the line segment of length a mark in reverse the length of b. To construct ab first construct a right triangle with legs of length 1 and a, then extend (or mark off on) the leg of length 1 a line segment of length b. Construct a line parallel to the hypothenuse of the original right triangle. Extend (or mark off on) the leg of length a to find the intersection of the parallel line constructed and the line containing the line segment of length a.*

Then one can show using similar triangles that the length of x in Figure 9.3 equals ab. A similar construction shows that a/b is constructible as well, which we leave as an exercise.

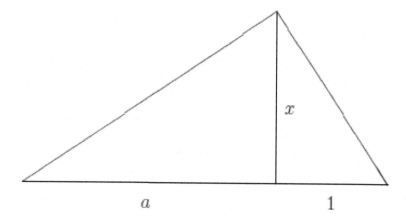

Figure 9.4 Diagram used for constructing $\sqrt{a}$.

Corollary 9.1 *If a, b are two constructible real numbers, then so are $a + b$, $a - b$, ab and a/b (when $b \neq 0$).*

Proof 9.6 *This corollary requires a proof by cases which relies heavily on the result just proved.*

Corollary 9.2 *The rational numbers are constructible. Furthermore, the collection of all constructible real numbers forms a subfield of $\mathbb{R}$.*

Proof 9.7 *First note that the integers are certainly constructible. The rest follows immediately from Corollary 9.1.*

Lemma 9.3 *If a is a constructible non-negative real number, then so is $\sqrt{a}$.*

Proof 9.8 *First construct a line segment of length a and of length 1 side by side on the same line. Construct the midpoint of this line and construct a semicircle with diameter the line segment of length $a + 1$. Construct a line perpendicular to the line segment of length $a + 1$ passing through the point where the line segment of length a and 1 meet. Find the point of intersection of this line and the semicircle. From geometry we know that this point together with the line segment of length $a + 1$ forms a right triangle. Furthermore, this altitude of the right triangle we just constructed indicated in Figure 9.4 has length $x = \sqrt{a}$ (use similar triangles).*

Corollary 9.3 *If F is a field of constructible real numbers and $a > 0$ constructible with $a \in F$ yet $\sqrt{a} \notin F$, then every element of the extension field $F(\sqrt{a})$ is constructible.*

Proof 9.9 *Since the field of constructible numbers contains F by assumption and $\sqrt{a}$, by Lemma 9.3, and $F(\sqrt{a})$ is the smallest field containing F and $\sqrt{a}$, this implies that $F(\sqrt{a})$ is contained in the field of constructible numbers.*

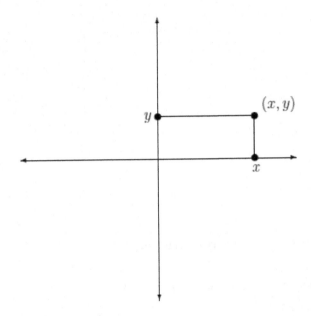

Figure 9.5 A point is constructible if its coordinates are constructible.

Remark 9.3 *The field $F(\sqrt{a})$ is called a* **quadratic** *extension of F since $x^2 - a \in$ $F[x]$ is the irreducible polynomial of $\sqrt{a}$ over F and so $[F(\sqrt{a}) : F] = 2$.*

Theorem 9.6 (Characterization of Constructible Numbers) *A real number b is constructible iff there exists a chain of fields of the form $\mathbb{Q} = F_0 \subset F_1 \subset F_2 \subset \cdots \subset F_n$ such that each F_i is a quadratic extension of F_{i-1} for $i = 1, 2, \ldots, n$ and $b \in F_n$.*

Proof 9.10 *First assume the existence of a chain of fields as in the statement of the theorem. We prove by induction on the length of the chain that b is constructible. For $n = 0$ we have $b \in \mathbb{Q}$ which we know is constructible. Assume now that $n > 0$. We are given that $b \in F_n = F_{n-1}(\sqrt{a_n})$ for some $a_n > 0$ in F_{n-1} yet $\sqrt{a_n} \notin F_{n-1}$. By induction, F_{n-1} is constructible and by Corollary 9.3, $F_n = F_{n-1}(\sqrt{a_n})$ is constructible.*

Now assume that b is constructible. Consider the xy-plane as the arena where all these constructions with straightedge and compass occur. Define a point (x, y) in the xy-plane **constructible** *if its coordinates x and y are constructible numbers. Indeed, it certainly is, since we can mark off on the x-and y-axis the values x and y and then drop two perpendiculars and find the point of intersection as in Figure 9.5.*

Because of the tools we have for constructing, it is clear that if P_1, P_2, P_3 and P_4 are constructible points, then so are (when they exist)

1. *The point(s) of intersection of lines $\overline{P_1 P_2}$ and $\overline{P_3 P_4}$.*

2. *The point(s) of intersection of line $\overline{P_1 P_2}$ and the circle with radius the line segment $\overline{P_3 P_4}$ and center P_3.*

3. *The point(s) of intersection of the circle with radius the line segment $\overline{P_1P_2}$ and center P_1 and the circle with radius the line segment $\overline{P_3P_4}$ and center P_3.*

In fact, these are the only ways of constructing new points with the tools available and hence the only way of producing constructible numbers. Hence, to finish the proof it is sufficient to show that all points constructed in this manner have coordinates which lie in at most a quadratic extension of the coordinates of the points P_1, P_2, P_3 and P_4.

Claim 9.1 *The standard equations of the lines and circles described above have co-efficients all of which are constructible numbers.*

In the case of a line passing through two constructible points $P_1(a, b)$ and $P_2(c, d)$, the equation of the line is

$$\frac{y - b}{x - a} = \frac{d - b}{c - a} \quad \Leftrightarrow \quad (y - b)(c - a) = (x - a)(d - b)$$

$$\Leftrightarrow \quad (d - b)x + (c - a)y + [b(c - a) - a(d - b)] = 0.$$

Note that $d - b$, $c - a$ and $b(c - a) - a(d - b)$ are all constructible numbers. In the case of a circle with radius $\overline{P_1P_2}$ and center P_1, the equation of the circle is

$$(x - a)^2 + (y - b)^2 = (c - a)^2 + (d - b)^2$$

$$\Leftrightarrow \quad x^2 + y^2 - 2ax - 2by + [a^2 + b^2 - (c - a)^2 - (d - b)^2] = 0.$$

Note that $-2a$, $-2b$ and $a^2 + b^2 - (c - a)^2 - (d - b)^2$ are all constructible numbers. Hence, the Claim is proved.

Let's call such a line or circle in Claim 9.1 a **constructible** *line or circle. As stated earlier, the proof will be complete if we show that the points of intersection of constructible lines and circles have coordinates which lie in at most a quadratic extension of the coefficients of the constructible lines and circles. To do this we will consider the three cases listed above.*

1. *Consider two constructible lines $ax + by = c$ and $dx + ey = f$. By Cramer's rule, the point of intersection of the two lines has coordinates*

$$x = \frac{\begin{vmatrix} c & b \\ f & e \end{vmatrix}}{\begin{vmatrix} a & b \\ d & e \end{vmatrix}} = \frac{ce - bf}{ae - bd}, \qquad y = \frac{\begin{vmatrix} a & c \\ d & f \end{vmatrix}}{\begin{vmatrix} a & b \\ d & e \end{vmatrix}} = \frac{af - cd}{ae - bd}.$$

Thus, x and y are constructible numbers.

2. *Consider the case of a constructible line $ax + by + c = 0$ and constructible circle $x^2 + y^2 + dx + ey + f = 0$. Solving the first equation for x or y (whichever is possible) and substituting into the second equation yields a quadratic equation with constructible coefficients. Then using the quadratic equation, the points of intersection have coordinates which lie in at most a quadratic extension of the coefficients of the constructible line and circle.*

3. *For the case of two circles, notice that we can reduce this case to the second case as follows: Let $x^2 + y^2 + ax + by + c = 0$ and $x^2 + y^2 + dx + ey + f = 0$ be two constructible circles. By subtracting the two equations one obtains a constructible line. Then the points of intersection on the two circles correspond to the intersection of this line with either of the two circles, which is the previous case already considered.*

Hence, the proof of the theorem is complete.

Remark 9.4 *We points out a couple of things which we will make use of in what follows.*

1. *Under the assumptions of Theorem 9.6, since each $F_i = F_{i-1}(\sqrt{a_i})$ for $i = 1, 2, \ldots, n$ it follows that $F_n = \mathbb{Q}(\sqrt{a_1}, \sqrt{a_2}, \ldots, \sqrt{a_n})$.*

2. *Since $[F_i : F_{i-1}] = 2$ for $i = 1, 2, \ldots, n$ it follows that $[F_n : \mathbb{Q}] = 2^n$.*

Corollary 9.4 *If b is a constructible number, then $[\mathbb{Q}(b) : \mathbb{Q}]$ is a power of 2. Hence, it is not possible to construct a real number b such that $[\mathbb{Q}(b) : \mathbb{Q}]$ is infinite or **not** a power of 2. In particular, b cannot be the root of an irreducible polynomial over $\mathbb{Q}$ of degree which is **not** a power of 2.*

Proof 9.11 *Suppose that b is constructible. By Theorem 9.6, there exists a chain of fields of the form $\mathbb{Q} = F_0 \subset F_1 \subset F_2 \subset \cdots \subset F_n$ such that each F_i is a quadratic extension of F_{i-1} for $i = 1, 2, \ldots, n$ and $b \in F_n$. As in the remarks above, since $b \in F_n$, it follows that*

$$2^n = [F_n : \mathbb{Q}] = [F_n : \mathbb{Q}(b)][\mathbb{Q}(b) : \mathbb{Q}].$$

Hence, $[\mathbb{Q}(b) : \mathbb{Q}]$ divides 2^n and so must be a power of 2.

9.3.1 Famous Impossibilities

We can now answer some questions about geometric constructions that were posed and not answered for many centuries.

Theorem 9.7 (Doubling the Cube) *Given a cube with side of one unit length, it is **not** possible to construct with straightedge and compass the side of a cube having volume twice that of the original one.*

Proof 9.12 *Since the original cube has side of length 1, then the cube with double the volume would have a side of length $\sqrt[3]{2}$. Hence, in order to construct this larger cube it would be necessary that $\sqrt[3]{2}$ be a constructible number. But $\sqrt[3]{2}$ is a root of the irreducible polynomial $x^3 - 2 \in \mathbb{Q}[x]$ and so by Corollary 9.4 cannot be constructible.*

Theorem 9.8 (Squaring the Circle) *Given a circle with radius of unit length, it is **not** possible with straightedge and compass to construct a square having the same area of the circle.*

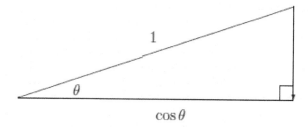

Figure 9.6 Diagram used for show that cosine of a constructible angle is also constructible.

Proof 9.13 *Since the circle has a radius of length 1, the square we wish to construct would have to have a side of length $\sqrt{\pi}$ and so it would be necessary that $\sqrt{\pi}$ be constructible. Suppose $\sqrt{\pi}$ were constructible. Then π would be constructible and thus, by Corollary 9.4, $[\mathbb{Q}(\sqrt{\pi}) : \mathbb{Q}] = 2^k$. In particular, π would be algebraic over $\mathbb{Q}$ contradicting that π is, in fact, transcendental over $\mathbb{Q}$.*

Before we present the next proof we point out that some angles can be trisected, like for instance a 180° angle can be trisected, since (as we shall see below) 60° is a constructible angle. What we mean by an angle θ being **constructible** is that one can construct with straightedge and compass two lines which intersect so that at least one of the angles between the lines is θ.

Theorem 9.9 (Trisecting the Angle) *It is **not always** possible to trisect an angle using straightedge and compass.*

Proof 9.14 *We prove this result by a series of claims.*

Claim 9.2 *If an angle θ is constructible, then the number $\cos\theta$ is constructible.*

Construct two lines which intersect to make an angle θ and mark off on one line from the point of intersection a line segment of length 1. From the other end of the line segment drop a perpendicular onto the other line to form a triangle (Figure 9.6). The resulting triangle has base of length $\cos\theta$.

Claim 9.3 *60° is a constructible angle.*

Construct two lines which intersect at right angles. Mark off a length 1 on one line and $\sqrt{3}$ (which is constructible) on the other line. Connect the two endpoints to form a triangle (Figure 9.7).

Claim 9.4 *The number $\cos 20°$ is a root of the polynomial $8x^3 - 6x - 1$.*

Notice that

$$\frac{1}{2} = \cos 60° = \cos(40° + 20°) = \cos 40° \cos 20° - \sin 40° \sin 20°$$

$$= (\cos^2 20° - \sin^2 20°) \cos 20° - (2 \sin 20° \cos 20°) \sin 20°$$

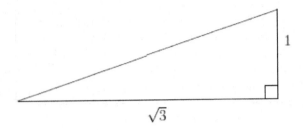

Figure 9.7 Diagram used for show that $\cos 60°$ is a constructible angle.

$$= \cos^3 20° - 3(\sin^2 20°)\cos 20° = \cos^3 20° - 3(1 - \cos^2 20°)\cos 20°$$

$$= 4\cos^3 20° - 3\cos 20°.$$

Multiplying both sides by 2 and throwing everything to one side yields $8\cos^3 20° - 6\cos 20° - 1 = 0$ *which proves the claim.*

Claim 9.5 $[\mathbb{Q}(\cos 20°) : \mathbb{Q}] = 3$

First note that $8x^3 - 6x - 1 \in \mathbb{Q}[x]$ *is irreducible in* $\mathbb{Q}[x]$, *for otherwise* $8x^3 - 6x - 1$ *would have a linear factor in* $\mathbb{Q}[x]$ *and so would have a rational root. By the Rational Root Theorem, the only candidates for rational roots for* $8x^3 - 6x - 1$ *are* ± 1, $\pm\frac{1}{2}$, $\pm\frac{1}{4}$, $\pm\frac{1}{8}$. *But none of these candidates are roots of* $8x^3 - 6x - 1$ *and so it must be irreducible. Therefore, by Claim 9.4,* $[\mathbb{Q}(\cos 20°) : \mathbb{Q}] = 3$.

We can now show that $60°$ *cannot be trisected using straightedge and compass. Suppose, to the contrary, that* $60°$ *could be trisected. Then, by Claim 9.3,* $20°$ *would be constructible. Then, by Claim 9.2,* $\cos 20°$ *would be a constructible number. However, Claim 9.5 and Corollary 9.4 are in direct contradiction.*

EXERCISES

1 Prove that if a, b are two positive constructible real numbers, then so is a/b (when $b \neq 0$).

2 Prove Corollary 9.1.

9.4 ALGEBRAIC EXTENSION & CLOSURE

We now introduce two additional and relevant concepts to the arena, namely *algebraic extension* and *algebraic closure*, which we will need in our future discussions. The similarity in the names of our terminology and in their definitions can easily lead to confusion and so the reader is advised to pay careful attention to the distinctions made between them.

Definition 9.6 *Let* $F \subseteq E$ *fields. The field* E *is* **algebraic over** F *if for every* $a \in E$ *is algebraic over* F. *The field* E *is then called an* **algebraic extension** *of* F.

Example 9.10 $\mathbb{Q}(\sqrt{2})$ *is algebraic over* $\mathbb{Q}$. *Take any* $a+b\sqrt{2} \in \mathbb{Q}(\sqrt{2})$. *Then* $a+b\sqrt{2}$ *is a zero of* $x^2 - 2ax + (a^2 - 2b^2) \in \mathbb{Q}[x]$.

Lemma 9.4 *Let* $F \subseteq E$ *be fields. If* $[E:F] < \infty$, *then* E *is algebraic over* F.

Proof 9.15 *Let* $[E:F] = n < \infty$ *and take any* $a \in E$. *Since the vector space dimension of* E *over* F *is* n, *the set of* $n+1$ *vectors* $1, a, a^2, \ldots, a^n$ *must be linearly dependent over* F. *Therefore, there are scalars* $c_0, c_1, c_2, \ldots, c_n \in F$ *not all zero such that* $c_0 1 + c_1 a + \cdots + c_n a^n = 0$. *But this says that* a *is algebraic over* F *since* a *is a zero of* $p(x) = c_0 + c_1 x + \cdots + c_n x^n \in F[x]$. *Since* a *was arbitrary in* E, *we have that* E *is algebraic over* F.

Theorem 9.10 *Let* $a_1, a_2, \ldots, a_n \in E \supseteq F$ *be fields with* $E = F(a_1, a_2, \ldots, a_n)$. *The following statements are equivalents:*

1. E *is algebraic over* F.

2. *Each* a_i *is algebraic over* F.

3. $[E:F] < \infty$.

Proof 9.16 *The first statement implies the second is immediate from the definition of* E *being algebraic over* F. *To prove the second statement implies the third, since each* a_i *is algebraic over* F *from previous work we know that* $[F(a_i) : F] = d_i < \infty$ *where* d_i *is the degree of the irreducible polynomial of* a_i *over* F. *Let's call that polynomial* $p_i(x) \in F[x]$. *We show the third statement is true by induction on* n, *the number of generators of* E *over* F. *For the* $n = 1$ *case,* $E = F(a_1)$ *and so* $[E:F] = [F(a_1) : F] = d_1 < \infty$. *For* $n > 1$, *set* $K = F(a_1, a_2, \ldots, a_{n-1})$ *so that* $F \subseteq K \subseteq E = K(a_n)$. *By the induction assumption, we know that* $[K:F] < \infty$. *We claim that* $[E:K] < \infty$ *and so by Theorem 9.3,* $[E:F] = [E:K][K:F] < \infty$ *and we are done. To see this, note that* $p_n(a_n) = 0$ *with* $p_n \in F[x] \subseteq K[x]$. *Hence,* a_n *is algebraic over* K *and so* $[E:K] = [K(a_n) : K] < \infty$.

The third statement implies the first follows immediately from Lemma 9.4.

Corollary 9.5 *If* $F \subseteq E \subseteq K$ *with* K *algebraic over* E *and* E *algebraic over* F, *then* K *is algebraic over* F.

Proof 9.17 *Take an* $a \in K$. *By assumption,* a *is algebraic over* E. *Let* $p(x) = c_0 + c_1 x + \cdots + c_n x^n \in E[x]$ *be the minimal polynomial of* a *over* E. *Consider the following chain of fields:*

$$F \subseteq E_1 = F(c_0, c_1, \ldots, c_n) \subseteq L = F(c_0, c_1, \ldots, c_n, a) = E_1(a).$$

By assumption, each c_i *is algebraic over* F *and so by Theorem 9.10,* $[E_1 : F] < \infty$. *Furthermore,* $[L:E_1] = n < \infty$. *Therefore,* $[L:E_1] < \infty$. *But then by Theorem 9.10 again, we have* a *is algebraic over* F.

Theorem 9.11 *Let E be algebraic over F. The following statements are equivalent:*

1. *E is finitely generated over F.*

2. *$[E : F] < \infty$.*

Proof 9.18 *The first statement implies second follows immediately from Theorem 9.10, since by assumption each of the generators is algebraic over F. To show the second statement implies the first, we employ induction on $n = [E : F] < \infty$. For $n = 1$, we've seen that $E = F$ and so $E = F(1)$ is finitely generated by 1 over F. For $n > 1$, take any $a \in E \setminus F$ which we know by assumption is algebraic over F and so $1 < [F(a) : F] < \infty$. Since $[E : F] < \infty$ then $[E : F(a)] < [E : F]$. Since E is algebraic over F, it is certainly algebraic over $F(a)$, thus we can invoke induction to get that $E = F(a)(a_1, a_2, \dots, a_k)$ for some $a_1, a_2, \dots, a_k \in E$. But then $E = F(a, a_1, \dots, a_k)$ and we are done.*

Definition 9.7 *Let $F \subseteq E$ be fields. The **algebraic closure of F in E**, written*

$$acl_E(F) = \{a \in E : a \text{ is algebraic over } F\}.$$

*We say that F is **algebraically closed in E** if $acl_E(F) = F$, i.e. If $a \in E$ is algebraic over F, then $a \in F$. We say that F is **algebraically closed** if $acl_E(F) = F$ for all fields $E \supseteq F$. We say that E is an **algebraic closure** of F if E is algebraically closed and E is algebraic over F. We will denote an algebraic closure of F by $\overline{F}$.*

Remark 9.5 *We make several remarks concerning $acl_E(F)$.*

1. *It is immediate from the definition that if E is algebraic over F, then $acl_E(F) = E$.*

2. *For any field $E \supseteq F$, $acl_E(F)$ is a subfield of E. To see this, Take any $a, b \in alg_E(F)$. By Theorem 9.10, since a and b are algebraic over F, we have that $F(a, b)$ is algebraic over F, i.e. every element in $F(a, b)$ is algebraic over F. Since $a, b \in F(a, b)$ a field, then so are $a - b, ab^{-1} \in F(a, b)$ and hence they are algebraic over F, i.e. $a - b, ab^{-1} \in acl_E(F)$.*

3. *It's immediate from its definition that $acl_E(F)$ is algebraic over F.*

4. *For any field $E \supseteq F$, we have $F \subseteq acl_E(F)$, since for $a \in F$ the linear polynomial $p(x) = x - a \in F[x]$ has a as a zero.*

5. *It is always the case that $acl_E(F)$ is algebraically closed in E, and so $acl_E(acl_E(F)) = acl_E(F)$. One proves this by showing that every $a \in E$ algebraic over $acl_E(F)$ is, in fact, algebraic over F. To see this, first set $K = acl_E(F)$ and take $a \in E$ algebraic over K. This means there is an $f(x) = a_n x^n + \dots + a_1 x + a_0 \in K[x]$ which has the element a as a zero. Since each $a_i \in K$, this means that each a_i is algebraic over F. But then by Theorem 9.10, $[F(a_0, a_1, \dots, a_n) : F] < \infty$. Now certainly a is algebraic over $F(a_0, a_1, \dots, a_n)$*

via the polynomial $f(x)$ and so $[F(a_0, a_1, \ldots, a_n)(a) : F(a_0, a_1, \ldots, a_n)] < \infty$. Therefore,

$$[F(a_0, a_1, \ldots, a_n, a) : F]$$
$$= [F(a_0, a_1, \ldots, a_n)(a) : F(a_0, a_1, \ldots, a_n)][F(a_0, a_1, \ldots, a_n) : F] < \infty.$$

Now, by Theorem 9.10, a is algebraic over F.

6. *Unlike Theorem 9.10, it is **not** true for an arbitrary field E algebraic over F that we have $[E : F] < \infty$, i.e. the converse of Lemma 9.4 is false. For example, set $F = \mathbb{Q}$ and $E = acl_{\mathbb{C}}(\mathbb{Q})$ a field, by the first remark, algebraic over $\mathbb{Q}$ (called the **field of algebraic numbers**). However, we show that $[E : F] = \infty$. To see this, consider the polynomial $p(x) = x^n - 2 \in \mathbb{Q}[x]$ for any odd number n. Using Eisenstein's Criterion, $p(x)$ is irreducible in $\mathbb{Q}[x]$. Since $p(x)$ is a polynomial of odd degree, it must have a real zero, say $a \in \mathbb{R}$ (and so $a \in E$). Hence, since $p(x)$ irreducible with a zero of $p(x)$ being the element a, we have that $p(x)$ is the irreducible polynomial of a over $\mathbb{Q}$, and so $[\mathbb{Q}(a) : \mathbb{Q}] = n$. Certainly, since $E \supseteq \mathbb{Q}(a)$, it's true that $[E : \mathbb{Q}] \geq [\mathbb{Q}(a) : \mathbb{Q}] = n$ for any odd number n. But then one must conclude that $[E : \mathbb{Q}] = \infty$. Note that by Theorem 9.11, that it must be the case that $acl_{\mathbb{C}}(\mathbb{Q})$ is not finitely generated over $\mathbb{Q}$.*

Example 9.11 *Here, we give several examples to illustrate the myriad of definitions we just introduced.*

1. *$\mathbb{Q}$ is not algebraically closed in $\mathbb{Q}(\sqrt{2})$ for among other reasons $\sqrt{2} \in \mathbb{Q}(\sqrt{2}) \setminus \mathbb{Q}$ is algebraic over $\mathbb{Q}$ (via the polynomial $x^2 - 2 \in \mathbb{Q}[x]$).*

2. *The field, $acl_{\mathbb{R}}(\mathbb{Q}) = \{a \in \mathbb{R} : a \text{ algebraic over } \mathbb{Q}\}$ although algebraically closed in $\mathbb{R}$ (see Remark 9.5.5) is not algebraically closed in $\mathbb{C}$, since i is algebraic over $acl_{\mathbb{R}}(\mathbb{Q})$ via $p(x) = x^2 + 1$, yet $i \notin acl_{\mathbb{R}}(\mathbb{Q})$. Hence, we conclude that $acl_{\mathbb{R}}(\mathbb{Q})$ is not algebraically closed.*

3. *$\mathbb{C}$ is not algebraic over $\mathbb{Q}$. The reason for this is because every algebraic extension of $\mathbb{Q}$ must be countable. Indeed, if E is algebraic over $\mathbb{Q}$ then every $a \in E$ is algebraic over $\mathbb{Q}$ and so every $a \in E$ is a zero of some polynomial in $\mathbb{Q}[x]$. This defines a map from E to $\mathbb{Q}[x]$ which is finite-to-one (i.e. for each $f(x) \in \mathbb{Q}[x]$ there are a finite number of $a \in E$ which are zeros of f) and so the cardinality of E is the same as the cardinality of $\mathbb{Q}[x]$ which is countable.*

EXERCISES

1 Consider the fields $F \subseteq K \subseteq E$. Prove that E is algebraic over F iff E is algebraic over K and K is algebraic over F.

2 Let $F \subseteq E$ be fields with E algebraically closed. Prove that $acl_E(F)$ is algebraically closed as well.

9.5 EXISTENCE THEOREMS

In Section 9.4, we defined some new structures one of which was the algebraic closure of a field. Now there is no reason a priori to assume such structures must exist. In this section, we give proofs of the existence of some of the concepts we defined in the previous sections. For instance, when we formed the finite field $\mathbb{Z}_2(a)$ we made one assumption which should really be justified, namely that the polynomial $x^2 + x + 1$ does indeed have a zero (which we called a) in some larger field containing $\mathbb{Z}_2$. The theorem below puts this issue to rest.

Theorem 9.12 *Let F be a field. For any $f(x) \in F[x]$ of degree at least 1, there exists a field $E \supseteq F$ with an $a \in E$ a zero of $f(x)$.*

Proof 9.19 *Since $F[x]$ is a UFD, we can factor $f(x)$ as a product of irreducibles. Let $p(x) = c_d x^d + \cdots + c_1 x + c_0$ be one of the irreducible factors of $f(x)$. It suffices to show that there exists a field $E \supseteq F$ with an $a \in E$ a zero of $p(x)$. Set $\overline{E} = F[x]/(p(x))$ which is a field, since $p(x)$ is irreducible. Consider the canonical map $\phi : F \to \overline{E}$ by $\phi(c) = \overline{c} = c + (p(x))$. One can check that this homomorphism is one-to-one (use the fact that $p(x)$ has degree at least 1). Therefore, $\overline{E}$ contains an isomorphic copy of F, namely $\overline{F} = \phi(F)$. Therefore, we will switch to this new base field $\overline{F}$ and it suffices to find a zero of $\overline{p}(x) = \overline{c_d} x^d + \cdots + \overline{c_1} x + \overline{c_0} \in \overline{F}[x]$ in an extension field of $\overline{F}$. It just so happens that $\overline{E}$ is the field we seek. Indeed, consider $\overline{x} = x + (p(x)) \in \overline{E}$. Notice that*

$$\overline{p}(\overline{x}) = \overline{c_d}\, \overline{x}^d + \cdots + \overline{c_1}\, \overline{x} + \overline{c_0} = \overline{p(x)} = \overline{0},$$

so that $\overline{x}$ is a zero of $\overline{p}(x)$ and the theorem is proved.

Corollary 9.6 *Let F be a field. For any $f_1(x), \ldots, f_n(x) \in F[x]$ all of degree at least 1, there exists a field $E \supseteq F$ with $a_1, \ldots, a_n \in E$ zeros of $f_1(x), \ldots, f_n(x)$, respectively.*

Proof 9.20 *Simply apply Theorem 9.12 n times to form a chain of extensions the largest of which is the field E we seek.*

Corollary 9.7 *The following statements are equivalent for a field F.*

1. *F is algebraically closed*

2. *Every $f \in F[x]$ of degree at least one has a zero in F.*

3. *Every $f \in F[x]$ of degree at least one factors completely as a product of linear factors in $F[x]$.*

4. *Every irreducible polynomial in $F[x]$ has degree exactly one.*

Proof 9.21 *The first statement implies the second follows immediately from the previous theorem and the fact that F is algebraically closed. The second statement implies the third follows by a simple induction on the degree of f. Indeed, if $\deg(f) = 1$ the* *result follows immediately without appealing to the second statement. If $deg(f) > 1$ and $c \in F$ is the given root of f, then by Corollary 7.3, $f(x) = (x-c)g(x)$. Since the degree of g is less than that of f, by induction, g factors completely into linear factors and hence so does f. The third statement implies the fourth is immediate. Finally, the fourth statement implies the first, for suppose $F \subseteq E$ a field and $a \in E$ is algebraic over F. Set $p(x)$ to be the irreducible polynomial of a over F. By assumption, $p(x) = bx + c$ for some $b, c \in F$. But then $0 = p(a) = ba + c$ and so $a = -b^{-1}c \in F$.*

We shall postpone the proof of the existence of an extension field of an arbitrary field which is algebraically closed until we discuss Zorn's Lemma in a later section. However, under the assumption it is true, we can prove the existence of the algebraic closure of an arbitrary field.

Theorem 9.13 *For any field F there exists an algebraic closure of F.*

Proof 9.22 *Let E be an algebraically closed field containing F and set $K = acl_E(F)$ a subfield of E containing F and algebraic over F and algebraically closed in E (see Remark 9.5 on $acl_E(F)$). We claim that K is the algebraic closure of F which we seek. It remains to prove that K is algebraically closed. To show this we prove every polynomial over K of degree at least one has a zero in K and then appeal to Corollary 9.7. Take $f(x) \in K[x]$ with $deg(f) \geq 1$. Since E is algebraically closed we know, by Corollary 9.12, there is an $a \in E$ a zero for $f(x)$. Therefore,*

$$a \in acl_E(K) = acl_E(acl_E(F)) = acl_E(F) = K.$$

Definition 9.8 *Let $F \subseteq E$ be fields and $f(x) \in F[x]$ of degree at least one. Then E is called the **splitting field of** $f(x)$ **over** F if for some $a_1, a_2, \ldots, a_n \in E$ we have $E = F(a_1, a_2, \ldots, a_n)$ and $f(x) = c(x - a_1)(x - a_2) \cdots (x - a_n)$, for some $c \in F$. We say that $f(x)$ **splits in** E.*

Example 9.12 *Let $F = \mathbb{Q}$ and $f(x) = x^3 - x^2 + x - 1 \in \mathbb{Q}[x]$ which has zeros $1, \pm i$. Then $E = \mathbb{Q}(1, i, -i)$ is a splitting field of $f(x)$ over $\mathbb{Q}$. Note that $E = \mathbb{Q}(i)$ and $[E : \mathbb{Q}] = 2$.*

Theorem 9.14 *For every $f(x) \in F[x]$ of degree at least one, there is a splitting field of $f(x)$ over F.*

Proof 9.23 *We prove this by induction on the degree of $f(x)$. When $deg(f) = 1$, the result is immediate. when $deg(f) > 1$, by Theorem 9.12, there is an extension field $E_1 \supseteq F$ and an $a_1 \in E_1$ a zero of $f(x)$. So write $f(x) = (x - a_1)f_1(x)$, where $f_1(x) \in E_1[x]$ and $deg(f_1) < deg(f)$. By induction, there exists a splitting field E_2 of $f_1(x)$ over E_1, i.e. $E_2 = E_1(a_2, \ldots, a_n)$ with $f_1(x) = c(x - a_2) \cdots (x - c_n)$, $c_2, \ldots, c_n \in E_2$ and $c \in F$. Then $E = F(a_1, a_2, \ldots, a_n)$ is the desired splitting field.*

Remark 9.6 *We make several remarks about splitting fields.*

1. *One can show by induction that if E is a splitting field for $f(x)$ over F and $deg(f) = n$, then $[E : F] \leq n!$.*

2. If E is a splitting field of $f(x)$ over F, then E is algebraic over F since each a_i is algebraic over F each being a zero of $f(x)$ (see Theorem 9.10).

3. By Theorem 9.7, if E is an algebraically closed field containing F, then every polynomial in F of degree at least one splits in E.

EXERCISES

1 Prove that the canonical map $\phi : F \to \overline{E}$ by $\phi(c) = \overline{c} = c + (p(x))$ in the proof of Theorem 9.12 is one-to-one.

2 Prove by induction that if E is a splitting field for $f(x)$ over F and $deg(f) = n$, then $[E : F] \leq n!$.

9.6   FINITE FIELDS

In this section, we classify completely the fields of finite order. Namely, we show that every finite field is of prime power order and for every prime to a power there is exists a unique field of that order. We first show that every field is of prime power order and for every prime power there exists a field of that order. We prove uniqueness at the end of the section.

Theorem 9.15 If E is a finite field, then $|E| = p^n$ where p is a prime and n is a positive integer. Furthermore, the characteristic of E is p and $[E : F] = n$ where $F \cong \mathbb{Z}_p$ is the prime subfield of E.

Proof 9.24 First note that $char(E) \neq 0$, for otherwise E would contain an isomorphic copy of $\mathbb{Q}$ making it infinite. Therefore, $char(E) = p$ for some prime p and contains an isomorphic copy of $\mathbb{Z}_p$, which we will designate by F. Since E is finite, then certainly so is the degree, $[E : F]$. Set $n = [E : F]$ and let $a_1, a_2, \ldots, a_n$ be a basis for E over F. Thus, every element of E can be written uniquely as a linear combination of $a_1, a_2, \ldots, a_n$ over F. This allows us to count the elements of E, namely there are exactly

$$\underbrace{|F| \cdot |F| \cdot \cdots \cdot |F|}_{n \text{ times}} = |F|^n = p^n.$$

We now know that any finite field is of prime power order, however we still need to show the converse that for every prime power there is a finite field of that given order. Before we do this we need to derive some results which are invoked in the existence proof.

Lemma 9.5 For any field F and irreducible polynomial $p(x) \in F[x]$, if $f(x) \in F[x]$ has a zero in common with $p(x)$ in some extension field of F, then $p(x)$ divides $f(x)$ in $F[x]$.

Proof 9.25 Suppose under the assumptions given in this lemma that $p(x)$ did not divide $f(x)$. Since $p(x)$ is irreducible, it must then be the case that $p(x)$ and $f(x)$ are *relatively prime. Hence, there are polynomials* $g(x), h(x) \in F[x]$ *such that* $g(x)p(x) +$ $h(x)f(x) = 1$. *Let* $a \in E \supseteq F$ *be the common zero of* $p(x)$ *and* $f(x)$. *Plugging this in we get*

$$1 = g(a)p(a) + h(a)f(a) = g(a) \cdot 0 + h(a) \cdot 0 = 0,$$

a contradiction.

Lemma 9.6 *If* $f(x), g(x) \in F[x]$ *are relatively prime in* $F[x]$, *then they are also relatively prime in* $E[x]$ *for any extension field* $E \supseteq F$.

Proof 9.26 *We prove the contrapositive statement. If* $f(x), g(x) \in F[x]$ *are not relatively prime in* $E[x]$, *then there is an irreducible polynomial* $q(x) \in E[x]$ *which divides both* $f(x)$ *and* $g(x)$. *Let* a *be a zero of* $q(x)$ *(which we know exists by Theorem 9.12 of Section 9.5). Then* a *is also a common zero of* $f(x)$ *and* $g(x)$. *Let* $p(x) \in F[x]$ *be an irreducible factor of* $g(x)$ *having* a *as a zero. Then, by Lemma 9.5,* $p(x)$ *divides* $f(x)$ *and so* $p(x)$ *is a common divisor of* $f(x)$ *and* $g(x)$, *i.e.* $f(x)$ *and* $g(x)$ *are not relatively prime in* $F[x]$.

We will need the concept of the *formal* derivative of a polynomial.

Definition 9.9 *Let* F *be a field and* $f(x) = a_0 + a_1 x + \cdots + a_n x^n \in F[x]$. *The* **formal derivative of** $f(x)$, *written* $f'(x) = a_1 + 2a_2 x + \cdots + na_n x^{n-1}$.

Notice that this definition of derivative conforms to the usual definition of derivative in Calculus and as such shares the same properties that the Calculus derivative enjoys, namely it's a linear operator with the same product rule, quotient rule, etc. Of course all these statements require verification, however for the sake of brevity we omit them (the ambitious reader can easily verify these statements).

Remark 9.7

By Corollary 9.7 of Section 9.5 we know that we can express

$$f(x) = c(x - a_1) \cdots (x - a_n) = c \prod_{i=1}^{n} (x - a_i),$$

for some $c \in F$ *and* $a_i \in E$ *the zeros of* $f(x)$. *Then by the product rule,*

$$f'(x) = c \sum_{j=1}^{n} \prod_{\substack{i=1 \\ i \neq j}}^{n} (x - a_i).$$

Hence, in the UFD $E[x]$, $(x - a_j)$ *divides* $f'(x)$ *for some* j *iff* $(x - a_j)$ *divides* $\prod_{\substack{i=1 \\ i \neq j}}^{n} (x - a_i)$. *And this is true iff* $a_i = a_j$ *for some* $i \neq j$, *or equivalently* $f(x)$ *does not have distinct zeros.*

Lemma 9.7 *Let* $F \subseteq E$ *be fields and* E *an algebraic closure of* F. *If* $f(x) \in F[x]$ *is a polynomial with* $\deg(f) = n \geq 1$, *then* f *has* n *distinct zeros in* E *iff* $f'(x) \neq 0$ *and* $\gcd(f(x), f'(x)) = 1$ *in* $F[x]$.

Proof 9.27 *Assume first that $f'(x) \neq 0$ and $\gcd(f(x), f'(x)) = 1$ in $F[x]$. By Lemma 9.6, $\gcd(f(x), f'(x)) = 1$ in $E[x]$ as well. So in particular, for all j, we have that $(x - a_j)$ does not divide $f'(x)$. By Remark 9.7, $f(x)$ has n distinct zeros. To show the reverse direction we prove the contrapositive statement. Suppose that either $f'(x) = 0$ or $\gcd(f(x), f'(x)) \neq 1$ in $F[x]$. If $f'(x) = 0$, then certainly, for all j, we have $(x - a_j)$ divides $f'(x)$ which, by Remark 9.7, implies that $f(x)$ does not have n distinct zeros. If $\gcd(f(x), f'(x)) \neq 1$ in $F[x]$, then for some j we have that $(x - a_j)$ divides $f'(x)$ which again, by Remark 9.7, implies that $f(x)$ does not have n distinct zeros.*

Example 9.13 *We apply Lemma 9.7 to several examples.*

1. *Consider the polynomial $f(x) = x^3 + 1 \in \mathbb{Q}[x]$ which has three distinct zeros in $\mathbb{C}$ (which later we show is an algebraic closure of $\mathbb{Q}$), since $f'(x) = 3x^2 \neq 0$ and $\gcd(f(x), f'(x)) = 1$. Indeed, the three distinct zeros are -1 and $\frac{1}{2} \pm \frac{\sqrt{3}}{2}i$.*

2. *Consider the polynomial $f(x) = x^2 + 2x + 1 \in \mathbb{Q}[x]$ which on the other hand does not have two distinct zeros, since $f'(x) = 2x + 2$ and $\gcd(f(x), f'(x)) = x + 1 \neq 1$. Indeed, -1 is a double root of $f(x)$.*

3. *Consider the polynomial $f(x) = x^3 + 1 \in \mathbb{Z}_3[x]$. One might argue that as in the first example $f(x)$ has three distinct zeros, however $\mathbb{C}$ is not an extension field of $\mathbb{Z}_3$. Indeed any extension field of $\mathbb{Z}_3$ would have to have characteristic 3 in order to have $\mathbb{Z}_3$ as a subfield. So we have to look at this polynomial afresh. Notice that $f'(x) = 0$ in $\mathbb{Z}_3[x]$, so in fact, by the lemma, $f(x)$ does **not** have distinct zeros in an algebraic closure. Moreover we need not look any further than $\mathbb{Z}_3$ to find all of its zeros. Notice that in $\mathbb{Z}_3[x]$, $x^3 + 1 = (x + 1)^3$. Thus, the element -1 or $2 \in \mathbb{Z}_3$ is a triple zero of $x^3 + 1$.*

Having proved Lemma 9.7, we are now in a position to give the existence proof of a finite field for any prime power order.

Theorem 9.16 *Given a prime p and a positive integer n, there is a field of order p^n.*

Proof 9.28 *Let E be an algebraic closure of $\mathbb{Z}_p$ and consider the polynomial $f(x) = x^{p^n} - x \in \mathbb{Z}_p[x]$. Since $f'(x) = -1 = p - 1 \neq 0$ and $\gcd(f(x), f'(x)) = 1$ in $\mathbb{Z}_p[x]$, by Lemma 9.7, we know that $f(x)$ has p^n distinct zeros in E. Set F_{p^n} to be the collection of p^n zeros of $f(x)$ in E. Note that $a \in F_{p^n}$ is a zero of $f(x)$ iff $a^{p^n} = a$. To complete the proof we need to show that F_{p^n} is a subfield of E. We will rely on a straightforward induction result (which we leave to the reader) which says that in a field E of characteristic p, for all $a, b \in E$ we have that $(a \pm b)^{p^n} = a^{p^n} \pm b^{p^n}$. Indeed, E is a field of characteristic p since it contains $\mathbb{Z}_p$ as a subfield. So take any $a, b \in F_{p^n}$ and notice that $(a + b)^{p^n} = a^{p^n} + b^{p^n} = a + b$, so that $a + b \in F_{p^n}$ and $(ab)^{p^n} = a^{p^n} b^{p^n} = ab$, so that $ab \in F_{p^n}$. Finally, for $a \in F_{p^n}$ and p odd, $(-a)^{p^n} = -a^{p^n} = -a$, so that $-a \in F_{p^n}$ and for $p = 2$, $(-a)^{p^n} = a^{p^n} = a = -a$, so that again $-a \in F_{p^n}$.*

To prove the uniqueness of a finite field for a given prime power order, we will first need to prove several lemmas. The first lemma is a fact about fields that perhaps you would not guess. The second lemma is reassuring in the sense that it says we can always construct finite fields by adjoining a zero of an irreducible polynomial to the base field. For instance, in Section 9.4, we constructed a field of order 4 by adjoining a zero of $x^2 + x + 1$ to the base field $\mathbb{Z}_2$.

Lemma 9.8 *Any finite subgroup of the multiplicative group of a field is cyclic.*

Proof 9.29 *This proof relies on the Classification of Finite Abelian Groups and the Fundamental Theorem of Algebra. Let $G \leq F^*$ where F is a field and G is finite. Since G is a finite abelian group, G is isomorphic to a direct sum of non-trivial cyclic subgroups, $G = H_1 \oplus H_2 \oplus \cdots \oplus H_n$, where $|H_i|$ divides $|H_{i-1}|$ for $i = 2, 3, \ldots, n$. Set $k_i = |H_i|$ for $i = 1, 2, \ldots, n$. Since the k_i successively divide each other this implies $lcm(k_1, k_2, \ldots, k_n) = k_n$. Therefore, for all $g \in G$ we have $g^{k_n} = 1$. In other words, every element of G is a root of the polynomial $f(x) = x^{k_n} - 1 \in F[x]$. By the Fundamental Theorem of Algebra, $f(x)$ has at most k_n distinct roots and so $|G| \leq k_n$. But $k_n = |H_n| \leq |G|$ and thus $G = H_n$ and is cyclic.*

Definition 9.10 *The generator of the multiplicative group of a finite field is called a **primitive element**.*

Example 9.14 *Consider the finite field with four elements computed in Example 9.9.3,*

$$\mathbb{Z}_2(a) = \{0, 1, a, 1 + a\}.$$

It's easy to check that a and $1 + a$ are primitive elements.

Lemma 9.9 *If $F \subseteq E$ are finite fields, then $E = F(a)$ for some $a \in E$ algebraic over F.*

Proof 9.30 *We have just seen that the multiplicative group of a finite field is cyclic so set $E^* = <a>$ for some $a \in E$. Then $E = F(a)$, since E is the smallest field containing F and a. To see this, suppose K is a field containing F and a. If $b \in E^*$, then for some positive integer k, we have $b = a^k \in K$, by closure. Since E and F are finite, then so is $[E : F] < \infty$ which, by Lemma 9.1 of Section 9.2, implies that a is algebraic over F.*

Example 9.15 *Let's form a finite field with nine elements and find generators of the multiplicative group associated with the field. We will start with $\mathbb{Z}_3$ and the polynomial $p(x) = x^2 + 1$ which is irreducible over $\mathbb{Z}_3$ (simply verify that no element of $\mathbb{Z}_3$ is a zero of $p(x)$). Let a be a zero of $p(x)$ in some extension field (which we know exists by Theorem 9.12 of Section 9.5). Then $\mathbb{Z}_3(a)$ is the finite field we seek. Indeed,*

$$\mathbb{Z}(a) = \mathbb{Z}[a] = \{0,\ 1,\ 2,\ a,\ 1 + a,\ 2 + a,\ 2a,\ 1 + 2a,\ 2 + 2a\}.$$

Since the order of $\mathbb{Z}(a)^*$ is eight, we know that there are $\phi(8) = 4$ generators of $\mathbb{Z}(a)^*$ (i.e. primitive elements). Let's find them. Notice that

$$\langle 1 \rangle = \{1\}, \ \langle 2 \rangle = \{1, \ 2\}, \ \langle a \rangle = \{1, \ 2, \ a, \ 2a\} = \langle 2a \rangle.$$

Therefore, the remaining four elements $1 + a, \ 2 + a, \ 1 + 2a, \ 2 + 2a$ must each be a generator of $\mathbb{Z}(a)^*$.

In the proof of Theorem 9.6, we proved the existence of a finite field of order p^n in a constructive way as the zeros of the polynomial $x^{p^n} - x$. The next Lemma shows that this is true for any finite field of order p^n and so we will be closer to our desired uniqueness result.

Lemma 9.10 *If E is a finite field of order p^n for p prime and n a positive integer, then the elements of E are exactly the zeros of the polynomial $x^{p^n} - x$ and so $x^{p^n} - x$ splits in E.*

Proof 9.31 *Since E^* is a group under multiplication, every element raised to the order of that group equals the identity, i.e. for all $a \in E^*$ we have $a^{p^n - 1} = 1$ or equivalently for all $a \in E^*$ we have $a^{p^n} = a$. Therefore, for all $a \in E$ we have $a^{p^n} - a = 0$ which shows that every element of E is indeed a zero of $x^{p^n} - x$. Furthermore, it must be the case that $x^{p^n} - x = (x - a_1)(x - a_2) \cdots (x - a_{p^n})$ where $E = \{a_1, a_2, \ldots, a_{p^n}\}$.*

Theorem 9.17 *Two finite fields of the same order are isomorphic.*

Proof 9.32 *Let E_1, E_2 be two finite fields of the same order. By Theorem 9.15, $|E_1| = |E_2| = p^n$ for some prime p and positive element n and E_1 and E_2 each contain an isomorphic copy of $\mathbb{Z}_p$, call it F_i, with $[E_i : F_i] = n$ for $i = 1, 2$. Without loss of generality, we can assume that E_1 and E_2 are, in fact, extension fields of $\mathbb{Z}_p$. By Lemma 9.9, $E_i = \mathbb{Z}_p(a_i)$ with a_i algebraic over $\mathbb{Z}_p$ for some $a_i \in E_i$ for $i = 1, 2$. Let $p(x) \in \mathbb{Z}_p[x]$ be the irreducible polynomial for a_1 over $\mathbb{Z}_p$. As usual, $E_1 = \mathbb{Z}_p(a_1) \cong \mathbb{Z}_p[x]/(p(x))$. By Lemma 9.10 and Lemma 9.5, $p(x)$ divides $x^{p^n} - x$. By Lemma 9.10, the elements of E_2 are exactly the zeros of the polynomial $x^{p^n} - x$. Therefore, some $b_j \in E_2$ is a zero of $p(x)$. This makes $p(x)$ the irreducible polynomial of $b_j \in E_2$ over $\mathbb{Z}_p$ so that $\mathbb{Z}_p(b_j) \cong \mathbb{Z}_p[x]/(p(x)) \cong E_1$. Since $|E_1| = |E_2|$ and $\mathbb{Z}_p(b_j) \subseteq E_2$ we must have $E_2 = \mathbb{Z}_p(b_j) \cong E_1$.*

For a given prime p and a positive integer n, the notation F_{p^n} shall denote the unique (up to isomorphism) finite field of order p^n called the **Galois field of order p^n**.

We will finish this section with further results regarding finite fields.

Lemma 9.11 *If $F \subseteq E$ finite fields, then $|E| = |F|^n$ for some positive integer n.*

Proof 9.33 *We've seen that $E = F(a)$ for some $a \in E$ algebraic over F and $[E : F] = n < \infty$. Let $a_1, a_2, \ldots, a_n$ be a basis for E over F. Since every element of E can* be written uniquely as linear combinations of $a_1, a_2, \ldots, a_n$ over F, we can therefore count the elements of E, namely

$$|E| = \underbrace{|F| \cdot |F| \cdot \cdots \cdot |F|}_{n \ times} = |F|^n.$$

Theorem 9.18 *For p prime and k, m positive integers, $F_{p^k} \subseteq F_{p^m}$ iff $k|m$.*

Proof 9.34 *If $F_{p^k} \subseteq F_{p^m}$, then by Lemma 9.11, $p^m = |F_{p^m}| = |F_{p^k}|^n = p^{kn}$, for some positive integer n, which implies that $m = kn$ or $k|m$. Now suppose that $k|m$ so that $m = kn$ for some positive integer n. By Lemma 9.10, every element of $a \in F_{p^m}$ satisfies $a^{p^m} = a$ and every element of $b \in F_{p^k}$ satisfies $b^{p^k} = b$. Notice that*

$$b^{p^m} = b^{p^{nk}} = b^{\overbrace{p^k p^k \cdots p^k}^{n \ times}} = (\cdots((b^{p^k})^{p^k})\cdots)^{p^k} = b,$$

so that $b \in F_{p^m}$ as well. Hence, $F_{p^k} \subseteq F_{p^m}$.

EXERCISES

---

1 Check that a and $1 + a$ are primitive elements of $\mathbb{Z}_2(a)$ in Example 9.9.3.

2 Verify for polynomials that the formal derivative satisfies the sum, product and quotient rule.

3 Let E be a field of characteristic p.

   a. Prove that p divides $\binom{p}{i}$ for $i = 1, 2, \ldots p - 1$.

   b. Prove that $(a \pm b)^p = a^p \pm b^p$ using the binomial theorem.

   c. Prove by induction on n that for all $a, b \in E$ we have that $(a \pm b)^{p^n} = a^{p^n} \pm b^{p^n}$.

CHAPTER 10

Galois Theory

IN THIS CHAPTER, we make a big connection between field theory and the theory of groups called Galois theory. Such connections are ways in which powerful tools can be created for proving difficult mathematical results. The result which motivated Evariste Galois to develop this theory and its connections is the investigation of solvability by radicals. For the reader who is interested in the history of mathematics, Galois' biography is quite interesting and dramatic.

In Section 10.1, we relate fields and groups via field homomorphisms of an extension field which fix the base field, thus arriving at Galois groups. We need Section 10.2 so that we can compute these Galois groups in a practical way. Section 10.3 is a bit off the beaten path, but at the very least the reader should be aware of the important results arrived at in this section. In Section 10.4, we introduce the notion of a splitting field and prove two important results linking the Galois group and the corresponding field extension. In Section 10.5, we introduce separable degree and link this concept to the size of the corresponding Galois group. In Section 10.6, introduce the notion of a Galois extension as we compare the lattice of subgroups of the Galois group and the corresponding lattice of intermediate fields in the corresponding field extension. In Section 10.7, we give the promised proof of Theorem 10.7 presented in Section 10.6 as well as prove a foundational result of Artin. In Section 10.8, we summarize our investigation in this chapter and prove the Fundamental Theorem of Galois Theory. In Section 10.9, we review Chapter 6 emphasizing the important concepts and results given therein regarding solvable groups. Finally, in Section 10.10, we investigate the notion of solvability by radicals.

10.1 FIELD HOMOMORPHISMS

In this section, we can begin to make an important connection between fields and groups. Namely, that every pair $F \subseteq E$ of fields can be related to a certain collection of field homomorphisms which form a group under composition. Recall that a non-trivial field homomorphism is necessarily one-to-one and therefore any non-trivial finite field endomorphism is an automorphism, i.e. bijective homomorphism from a field to itself. We will be looking at a certain subset of field homomorphisms.

DOI: 10.1201/9781003335283-10

Definition 10.1 *Let E_1 and E_2 be two fields containing the same subfield F. A field homomorphism $\phi : E_1 \to E_2$ is an F-**homomorphism** if it is the identity map on F, i.e. for all $c \in F$ we have $\phi(c) = c$.*

Remark 10.1 *Here, we make some quick observations about F-homomorphisms.*

1. *We point out that any F-homomorphism between fields is also a F-vector space homomorphism (i.e. linear transformation) between the two fields. We have actually employed this fact several times already in earlier lessons. We now give a formal argument that this is so. Suppose $\phi : E_1 \to E_2$ is an F-homomorphism. Then for vectors $a, b \in E_1$ we have $\phi(a+b) = \phi(a)+\phi(b)$, since ϕ preserves field addition. And for a scalar $c \in F$ and vector $a \in E_1$, $\phi(ca) = \phi(c)\phi(a) = c\phi(a)$, since ϕ preserves field multiplication and fixes F.*

2. *The reader should verify that the collection of F-homomorphisms from E_1 to E_2 with composition forms a group.*

Example 10.1 *An important example of an $\mathbb{Z}_p$-automorphism of F_{p^n} is the **Frobenius** automorphism defined by $\sigma_p(a) = a^p$. The collection of maps $\sigma_{p^k} : F_{p^n} \to F_{p^n}$ by $\sigma_{p^k}(a) = a^{p^k}$ (for $k = 1, 2, \ldots$) are all $\mathbb{Z}_p$-automorphisms of F_{p^n}.*

Remark 10.2 *We make several remarks regarding the Frobenius automorphism.*

1. *For any positive integer m we have that $(\sigma_{p^k})^m = \sigma_{p^{mk}}$, since for any $a \in F_{p^n}$,*

$$(\sigma_{p^k})^m(a) = \underbrace{\sigma_{p^k}\sigma_{p^k}\cdots\sigma_{p^k}}_{m \text{ times}}(a) = (\cdots((a^{p^k})^{p^k}\cdots)^{p^k} = a^{\overbrace{p^k p^k \cdots p^k}^{m \text{ times}}} = a^{p^{mk}}$$
$$= \sigma_{p^{mk}}(a).$$

2. *The homomorphism $\sigma_{p^n} : F_{p^n} \to F_{p^n}$ is the identity, since each element of F_{p^n} is a zero of $x^{p^n} - x$.*

3. *The inverse of σ_{p^k} is $\sigma_{p^{-k}}$, since for any $a \in F_{p^n}$,*

$$\sigma_{p^k}(\sigma_{p^{-k}}(a)) = \sigma_{p^k}(a^{p^{-k}}) = (a^{p^{-k}})^{p^k} = a.$$

4. *The order of $\sigma_p : F_{p^n} \to F_{p^n}$ is n, since $(\sigma_p)^n = \sigma_{p^n}$ which we know to be the identity and if it were the case that $(\sigma_p)^k$ were the identity for $0 < k \leq n$, then that would mean every element of F_{p^n} is a zero of $x^{p^k} - x$ and so by Lemma 9.10, $F_{p^n} \cong F_{p^k}$ and $k = n$.*

5. *For $k \geq m$, the homomorphisms $\sigma_{p^k}, \sigma_{p^m} : F_{p^n} \to F_{p^n}$ are equal iff $k \equiv m \pmod{n}$, since $\sigma_{p^k} = \sigma_{p^m}$ iff $\sigma_{p^k}(\sigma_{p^m})^{-1}$ is the identity iff $(\sigma_p)^{k-m}$ is the identity iff $n | (k-m)$ iff $k \equiv m \pmod{n}$.*

Definition 10.2 *Let $F \subseteq E$ be fields. The set of all F-automorphisms of E with composition, denoted by $Gal(E/F)$, is called the **Galois group** of E over F.*

Example 10.2 *We can already determine some instances of elements in Galois groups.*

1. *From our earlier remarks we see that $\sigma_p, \sigma_{p^2}, \ldots, \sigma_{p^n} \in Gal(F_{p^n}/F_p)$. with $\langle \sigma_p \rangle = \{\sigma_p, \sigma_{p^2}, \ldots, \sigma_{p^n}\}$.*

2. *More generally, σ_{p^k} is an element of $Gal(F_{p^n}/F_{p^k})$ of order n/k. Indeed, $(\sigma_{p^k})^{(n/k)} = \sigma_{p^n}$ which is the identity in $Gal(F_{p^n}/F_{p^k})$ and $(\sigma_{p^k})^m$ is the identity in $Gal(F_{p^n}/F_{p^k})$ iff $n|mk$ iff $(n/k)|m$. Hence, $Gal(F_{p^n}/F_{p^k})$ contains the cyclic subgroup $\langle \sigma_{p^k} \rangle = \{\sigma_{p^k}, \sigma_{p^{2k}}, \ldots, \sigma_{p^n}\}$ with $|\langle \sigma_{p^k} \rangle| = n/k$.*

What we especially want to investigate about these F-homomorphisms is what they do to the zeros of a polynomial.

Definition 10.3 *Let $a_1 \in E_1$ and $a_2 \in E_2$ be two fields containing the same subfield F with a_1 and a_2 both algebraic over F. We say that a_1 and a_2 are **conjugate over** F if they have the same irreducible polynomial over F.*

Example 10.3 *We illustrate conjugacy with several examples. This type of conjugacy is not equivalent to conjugacy of complex numbers.*

1. *Set $E_1 = E_2 = \mathbb{C}$ and $F = \mathbb{Q}$. The elements $1 + i$ and $1 - i$ are conjugate over $\mathbb{Q}$ since they are both zeros of the same irreducible polynomial $x^2 - 2x + 2$ over $\mathbb{Q}$.*

2. *Set $E_1 = E_2 = \mathbb{C}$ and consider the elements $\sqrt[4]{2}$ and $i\sqrt[4]{2}$ which are conjugate over $\mathbb{Q}$ since the irreducible polynomial of $\sqrt[4]{2}$ and $i\sqrt[4]{2}$ over $\mathbb{Q}$ is $x^4 - 2$, yet **not** conjugate over $\mathbb{Q}(\sqrt{2})$ since the irreducible polynomial of $\sqrt[4]{2}$ over $\mathbb{Q}(\sqrt{2})$ is $x^2 - \sqrt{2}$ while the irreducible polynomial of $i\sqrt[4]{2}$ over $\mathbb{Q}(\sqrt{2})$ is $x^2 + \sqrt{2}$. In fact, all four elements $\pm\sqrt[4]{2}, \pm i\sqrt[4]{2}$ are all conjugate over $\mathbb{Q}$.*

Theorem 10.1 *Let $a_1 \in E_1$ and $a_2 \in E_2$ be two fields containing the same subfield F with a_1 and a_2 both algebraic over F. The following statements are equivalent:*

1. *a_1 and a_2 are conjugate over F.*

2. *There exists an F-homomorphism $\phi : F(a_1) \to F(a_2)$ with $\phi(a_1) = a_2$.*

3. *There exists an F-isomorphism $\phi : F(a_1) \to F(a_2)$ with $\phi(a_1) = a_2$.*

Proof 10.1 *To show the first statement implies the third, let $p(x)$ be the irreducible polynomial over F for which both a_1 and a_2 are zeros. Consider the evaluation epimorphisms $\Psi_{a_i} : F[x] \to F[a_i] \cong F(a_i)$ by $\Psi_{a_i}(f(x)) = f(a_i)$ for $i = 1, 2$. Both maps have kernel $(p(x))$. Let ϕ_i be the map which makes $F[x]/(p(x))$ and $F(a_i)$ isomorphic defined as $\phi_i(\overline{f(x)}) = f(a_i)$ for $i = 1, 2$. Then the composition $\phi_2 \circ \phi_1^{-1}$ is an F-isomorphism from $F(a_1)$ to $F(a_2)$ with $\phi_2 \circ \phi_1^{-1}(a_1) = \phi_2(\overline{x}) = a_2$.*

The third statement implies the second is immediate.

To show the second statement implies the first, let $\phi : F(a_1) \to F(a_2)$ *be an F-homomorphism with* $\phi(a_1) = a_2$. *Since* a_1 *is algebraic over* F *there exists an irreducible polynomial* $p(x) \in F[x]$ *for which* a_1 *is a zero. Write* $p(x) = c_0 + c_1 x + \cdots + c_d x^d$. *Notice that*

$$0 = \phi(0) = \phi(p(a_1)) = \phi(c_0 + c_1 a_1 + \cdots + c_d a_1^d) =$$

$$\phi(c_0) + \phi(c_1)\phi(a_1) + \cdots + \phi(c_d)\phi(a_1)^d = c_0 + c_1 a_2 + \cdots + c_d a_2^d = p(a_2).$$

Hence, a_2 *is the zero of the same polynomial irreducible polynomial over* F *as* a_1.

EXERCISES

1 Given the fields E_1 and E_2 below each containing the base field F decide if a and b are conjugate over F.

  a. $E_1 = \mathbb{C} = E_1$, $F = \mathbb{Q}$, $a = i$ and $b = -i$.

  b. $E_1 = \mathbb{R} = E_1$, $F = \mathbb{Q}$, $a = \sqrt{2}$ and $b = -\sqrt{2}$.

  c. $E_1 = \mathbb{R} = E_1$, $F = \mathbb{Q}(\sqrt{2})$, $a = \sqrt{2}$ and $b = -\sqrt{2}$.

2 Verify that the collection of F-homomorphisms from E_1 to E_2 with composition forms a group.

3 Verify that the collection of maps $\sigma_{p^k} : F_{p^n} \to F_{p^n}$ by $\sigma_{p^k}(a) = a^{p^k}$ $(k = 1, 2, \ldots)$ are all $\mathbb{Z}_p$-automorphisms of F_{p^n}.

10.2 COMPUTING GALOIS GROUPS

In order to compute concrete Galois groups we first need some results which will help us characterize the elements of a Galois group. Two theorems are presented followed by some examples which apply these results.

Theorem 10.2 *Consider the fields* $F \subseteq E$ *and let* $a_1, a_2, \ldots, a_n \in E$ *be algebraic over* F *and set* $K = F(a_1, a_2, \ldots, a_n)$. *Then each element* $\phi \in Gal(K/F)$ *is completely determined by where it sends the generators of* K *over* F, *i.e. it is sufficient to know the values of* $\phi(a_1), \phi(a_2), \ldots, \phi(a_n)$ *in order to compute* $\phi(a)$ *for any* $a \in K$ *and should there be another* $\psi \in Gal(K/F)$ *with* $\psi(a_i) = \phi(a_i)$, $1 \leq i \leq n$, *then* $\psi(a) = \phi(a)$ *for all* $a \in K$.

Proof 10.2 *The proof is by induction on the number of generators. For the base case we have* $K = F(a_1) \cong F[a_1]$. *Take any* $f(a_1) \in F(a_1)$ *and write* $f(a_1) = c_0 + c_1 a_1 + \cdots + c_k a_1^k$ *with each* $c_i \in F$. *Then for* $\phi \in Gal(K/F)$ *we have that* $\phi(f(a_1)) = c_0 + c_1\phi(a_1) + \cdots + c_k\phi(a_1)^k$ *and thus the value of* $\phi(f(a_1))$ *is completely determined by the value of* $\phi(a_1)$. *For* $n > 1$, *set* $K_1 = F(a_1, a_2, \ldots, a_{n-1})$ *so that* $K = K_1(a_n) \cong K_1[a_n]$ *(since* a_n *is algebraic over* F *it is certainly algebraic over* $K_1 \supseteq F$ *as well). Take any* $f(a_n) \in K_1(a_n)$ *and write* $f(a_n) = b_0 + b_1 a_n + \cdots + b_k a_n^k$ *with* each $b_i \in K_1$. Then for $\phi \in Gal(K/F)$ we have that $\phi(f(a_n)) = \phi(b_0) + \phi(b_1)\phi(a_n) + \cdots + \phi(c_k)\phi(a_n)^k$. By induction, each $\phi(b_i)$ is completely determined by the values of $\phi(a_1), \phi(a_2), \ldots, \phi(a_{n-1})$ so that $\phi(f(a_n))$ is completely determined by the values of $\phi(a_1), \phi(a_2), \ldots, \phi(a_n)$.

Theorem 10.3 Let $F \subseteq E$ be fields with $f(x) \in F[x]$ and suppose that $a_1, a_2, \ldots, a_n$ are the zeros of $f(x)$ contained in E. Then each $\phi \in Gal(E/F)$ defines a unique permutation of $a_1, a_2, \ldots, a_n$.

Proof 10.3 Take $\phi \in Gal(E/F)$. For each a_i notice that ϕ maps $F(a_i)$ into $F(\phi(a_i))$, since $\phi(c_0 + c_1 a_i + \cdots + c_n a_i^n) = c_0 + c_1 \phi(a_i) + \cdots + c_n \phi(a_i)^n$. Hence, we have the restriction $\phi : F(a_i) \to F(\phi(a_i))$. By Theorem 10.1, a_i and $\phi(a_i)$ are conjugate and this means that $\phi(a_i) = a_j$ for some $j \in \{1, 2, \ldots, n\}$. Now since ϕ is one-to-one and the set $\{a_1, a_2, \ldots, a_n\}$ is finite, we see that the restriction $\phi : F(a_1, a_2, \ldots, a_n) \to F(a_1, a_2, \ldots, a_n)$ defines a permutation of $a_1, a_2, \ldots, a_n$ (one should check that ϕ maps into $F(a_1, a_2, \ldots, a_n)$). By Theorem 10.2, each element of $Gal(E/F)$ must define a distinct permutation of $a_1, a_2, \ldots, a_n$.

Example 10.4 Using the two theorem just proved we now compute some Galois groups.

1. $Gal(\mathbb{Q}(\sqrt[3]{2})/\mathbb{Q})$ is the trivial group. Indeed, $\sqrt[3]{2}$ is a zero of $f(x) = x^3 - 2 \in \mathbb{Q}$ and so for any $\phi \in Gal(\mathbb{Q}(\sqrt[3]{2})/\mathbb{Q})$, by Theorem 10.3, $\phi(\sqrt[3]{2})$ must be a zero of $f(x)$ contained in $\mathbb{Q}(\sqrt[3]{2})$. However, $\sqrt[3]{2}$ is the only zero of $f(x)$ contained in $\mathbb{Q}(\sqrt[3]{2})$ (exercise). Therefore, it must be the case that $\phi(\sqrt[3]{2}) = \sqrt[3]{2}$. Since the identity map also has this property, by Theorem 10.2, ϕ must be the identity automorphism.

2. $Gal(\mathbb{Q}(i)/\mathbb{Q}) \cong \mathbb{Z}_2$. Indeed, i is a zero of $f(x) = x^2 + 1 \in \mathbb{Q}[x]$ and the other zero of $f(x)$ is $-i$ also contained in $\mathbb{Q}(i)$ so that there are potentially two $\mathbb{Q}$-automorphisms of $\mathbb{Q}(i)$: one sending i to i and $-i$ to $-i$ and the other sending i to $-i$ and $-i$ to i. We can, in fact, describe two such $\mathbb{Q}$-automorphisms. Of course, the first would be the identity automorphism. For the second, first note that since i is algebraic over $\mathbb{Q}$ and $[\mathbb{Q}(i) : \mathbb{Q}] = 2$ that $\mathbb{Q}(i) = \{a + bi : a, b \in \mathbb{Q}\}$ (we know that $1, i$ forms a basis for $\mathbb{Q}(i)$ over $\mathbb{Q}$). Define $\phi : \mathbb{Q}(i) \to \mathbb{Q}(i)$ by $\phi(a + bi) = a - bi$. The reader should check that this indeed defines a automorphism of $\mathbb{Q}(i)$. Therefore, $Gal(\mathbb{Q}(i)/\mathbb{Q})$ is a group with exactly two elements and so is isomorphic to $\mathbb{Z}_2$.

3. $Gal(\mathbb{Q}(\sqrt{2}, \sqrt{3})/\mathbb{Q})$ is the Klein 4-group. Indeed, $\sqrt{2}$ is a zero of $f(x) = x^2 - 2 \in \mathbb{Q}[x]$ and the other zero of $f(x)$ is $-\sqrt{2}$ also contained in $\mathbb{Q}(\sqrt{2}, \sqrt{3})$ and $\sqrt{3}$ is a zero of $g(x) = x^2 - 3 \in \mathbb{Q}[x]$ and the other zero of $g(x)$ is $-\sqrt{3}$ also contained in $\mathbb{Q}(\sqrt{2}, \sqrt{3})$. By Theorem 10.3, each element of $Gal(\mathbb{Q}(\sqrt{2}, \sqrt{3})/\mathbb{Q})$ defines a permutation of the zeros of $f(x)$ and a permutation of the zeros of $g(x)$. Hence, there are potentially as many as four elements in $Gal(\mathbb{Q}(\sqrt{2}, \sqrt{3})/\mathbb{Q})$ corresponding to the permutations below:

$$\begin{pmatrix} \sqrt{2} & -\sqrt{2} & \sqrt{3} & -\sqrt{3} \\ \sqrt{2} & -\sqrt{2} & \sqrt{3} & -\sqrt{3} \end{pmatrix}, \qquad \begin{pmatrix} \sqrt{2} & -\sqrt{2} & \sqrt{3} & -\sqrt{3} \\ -\sqrt{2} & \sqrt{2} & \sqrt{3} & -\sqrt{3} \end{pmatrix},$$

$$\begin{pmatrix} \sqrt{2} & -\sqrt{2} & \sqrt{3} & -\sqrt{3} \\ \sqrt{2} & -\sqrt{2} & -\sqrt{3} & \sqrt{3} \end{pmatrix}, \qquad \begin{pmatrix} \sqrt{2} & -\sqrt{2} & \sqrt{3} & -\sqrt{3} \\ -\sqrt{2} & \sqrt{2} & -\sqrt{3} & \sqrt{3} \end{pmatrix}.$$

We can, in fact, describe four such $\mathbb{Q}$-automorphisms of $\mathbb{Q}(\sqrt{2}, \sqrt{3})$. The first automorphism is, of course, the identity automorphism. To describe the remaining three let's first get a better description of $\mathbb{Q}(\sqrt{2}, \sqrt{3})$. Consider the chain of fields

$$\mathbb{Q} \subseteq \mathbb{Q}(\sqrt{2}) \subseteq \mathbb{Q}(\sqrt{2}, \sqrt{3}).$$

Since $\sqrt{2}$ is a zero of the irreducible polynomial $f(x) = x^2 - 2$ over $\mathbb{Q}$ (Eisenstein), $[\mathbb{Q}(\sqrt{2}) : \mathbb{Q}] = 2$ and $1, \sqrt{2}$ forms a basis for $\mathbb{Q}(\sqrt{2})$ over $\mathbb{Q}$. Since $\sqrt{3}$ is a zero of the irreducible polynomial $g(x) = x^2 - 3$ over $\mathbb{Q}(\sqrt{2})$ (it must be irreducible, for otherwise $[\mathbb{Q}(\sqrt{2}, \sqrt{3}) : \mathbb{Q}(\sqrt{2})] = 1$ which puts $\sqrt{3}$ in $\mathbb{Q}(\sqrt{2})$, a contradiction), $[\mathbb{Q}(\sqrt{2}, \sqrt{3}) : \mathbb{Q}(\sqrt{2})] = 2$ and $1, \sqrt{3}$ forms a basis for $\mathbb{Q}(\sqrt{2}, \sqrt{3})$ over $\mathbb{Q}(\sqrt{2})$. Hence, by Theorem 9.3,

$$[\mathbb{Q}(\sqrt{2}, \sqrt{3}) : \mathbb{Q}] = [\mathbb{Q}(\sqrt{2}, \sqrt{3}) : \mathbb{Q}(\sqrt{2})][\mathbb{Q}(\sqrt{2}) : \mathbb{Q}] = 2 \cdot 2 = 4.$$

Furthermore, in the proof of Theorem 9.3 we see that the collection of products of the bases $1, \sqrt{2}$ and $1, \sqrt{3}$ forms a basis for $\mathbb{Q}(\sqrt{2}, \sqrt{3})$ over $\mathbb{Q}$. Namely, the basis is $1, \sqrt{2}, \sqrt{3}, \sqrt{6}$. Therefore,

$$\mathbb{Q}(\sqrt{2}, \sqrt{3}) = \{a + b\sqrt{2} + c\sqrt{3} + d\sqrt{6} \ : \ a, b, c, d \in \mathbb{Q}\}.$$

Consider the following maps from $\mathbb{Q}(\sqrt{2}, \sqrt{3})$ to itself defined by

$$a + b\sqrt{2} + c\sqrt{3} + d\sqrt{6} \ \mapsto \ a + b\sqrt{2} + c\sqrt{3} + d\sqrt{6},$$

$$a + b\sqrt{2} + c\sqrt{3} + d\sqrt{6} \ \mapsto \ a - b\sqrt{2} + c\sqrt{3} - d\sqrt{6},$$

$$a + b\sqrt{2} + c\sqrt{3} + d\sqrt{6} \ \mapsto \ a + b\sqrt{2} - c\sqrt{3} - d\sqrt{6},$$

$$a + b\sqrt{2} + c\sqrt{3} + d\sqrt{6} \ \mapsto \ a - b\sqrt{2} - c\sqrt{3} + d\sqrt{6}.$$

The reader should check that these maps are indeed automorphisms which preserve addition and multiplication. Notice that these maps are permuting the zeros of $f(x)$ and $g(x)$ as required. Therefore, we know that $|Gal(\mathbb{Q}(\sqrt{2}, \sqrt{3})/\mathbb{Q})| = 4$. It remains to determine which of the order four groups this Galois group is, but this is simple. Observe that if you square each of the automorphisms (under composition) you get the identity automorphism and this property characterizes the order four group to be the Klein 4-group.

Our next objective in the coming sections is to put a bound on the number of possible elements in a given Galois group. Perhaps the reader has already noticed from the examples that there is a relationship between the size of $Gal(E/F)$ and the value of $[E : F]$. We will show that, in general, $|Gal(E/F)| \leq [E : F]$. We will also discover conditions for when $|Gal(E/F)| = [E : F]$.

Example 10.5 *Take $E = \mathbb{Q}(\sqrt[3]{2})$ and $F = \mathbb{Q}$. In Example 10.4.1, we found that $|Gal(E/F)| = 1$ while in Example 9.8.2 we found that $[E : F] = 3$. Hence, in this example $|Gal(E/F)| < [E : F]$.*

EXERCISES

1 In each of the problems below, compute $Gal(E/F)$ as we did in Example 10.4.

   a. $E = \mathbb{Q}(\sqrt{2})$ and $F = \mathbb{Q}$.

   b. $E = \mathbb{Q}(\sqrt[4]{2})$ and $F = \mathbb{Q}$.

   c. $E = \mathbb{Q}(\sqrt[4]{2})$ and $F = \mathbb{Q}(\sqrt{2})$.

2 In the proof of Theorem 10.3, verify that ϕ maps into $F(a_1, a_2, \ldots, a_n)$.

3 Verify that $\sqrt[3]{2}$ is the only zero of $f(x) = x^3 - 2$ contained in $\mathbb{Q}(\sqrt[3]{2})$.

4 In Example 10.4.2, verify that $\phi : \mathbb{Q}(i) \to \mathbb{Q}(i)$ by $\phi(a + bi) = a - bi$ defines a automorphism of $\mathbb{Q}(i)$.

5 In Example 10.4.3, verify that the four defined maps are indeed automorphisms which preserve addition and multiplication.

10.3   APPLICATIONS OF ZORN'S LEMMA

The main result of the next two sections, namely that the size of the Galois group $Gal(E/F)$ is bounded by the index $[E : F]$ requires the use of Zorn's Lemma. We will assume the reader has no prior knowledge of Zorn's Lemma and start from there. Since this is not a course on Set Theory, we will be brief. One of the foundational axioms which mathematicians assume about sets is the Axiom of Choice which basically states that given any infinite collection of sets one can define a rule which selects an element from each of those sets. It turns out that the Axiom of Choice is equivalent to Zorn's Lemma, but before we can state it we need to define some terminology some of which you have already seen in this text.

Definition 10.4 *A relation $\leq$ on a set X is a* **partial ordering** *if it satisfies the reflexive, anti-symmetric and transitivity properties, i.e.*

   1. For all $x \in X$, we have $x \leq x$.

   2. For all $x, y \in X$ if $x \leq y$ and $y \leq x$, then $x = y$.

   3. For all $x, y, z \in X$ if $x \leq y$ and $y \leq z$, then $x \leq z$.

The pair $(X, \leq)$ is then called a **poset** *(i.e. a partially ordered set). The set is called* **linearly ordered** *if in addition we have the following property: For all $x, y \in X$ either $x \leq y$ or $y \leq x$.*

Example 10.6 *Some examples of posets are the integers with $\leq$ (hence, the notation) or the collection of subsets of a set together with $\subseteq$. Note that the first example is, in fact, a linear ordering while the second, in general, is not.*

Definition 10.5 *Let $(X, \leq)$ be a poset.*

1. *An element $x \in X$ is a* **maximal** *element of X if for all $y \in X$, whenever we have $x \leq y$ it must be the case that $y = x$.*

2. *An element $x \in X$ is an* **upper bound** *of a subset $Y \subseteq X$ if for all $y \in Y$, we have $y \leq x$.*

3. *A* **chain** *in X is a collection of subsets of X which is linearly ordered by $\subseteq$.*

4. *X is* **inductively ordered** *if every chain in X has an upperbound in X.*

We can now state Zorn's Lemma which as we have stated is a consequence of the Axiom of Choice and the proof shall be omitted.

Lemma 10.1 (Zorn) *A non-empty inductively ordered poset has a maximal element.*

Example 10.7 *We illustrate in several examples the use of Zorn's lemma.*

1. *We will use Zorn's Lemma to show that every ring with $0 \neq 1$ has a maximal ideal. Let X be the collection of all proper ideals of R and partially order them by inclusion. Note that X is non-empty, since it contains the trivial ideal (here we need $0 \neq 1$). If we can show X has a maximal element, then we will be done. By Zorn's Lemma, it's enough that we show every chain in X has an upperbound. Let $\mathcal{C}$ be a chain of elements in X. We show that the union of all the elements in the chain, let's call it J, is the upperbound we seek. First note that J is indeed a proper ideal of R (and thus in X). To see this take $a, b \in J$. It must be that $a \in I_1$ and $b \in I_2$ for some $I_1, I_2 \in \mathcal{C}$. Since $\mathcal{C}$ is linearly ordered, without loss of generality, let's suppose that $I_1 \subseteq I_2$. Therefore $a, b \in I_2$ and so $a - b \in I_2 \subseteq J$. Now take $a \in J$ and $r \in R$. Again $a \in I$ for some $I \in \mathcal{C}$ and so $ra, ar \in I \subseteq J$. What makes J proper is the fact that 1 is not contained in any of the elements of $\mathcal{C}$ (exercise) and thus is not in J as well. Furthermore, $I \subseteq J$ for all $I \in \mathcal{C}$ and so J is the upperbound we seek.*

2. *We will use Zorn's Lemma to show that every non-trivial vector space over a field has a basis. Let X be the collection of all linearly independent subsets of a non-trivial vector space V. Note that X is non-empty, since X contains the singletons sets consisting of the non-zero elements of V. If we show X has a maximal element, then this element is the basis we seek (in linear algebra one can prove that a maximal linearly independent set of vectors form a basis for the vector space). By Zorn's Lemma, it's enough that we show every chain in X has an upperbound. Let $\mathcal{C}$ be a chain of elements in X. We show that the union of all the elements in the chain, let's call it Y, is the upperbound we seek. We need to show then that Y is a linearly independent subset of V. Suppose there are scalars $a_1, a_2, \ldots, a_n$ in the field and vectors $v_1, v_2, \ldots, v_n \in V$ such that $a_1 v_1 + a_2 v_2 + \cdots + a_n v_n = 0$. Now each v_i is in some linearly independent subset in $\mathcal{C}$ and since the elements of $\mathcal{C}$ are linearly ordered there must be some element of $\mathcal{C}$, call it S, where all the vectors $v_1, v_2, \ldots, v_n$ reside. Since S is a*

linearly independent subset of V, it follows that $a_1 = a_2 = \cdots = a_n = 0$, and so Y is a linearly independent subset of V and hence is an element of X. It is certainly an upperbound of C being the union of all the elements of C.

3. We can now prove using Zorn's Lemma that every field F has an extension which is algebraically closed. First, we show that F has a field extension E such that every polynomial $f(x) \in F[x]$ of degree at least one has a root in E. Let's index the polynomials in $F[x]$ of degree at least one by an indexing set I. In other words, $P = \{f_i(x) \; : \; i \in I\}$. Let $X = \{x_i \; : \; i \in I\}$ be a set of indeterminates indexed by the same set I. Define $R = F[X]$ be the collection of polynomial in a finite number of indeterminates coming from X. We leave it as an exercise for the reader to verify that R is an integral domain. Set J equal to the ideal in R generated by P.

Claim 10.1 J is a proper ideal in R.

Suppose to the contrary that $J = R$. Then, in particular, $1 \in J$ and so

$$1 = \sum_{i=1}^{n} g_i(x_{j_1}, \ldots, x_{j_{m_i}}) f_i(x_i), \quad \text{where each} \quad g_i \in R.$$

By Corollary 9.6, there is a field $K \supseteq F$ with $a_1, \ldots, a_n \in K$ zeros of $f_1(x_1), \ldots, f_n(x_1)$, respectively. In the expression representing 1 above evaluate each x_i at a_i and any other variables evaluate at 0 to get

$$1 = \sum_{i=1}^{n} c_i f_i(a_i) = \sum_{i=1}^{n} c_i \cdot 0 = 0, \quad \text{a contradiction.}$$

Thus, having proved the Claim, using Zorn's Lemma in a similar manner to the first example there exists a maximal ideal M of R containing J. Now $\overline{E} = R/M$ is a field and the map $\phi : F \to R/M$ by $\phi(c) = \overline{c} = c + M$ is a field monomorphism (recall all non-trivial field homomorphisms are monomorphisms). Therefore, $\overline{E}$ contains an isomorphic copy of F, namely $\overline{F} = \phi(F)$. Therefore, we will switch to this new base field $\overline{F}$ and look for zeros as we did in Theorem 9.12. Note that since each $f_i(x_i) \in M$ this implies that $\overline{x_i}$ is a root of $\overline{f_i}(x_i)$, where $\overline{f_i}(x_i) = \overline{c_d} x_i^d + \cdots + \overline{c_1} x_i + \overline{c_0}$ when $f_i(x_i) = c_d x_i^d + \cdots + c_1 x_i + c_0$. Hence, $\overline{E}$ is the field we seek.

To show the existence of an algebraically closed extension of F, by Corollary 9.7, it's enough to construct a field extension E of F such that every polynomial $f(x) \in E[x]$ of degree at least one has a root in E. To do the, define a chain of fields $F = F_0 \subseteq F_1 \subseteq F_2 \subseteq \cdots$ such that for each $n \in \mathbb{N}$ and every $f_n(x) \in F_n[x]$ there is a root of $f_n(x)$ in F_{n+1} (which we know we can define by the work just completed). Now set $E = \bigcup_{n \in \mathbb{N}} F_n$ and E is the field we seek. Indeed, take any $f(x) \in E[x]$. Then the finite number of coefficients in $f(x)$ must lie in some F_n and so $f(x) \in F_n[x]$. By our construction, $f(x)$ has a root in $F_{n+1} \subseteq E$.

In preparation for our next result, we will require Zorn's Lemma to prove a preliminary result on the way toward this result. But before we do this preliminary result, we need a more preliminary result in order to prove the preliminary result! First, we introduce some terminology and notation.

Definition 10.6 *Let X_1, X_2, Y_1, Y_2 be sets and let $f : X_1 \to Y_1$ and $g : X_2 \to Y_2$ be functions. We say that g **extends** f (or g is an **extension of** f) if $X_1 \subseteq X_2$ and for all $x \in X_1$ we have that $g(x) = f(x)$.*

Let $\phi : F_1 \to F_2$ be a field homomorphism and $f(x) = c_0 + c_1 x + \cdots + c_n x^n \in F_1[x]$. We will denote $\phi(c_0) + \phi(c_1)x + \cdots + \phi(c_n)x^n$ by the notation $\phi(f(x))$.

Lemma 10.2 *Consider the fields $F_1 \subseteq E_1$ and $F_2 \subseteq E_2$ and let $\phi : F_1 \to F_2$ be a field isomorphism. Take an $a \in E_1$ algebraic over F_1 with irreducible polynomial $p(x) \in F_1[x]$ of a over F_1. Set $q(x) = \phi(p(x)) \in F_2[x]$. For each zero $b \in E_2$ of $q(x)$, there exists a unique isomorphic extension of ϕ from $F_1(a)$ to $F_2(b)$.*

Proof 10.4 *Let $d = deg(p) = [F_1(a) : F_1]$. By Theorem 9.4, $1, a, \ldots, a^{d-1}$ forms a basis for $F_1(a)$ over F_1. In other words, every element of $F_1(a)$ can be written uniquely as $c_0 + c_1 a + \cdots + c_{d-1} a^{d-1}$ for each $c_i \in F_1$. Since ϕ is an isomorphism $q(x)$ is irreducible of the same degree as $p(x)$ (check) and so $1, b, \ldots, b^{d-1}$ forms a basis for $F_2(b)$ over F_2. Define $\overline{\phi} : F_1(a) \to F_2(b)$ by*

$$\overline{\phi}(c_0 + c_1 a + \cdots + c_{d-1} a^{d-1}) = \phi(c_0) + \phi(c_1)b + \cdots + \phi(c_{d-1})b^{d-1}.$$

We leave it to the reader to check that this is the desired extension.

Corollary 10.1 *Let $F \subseteq E$ be fields with E algebraic over F. If $a \in E$ and $p(x) \in F[x]$ is the irreducible polynomial of a over F, then the number of F-homomorphisms of E into $\overline{F}$ equals the number of distinct zeros of $p(x)$.*

Proof 10.5 *By Lemma 10.2, the number of F-homomorphisms from E into $\overline{F}$ is at least as many as the number of distinct zeros of $p(x)$ and by Theorem 10.1 any F-homomorphism from E_1 into $\overline{F}$ must send a to a zero of $p(x)$. Hence, the number of F-homomorphisms of E_1 into $\overline{F}$ is exactly the number of distinct zeros of $p(x)$.*

Lemma 10.3 *Consider the fields $F \subseteq K \subseteq E_1$ and $F \subseteq E_2$ with E_1 algebraic over F and E_2 algebraically closed. Then any F-homomorphism $\phi : K \to E_2$ extends to an F-homomorphism from E_1 to E_2.*

Proof 10.6 *Set X equal to the collection of F-homomorphisms from L to E_2, where $K \subseteq L \subseteq E_1$ and order them by extension, i.e. $\phi \leq \psi$ iff ψ extends ϕ. It is easy to see that this is a partial ordering. The set X is non-empty, since the original ϕ in our assumptions is in X. Finally, to apply Zorn's Lemma, we need to show that X is inductively ordered. We will show that for any chain $\mathcal{C}$ in X, the union of the elements of $\mathcal{C}$, call it χ, is an upperbound for $\mathcal{C}$ in X. There are several things to check. First, we need to point out that χ is a well-defined map, since $\mathcal{C}$ by definition* *is linearly ordered. Second, χ is in X since its domain contains K and it extends the original map $\phi : K \to E_2$. Finally, χ is an upperbound for $\mathcal{C}$ since it certainly extends all the elements in $\mathcal{C}$.*

Therefore, by Zorn's Lemma, there is an F-homomorphism $\overline{\phi} : E \to E_2$ extending all the elements of X with $K \subseteq E \subseteq E_1$. We show now that, in fact, $E = E_1$. Suppose to the contrary we had an $a \in E_1 \setminus E$. Since E_1 is algebraic over F, we have that a is algebraic over F and so a is algebraic over E as well. Let $p(x) \in E[x]$ be the irreducible polynomial for a over E. Set $q(x) = \overline{\phi}(p(x)) \in E_2[x]$. Since $\phi(a)$ is a zero of $q(x)$, by Lemma 10.2, $\overline{\phi}$ can be extended to a F-homomorphism from $E(a)$ to $\overline{\phi}(E)(\overline{\phi}(a)) \subseteq E_2$. But this would contradict the maximality of $\overline{\phi}$.

EXERCISES

1 Explain why in Example 10.7.1 that 1 is not contained in any of the elements of $\mathcal{C}$.

2 Verify in Example 10.7.3 that R is an integral domain.

3 Verify in Lemma 10.2 that $q(x)$ is irreducible of the same degree as $p(x)$.

4 Verify in Lemma 10.2 that $\overline{\phi}$ is the desired extension.

5 Verify in Lemma 10.3 that $\leq$ defined in the proof is a partial ordering.

10.4 TWO IMPORTANT THEOREMS

In this section, we will achieve the goal of putting a bound on the size of the Galois group, namely that $|Gal(E/F)| \leq [E : F]$. We will also work toward the goal of characterizing when the inequality above is an equality.

Theorem 10.4 *Consider the fields $F \subseteq E \subseteq \overline{F}$ where $\overline{F}$ is an algebraic closure of F and $[E : F] < \infty$. Then the number of F-homomorphisms from E to $\overline{F}$ is no more than $[E : F]$.*

Proof 10.7 *The proof is by induction on $n = [E : F]$. If $n = 1$, then $E = F$ and so the only F-homomorphism is the identity map. For $n > 1$ pick any $a \in E \setminus F$ and set $E_1 = F(a)$. We first show that the number of F-homomorphisms from E_1 into $\overline{F}$ is no more than $[E_1 : F]$. To see this, first note that a is algebraic over F since $[E : F] < \infty$. Indeed, by Lemma 9.4, all of E is algebraic over F. Let $p(x) \in F[x]$ be the irreducible polynomial of a over F. By Corollary 10.1, the number of F-homomorphisms from E_1 into $\overline{F}$ equals the number of distinct zeros of $p(x) \leq deg(p) = [E_1 : F]$.*

Since $[E : E_1] = [E : F]/[E_1 : F] < [E : F]$, by induction, the number of E_1-homomorphisms from E into $\overline{E_1}$ (check that $\overline{E_1}$ is an algebraic closure of F) is no more than $[E : E_1]$. Let's enumerate all the homomorphisms that we have found so far. Denote $\phi_1, \phi_2, \ldots, \phi_k$ to be the F-homomorphisms from E_1 into $\overline{F}$ with $k \leq [E_1 : F]$. Denote $\psi_1, \psi_2, \ldots, \psi_m$ to be the E_1-homomorphisms from E into $\overline{F}$ with $m \leq [E : E_1]$. By Lemma 10.3, each ϕ_i extends to $\overline{\phi_i}$ an F-homomorphism *from $\overline{F}$ into $\overline{F}$. Notice that each composition $\overline{\phi}_i \circ \psi_j$ is an F-homomorphism from E into $\overline{F}$ and the number of such compositions is $k \cdot m \leq [E_1 : F][E : E_1] = [E : F]$. Therefore, to complete the proof it suffices to show that any F-homomorphism from E into $\overline{F}$ is equal to one of the compositions $\overline{\phi}_i \circ \psi_j$ for some i and j. To this end, let $\phi : E \to \overline{F}$ be any F-homomorphism. Notice that the restriction of ϕ to E_1, denoted as $\phi \upharpoonright E_1$, must equal ϕ_i for some $1 \leq i \leq k$. Consider the composition $(\overline{\phi}_i)^{-1} \circ \phi$ a homomorphism from E into $\overline{F}$. Since $\phi \upharpoonright E_1 = \phi_i$ and $\overline{\phi}_i \upharpoonright E_1 = \phi_i$, for all $b \in E_1$ we have $\phi(b) = \overline{\phi}_i(b)$ or $(\overline{\phi}_i)^{-1}(\phi(b)) = b$. Thus $(\overline{\phi}_i)^{-1} \circ \phi$ fixes E_1 and hence is an E_1-homomorphism. Therefore, $(\overline{\phi}_i)^{-1} \circ \phi = \psi_j$ for some $1 \leq j \leq m$, and so $\phi = \overline{\phi}_i \circ \psi_j$.*

Corollary 10.2 *For any fields $F \subseteq E$ with $[E : F] < \infty$ we have that $|Gal(E/F)| \leq [E : F]$.*

Proof 10.8 *This result follows immediately from the observation that $E \subseteq \overline{F}$ and so every F-automorphism from E onto E is an F-homomorphism from E into $\overline{F}$.*

Example 10.8 *Let's return to the case of finite fields. We saw that*

$$\langle \sigma_{p^k} \rangle = \{\sigma_{p^k}, \sigma_{p^{2k}}, \ldots, \sigma_{p^n}\} \subseteq Gal(F_{p^n}/F_{p^k})$$

with $|\langle \sigma_{p^k} \rangle| = n/k$ so that $|Gal(F_{p^n}/F_{p^k})| \geq n/k$. Furthermore, since $[F_{p^n} : F_{p^k}] = n/k$, by Corollary 10.2, $|Gal(F_{p^n}/F_{p^k})| \leq n/k$. Hence, $|Gal(F_{p^n}/F_{p^k})| = n/k$ with $Gal(F_{p^n}/F_{p^k}) = \langle \sigma_{p^k} \rangle$. One consequence of this example is that for $F \subseteq E$ finite fields, the Galois group $Gal(E/F)$ is always cyclic.

Our next goal is to characterize the case when $|Gal(E/F)| = [E : F]$. The next result will be helpful toward that goal.

Theorem 10.5 *Let $F \subseteq E \subseteq \overline{F}$ with E algebraic over F and $\overline{F}$ an algebraic closure of E and F. Consider the following statements:*

1. If $\phi : E \to \overline{F}$ is an F-homomorphism, then $\phi(E) \subseteq E$.

2. If $\phi : E \to \overline{F}$ is an F-homomorphism, then $\phi(E) = E$.

3. If $f(x) \in F[x]$ has a zero in E, then $f(x)$ has all its zeros in E.

4. E is a splitting field of some $f(x) \in F[x]$.

Then the first three statements are equivalent, while the fourth statement implies the first three statements. In addition, the first three statements imply the fourth statement under the condition that $[E : F] < \infty$.

Proof 10.9 *In order to show the first three statements are equivalent, we first assume that the first statement is true and we show the second follows. Take an F-homomorphism $\phi : E \to E$ and we need to show ϕ maps onto E. Take any $b \in E$ and set*

$$X = \{b' \in E \; : \; b \text{ and } b' \text{ are conjugate}\}.$$

Since b is algebraic over F it is the zero of some irreducible polynomial $p(x) \in F[x]$ with $|X| \leq \deg(p) = [F(b) : F] < \infty$. So we can enumerate $X = \{b_1, b_2, \ldots, b_n\}$ and set $E_1 = F(b_1, b_2, \ldots, b_n)$. Since each b_i is algebraic over F, by Theorem 9.10, $[E_1 : F] < \infty$. Since $\phi(X) \subseteq X$ this implies that $\phi(E_1) \subseteq E_1$ (see Theorem 10.2 and Theorem 10.3). Set ϕ_1 to be the restriction of ϕ to E_1. As mentioned earlier we can view an F-homomorphism like ϕ_1 as an F-vector space homomorphism. Since ϕ_1 is one-to-one with $\phi_1(E_1) \subseteq E_1$ and the vector space dimension of E_1 over F is finite, it follows that ϕ_1 maps onto E_1. This means there is an $a \in E_1 \subseteq E$ such that $\phi_1(a) = b$ and so $\phi(a) = b$ which makes ϕ map onto E.

Now assume that the second statement is true to prove the third is true. Without loss of generality, we can assume the polynomial $f(x) \in F[x]$ is irreducible over F. Let $a \in E$ be a zero of $f(x)$. Take any $b \in \overline{F}$ any other zero of $f(x)$. By Theorem 10.1, there is an F-homomorphism $\phi : F(a) \to F(b)$ with $\phi(a) = b$. By Lemma 10.3, we can extend ϕ to $\overline{\phi} : E \to \overline{F}$. By assumption, $\overline{\phi}(E) = E$ and so $b = \phi(a) = \overline{\phi}(a) \in E$.

Now assume the third statement is true to prove the first is true. Let $\phi : E \to \overline{F}$ be an F-homomorphism and take $a \in E$. Since E is algebraic over F this means that a is a zero of some irreducible polynomial $p(x) \in F[x]$. Since $\phi(a)$ is conjugate to a over F, by assumption, $\phi(a) \in E$ and so $\phi(E) \subseteq E$.

Now we show that the fourth statement implies the first (and thus, the second and third as well). By assumption, $E = F(a_1, a_2, \ldots, a_n)$ where $X = \{a_1, a_2, \ldots, a_n\}$ are the zeros of some polynomial $f(x) \in F[x]$. Let $\phi : E \to \overline{F}$ be any F-homomorphism. Since each $\phi(a_i)$ is a zero of $f(x)$ we have that $\phi(X) \subseteq X$ and so $\phi(E) \subseteq E$.

Finally, we show the third statement (and thus, the first two as well) imply the fourth statement in the case that $[E : F] < \infty$. Since $[E : F] < \infty$, by Theorem 9.11, $E = F(a_1, a_2, \ldots, a_n)$ for some $a_1, a_2, \ldots, a_n \in E$. Since each a_i is algebraic over F, each has an irreducible polynomial $p_i(x)$ over F. Set $f(x) = p_1(x)p_2(x) \cdots p_n(x)$ and let X denote the (finite number of) zeros of $f(x)$. By assumption, $X \subseteq E$ and since $F(X)$ is the smallest field containing F and X, we have $F(X) \subseteq E$. Since $E = F(a_1, a_2, \ldots, a_n)$ we also have $E \subseteq F(X)$, and so $E = F(X)$ the splitting field of $f(x)$ over F.

Definition 10.7 *Let $F \subseteq E \subseteq \overline{F}$ with E algebraic over F and $\overline{F}$ an algebraic closure of E and F. If the first three statements in Theorem 10.5 hold for E, we say E is a* **normal** *extension of F.*

As Theorem 10.5 attests, in the case that $[E : F] < \infty$, E being a splitting field of some polynomial in F is equivalent to E being a normal extension of F.

EXERCISES

1 In the proof of Theorem 10.4 verify that $\overline{E_1}$ is an algebraic closure of F.

2 Let $F \subseteq K \subseteq E$ be fields with E a finite normal extension of K and K a finite normal extension of F.

a. Prove that if $Gal(E/F)$ is abelian, then so are $Gal(E/K)$ and $Gal(K/F)$.

b. Prove that if $Gal(E/F)$ is cyclic, then so are $Gal(E/K)$ and $Gal(K/F)$.

10.5   SEPARABLE DEGREE

As we saw in Section 10.4, counting the number of F-homomorphisms from an extension field E of F into an algebraic closure of F is the main quantity we want to pin down in order to get results about the size of $Gal(E/F)$. For this reason we introduce the following special notation and terminology.

Definition 10.8 *Let $F \subseteq E$ be fields. The* **separable degree** *of E over F, written $[E : F]_s$, is the number of F-homomorphisms from E into $\overline{E}$.*

Remark 10.3 *We now tie separable degree into the previous notions we have thus far.*

1. *In general,* $|Gal(E/F)| \leq [E : F]_s$.

2. *For E algebraic over F, any algebraic closure $\overline{E}$ of E also serves as an algebraic closure of F, since $\overline{E}$ is algebraic over E and E is algebraic over F.*

3. *The algebraic closure of F is unique up to isomorphism. Indeed, suppose E_1 and E_2 are both algebraic closures of a field F. By Lemma 10.3, we can extend the inclusion map from F into E_2 to an F-homomorphism (necessarily one-to-one) ϕ from E_1 to E_2. Finally, we need to show that ϕ maps onto E_2 (and hence E_1 and E_2 are isomorphic). Set $E = \phi(E_1)$, an algebraically closed field (since E_1 is algebraically closed – exercise). Take any $a \in E_2$. Since E_2 is algebraic over F, the element a is algebraic over F. Since E is algebraically closed, it must be the case that $a \in E$. Hence, $E = E_2$ and ϕ maps onto E_2.*

4. *From the previous remark, it follows that the value of $[E : F]_s$ is the same regardless of what algebraic closure of F we use (exercise).*

5. *Theorem 10.4 can be rephrased as follows: If $[E : F] < \infty$, then $[E : F]_s \leq [E : F]$.*

6. *If E is a normal extension of F, then by Theorem 10.5.2, it follows that $|Gal(E/F)| = [E : F]_s$.*

7. *If $[E : F] < \infty$, then by Theorem 10.5, E is a normal extension of F iff $|Gal(E/F)| = [E : F]_s$. Note that for the direction in which we assume $|Gal(E/F)| = [E : F]_s$, we need $[E : F] < \infty$ in order to have E algebraic over F and so, by the first remark, $\overline{E}$ serves as an algebraic closure of F. Hence, considering F-homomorphisms from E into $\overline{E}$ is the same as considering them from E into $\overline{F}$.*

8. *The second paragraph in the proof of Theorem 10.4 (with a bit of thought) essentially gives the result that $[E : F]_s = [E : K]_s[K : F]_s$ for any fields $F \subseteq K \subseteq E$ under the assumption that $[E : F] < \infty$.*

Example 10.9 *We illustrate separable degree with an example.*

1. *In preparation for the second example we point out that outside of every field F there exists a transcendental element. Indeed, Set $E = F(x) \supset F$ where x is an indeterminate not mentioned in F. Then x is transcendental over F, since if it were algebraic over F there would be a polynomial $f(x) \in F[x]$ of degree at least one with $f(x) = 0$ (here we have evaluated $f(x)$ at $x = x$). But then $f(x)$ would be the zero-polynomial, a contradiction.*

2. *We give an example of the case when $[E : F]_s$ is strictly smaller than $[E : F]$. Consider the field $\mathbb{Z}_p$ and set $F = \mathbb{Z}_p(t)$ where t is transcendental over $\mathbb{Z}_p$. Notice that $p(x) = x^p - t \in F[x]$ has no zero in F, for otherwise t would be algebraic over $\mathbb{Z}_p$. Indeed, suppose that $f(t)/g(t) \in F$ were a zero of $p(x)$. Then $f(t)^p/g(t)^p - t = 0$ which implies that $f(t)^p - tg(t)^p = 0$ and so t would be a zero of the polynomial $f(x)^p - xg(x)^p \in \mathbb{Z}_p[x]$. Choose any zero $a \in \overline{F}$ of $p(x)$ and set $E = F(a) \subseteq \overline{F}$. Since $\mathrm{char}(F) = p$, we have that $(x - a)^p = x^p - a^p = x^p - t = p(x)$. Thus, a is a zero of $p(x)$ of multiplicity p and hence the only zero of $p(x)$. Now this implies that any F-homomorphism from E into $\overline{E}$ must send a to a and thus this map must be the identity homomorphism. In sum, $[E : F]_s = 1$. Since $p(x)$ is the irreducible polynomial for a over F ($p(x)$ has no zero in F), we have that $[E : F] = p$ and so $[E : F]_s < [E : F]$.*

We now introduce some more terminology using the same word **separable** and show how the two notions are connected.

Definition 10.9 *Let $f(x) \in F[x]$ and $a \in \overline{F}$ be a zero of $f(x)$.*

1. *The element a is a **simple** zero of $f(x)$ if $(x - a)| f(x)$, yet it's not the case that $(x - a)^2| f(x)$.*

2. *An irreducible polynomial in $F[x]$ is **separable over** F if all its zeros are simple.*

3. *An arbitrary polynomial in $F[x]$ is **separable over** F if all its irreducible factors are separable over F.*

4. *An element a algebraic over F is **separable over** F if its irreducible polynomial is separable over F.*

5. *A field E is **separable over** F if E is algebraic over F and each element of E is separable over F.*

Theorem 10.6 *Consider the field $F \subseteq E$ with $[E : F] < \infty$. E is separable over F iff $[E : F]_s = [E : F]$.*

Proof 10.10 *First assume that E is separable over F. We show that $[E : F]_s = [E : F]$ by induction on $n = [E : F]$. In the case that $n = 1$ we get that $E = F$ and so the only F-homomorphism from E into $\overline{E}$ is the identity homomorphism. Hence,*

$[E : F]_s = 1 = [E : F]$. *For $n > 1$, choose an $a \in E \backslash F$ and set $K = F(a)$ and let $p(x)$ be the irreducible polynomial for a over F. By induction, $[E : K]_s = [E : K]$ (since E separable over F implies that E separable over K – the irreducible polynomial of a over K divides the irreducible polynomial of a over F). Since E is separable over F the irreducible polynomial $p(x)$ has only simple roots and hence has as many zeros as its degree. As we have seen, for each zero of $p(x)$ there is an F-homomorphism from K into $\overline{K}$ which sends a to some zero of $p(x)$. Hence, $[K : F]_s$ equals the number of zeros of $p(x)$ which equals $deg(p) = [K : F]$ and so*

$$[E : F] = [E : K][K : F] = [E : K]_s[K : F]_s = [E : F]_s.$$

Now assume that $[E : F]_s = [E : F]$ and we show E is separable over F. Take any $a \in E$ and set $K = F(a)$. By our earlier remarks, we know that $[E : K]_s \leq [E : K]$ and $[K : F]_s \leq [K : F]$. But since

$$[E : F]_s = [E : K]_s[K : F]_s \leq [E : K][K : F] = [E : F],$$

and $[E : F]_s = [E : F]$, then it must be the case that, in fact, $[E : K]_s = [E : K]$ and $[K : F]_s = [K : F]$. In particular, the fact that $[K : F]_s = [K : F]$ says that the irreducible polynomial for a over F has as many conjugates as the degree of $p(x)$. But that could only mean that all the zeros of $p(x)$ are simple.

Hence, we have now reached the main goal of the last few sections.

Corollary 10.3 *Let $F \subseteq E$ be fields with $[E : F] < \infty$. Then $|Gal(E/F)| = [E : F]$ iff E is both a normal and separable extension of F.*

Proof 10.11 *Since $[E : F] < \infty$, we know that $|Gal(E/F)| \leq [E : F]_s \leq [E : F]$. First assume that $|Gal(E/F)| = [E : F]$. By our work above, we have that $[E : F]_s = [E : F]$, which by Theorem 10.6, implies that E is separable over F and $|Gal(E/F)| = [E : F]_s$, by an earlier remark implies that E is normal over F. Now assume that E is both normal and separable over F. Citing the same references in the previous two lines we get that $|Gal(E/F)| = [E : F]_s$ and $[E : F]_s = [E : F]$ which implies that $|Gal(E/F)| = [E : F]$.*

EXERCISES

1 In Remark 10.3.3, verify that E_1 is algebraically closed.

2 Verify the statement made in Remark 10.3.4.

3 Verify the statement made in Remark 10.3.8.

10.6 GALOIS EXTENSIONS

Our next major goal is the Fundamental Theorem of Galois Theory which relates the lattice of subgroups of $Gal(E/F)$ to a lattice of certain intermediate subfields between E and F. A key assumption of this result is that E be a Galois extension of F. In this section, we will define and explore this idea.

Definition 10.10 *Let $F \subseteq E$ be fields.*

1. *For any field E let $Aut(E)$ denote the group of automorphisms of E with composition.*

2. *For any field E and any subgroup $G \leq Aut(E)$, the G-invariant subfield of E, denoted by E^G, is the collection of all elements of E which are fixed by every element of G, i.e.*

$$E^G = \{a \in E \mid \phi(a) = a \ \text{ for all } \ \phi \in G\}.$$

Remark 10.4   1. *One can easily check that E^G is a subfield of E. (exercise)*

2. *If $G \leq Gal(E/F)$, then $F \subseteq E^G \subseteq E$. In general, if K is a subfield of E containing F, then K is called an* **intermediate subfield of E over F**.

Example 10.10 *Here, we give some examples of G-invariant subfields.*

1. *Set $E = \mathbb{Q}(\sqrt[4]{2})$ and $F = \mathbb{Q}$. Since, by Theorem 10.3, any element of $Gal(E/F)$ is completely determined by where it sends the zeros of $x^4 - 2$ contained in E, then $|Gal(E/F)| = 2$ and consists of the identity homomorphism and the F-homomorphism sending $\sqrt[4]{2}$ to $-\sqrt[4]{2}$. Call this second map ϕ. Since $x^4 - 2$ is the irreducible polynomial for $\sqrt[4]{2}$ over F (Eisenstein), by Theorem 9.4, a basis for E over F is $1, \sqrt[4]{2}, (\sqrt[4]{2})^2 = \sqrt{2}, (\sqrt[4]{2})^3$. Hence, the elements of E have the form*

$$c_0 + c_1 \sqrt[4]{2} + c_2 \sqrt{2} + c_3 (\sqrt[4]{2})^3,$$

where $c_0, c_1, c_2, c_3 \in F$. Let $G = Gal(E/F)$ and we compute E^G in this setting. Since the identity homomorphism fixes everything, to find E^G it suffices to find out what the second element of G fixes. First note that

$$\phi(\sqrt{2}) = \phi[(\sqrt[4]{2})^2] = \phi(\sqrt[4]{2})^2 = (-\sqrt[4]{2})^2 = \sqrt{2} \ \text{ and}$$

$$\phi[\sqrt[4]{2}^3] = \phi(\sqrt[4]{2})^3 = (-\sqrt[4]{2})^3 = -\sqrt[4]{2}^3.$$

Therefore, ϕ fixes an element of E iff

$$\phi(c_0 + c_1 \sqrt[4]{2} + c_2 \sqrt{2} + c_3 (\sqrt[4]{2})^3) = c_0 + c_1 \sqrt[4]{2} + c_2 \sqrt{2} + c_3 (\sqrt[4]{2})^3 \ \text{ iff}$$

$$c_0 + c_1 \phi(\sqrt[4]{2}) + c_2 \phi(\sqrt{2}) + c_3 \phi((\sqrt[4]{2})^3) = c_0 + c_1 \sqrt[4]{2} + c_2 \sqrt{2} + c_3 (\sqrt[4]{2})^3 \ \text{ iff}$$

$$c_0 - c_1 \sqrt[4]{2} + c_2 \sqrt{2} - c_3 \sqrt[4]{2}^3 = c_0 + c_1 \sqrt[4]{2} + c_2 \sqrt{2} + c_3 (\sqrt[4]{2})^3 \ \text{ iff}$$

$c_1 = -c_1$ and $c_3 = -c_3$, i.e. $c_1 = c_3 = 0$. Hence, the elements of E fixed by ϕ look like $c_0 + c_2 \sqrt{2}$, which is precisely the elements of $\mathbb{Q}(\sqrt{2})$. In other words, $E^G = \mathbb{Q}(\sqrt{2})$.

2. *Recall our work investigating the Galois group of $E = \mathbb{Q}(\sqrt{2}, \sqrt{3})$ over $F = \mathbb{Q}$. We described a typical element of E as having the form $a+b\sqrt{2}+c\sqrt{3}+d\sqrt{6}$ and we found $Gal(E/F)$ to be the Klein-4 group having the following four elements:*

$$a + b\sqrt{2} + c\sqrt{3} + d\sqrt{6} \mapsto a + b\sqrt{2} + c\sqrt{3} + d\sqrt{6},$$
$$a + b\sqrt{2} + c\sqrt{3} + d\sqrt{6} \mapsto a - b\sqrt{2} + c\sqrt{3} - d\sqrt{6},$$
$$a + b\sqrt{2} + c\sqrt{3} + d\sqrt{6} \mapsto a + b\sqrt{2} - c\sqrt{3} - d\sqrt{6},$$
$$a + b\sqrt{2} + c\sqrt{3} + d\sqrt{6} \mapsto a - b\sqrt{2} - c\sqrt{3} + d\sqrt{6}.$$

Let's name these homomorphisms as 1, ϕ_1, ϕ_2 and ϕ_3, respectively. Consider the following subgroups of $Gal(E/F)$: $G_0 = \{1\}$, $G_1 = \{1, \phi_1\}$, $G_2 = \{1, \phi_2\}$ and $G = Gal(E/F)$. First, let's compute E^{G_1} which is dependent on what ϕ_1 fixes. Notice that

$$\phi_1(a + b\sqrt{2} + c\sqrt{3} + d\sqrt{6}) = a + b\sqrt{2} + c\sqrt{3} + d\sqrt{6} \;\; iff$$
$$a - b\sqrt{2} + c\sqrt{3} - d\sqrt{6} = a + b\sqrt{2} + c\sqrt{3} + d\sqrt{6} \;\; iff$$

$b = d = 0$ and so $E^{G_1} = \mathbb{Q}(\sqrt{3})$. In a similar way we get that $E^{G_2} = \mathbb{Q}(\sqrt{2})$. Furthermore, since elements of E^G must be fixed by both ϕ_1 and ϕ_2 we get that $E^G = F$. Finally, E^{G_0} is certainly E.

3. *Refer to Example 10.9 where $F = \mathbb{Z}_p(t)$ with t transcendental over $\mathbb{Z}_p$ and $E = F(a)$ where $a \in \overline{F} \setminus F$ was the only zero of $x^p - t$. We saw that $[E : F]_s = 1$ and so $|Gal(E/F)| = 1$ which means $Gal(E/F)$ has only the identity homomorphism. Thus, if we set $G = Gal(E/F)$, then $E^G = E$, the whole field.*

Amazingly enough, there is a connection between the $Gal(E/F)$-invariant subfields of E and the concepts of normality and separability. Before we can state this result we need another definition.

Definition 10.11 *Let $F \subseteq E$ be fields and set $G = Gal(E/F)$. The field E is said to be **Galois over** F if $E^G = F$. We also say that E is a **Galois extension** of F.*

Example 10.11 *Referring to the previous three examples just presented,*

1. *$E = \mathbb{Q}(\sqrt[4]{2})$ is not Galois over $F = \mathbb{Q}$, since $E^G = \mathbb{Q}(\sqrt{2}) \neq F$.*

2. *$E = \mathbb{Q}(\sqrt{2}, \sqrt{3})$ is indeed Galois over $F = \mathbb{Q}$, since $E^G = F$.*

3. *$E = \mathbb{Z}_p(t)(a)$ is not Galois over $F = \mathbb{Z}_p(t)$, since $E^G = E \neq F$.*

So the question is what makes some field extension Galois while others are not. The following theorem, whose proof we present in the next section, gives the reason in the nutshell:

Theorem 10.7 *Let $F \subseteq E$ be fields with $[E : F] < \infty$. The field E is Galois over F iff E is both normal and separable over F.*

Example 10.12 *Referring again to the previous three examples,*

1. *The reason why $E = \mathbb{Q}(\sqrt[4]{2})$ fails to be Galois over $F = \mathbb{Q}$ is that E is not normal over F. Indeed, not every zero of $x^4 - 2$ is contained in E as Theorem 10.5.3 requires.*

2. *The above theorem is confirmed by the fact that $E = \mathbb{Q}(\sqrt{2}, \sqrt{3})$ is Galois over $F = \mathbb{Q}$. Indeed, since $|Gal(E/F)| = 4 = [E : F]$, by Corollary 10.3, this implies that E is both normal and separable over F*

3. *The reason why $E = \mathbb{Z}_p(t)(a)$ fails to be Galois over $F = \mathbb{Z}_p(t)$ is that E is not separable over F. Indeed, we computed $[E : F]_s = 1 \neq p = [E : F]$ which by Theorem 10.6 shows that E is not separable over F.*

EXERCISES

1 Let $E = \mathbb{Q}(\sqrt[3]{2}, \zeta)$ where $\zeta = -\frac{1}{2} + \frac{\sqrt{3}}{2}i$. Let $F = \mathbb{Q}$ and $G = Gal(E/F)$.

   a. Verify that $\zeta^3 = 1$ and the three roots of $x^3 - 2$ are $\sqrt[3]{2}$, $\zeta\sqrt[3]{2}$ and $\zeta^2\sqrt[3]{2}$.

   b. Find an irreducible polynomial having ζ as a root.

   c. Compute $[E : F]$ and use this to put a bound on the number of elements in G.

   d. Now compute the size of G by listing an associated permutation for each element of G.

   e. Explain why G is **not** abelian.

   f. Give a nice description of the elements of E as F-linear combinations of a specific basis.

   g. Compute the invariant subfield E^H for any subgroup H of order two of your choosing.

2 Verify Remark 10.4.1.

10.7   SOME PRELIMINARY THEOREMS

There are two results to prove in this section. The second result is Artin's Lemma and has a surprisingly simple proof based on elementary ideas from linear algebra. But first, as promised, we provide the proof of Theorem 10.7 mentioned in the Section 10.6.

Proof 10.12 *First assume that E is both normal and separable over F with $[E : F] < \infty$. Set $G = Gal(E/F)$ and $K = E^G$ and we will show that $K = F$ and so E is Galois over F. To see this, note that by Theorem 10.5.4, E is the splitting field over F of some polynomial $f(x) \in F[x]$ and by Corollary 10.3, $|Gal(E/F)| = [E : F]$. Now since $F \subseteq K$, it is also the case that E is the splitting field over K of the same $f(x) \in K[x]$. Again, by Theorem 10.5, E is normal over K as well.*

Since E is separable over F it is separable over the intermediate field K as well (the justification is identical to a portion of second half of the proof of Theorem 10.6). Now, by Corollary 10.3, we have that $|\text{Gal}(E/K)| = [E : K]$. Observe since $K = E^G$ it is the case that $\text{Gal}(E/F) = \text{Gal}(E/K)$ and so $[E : F] = [E : K]$. But this implies that $[K : F] = 1$ and so $K = F$.

Now we assume $E^G = F$ with $[E : F] < \infty$. Take any $a \in E$ and let $p(x) \in F[x]$ be the irreducible polynomial of a over F. To show that E is both normal and separable over F, it is enough to show that $p(x)$ has all its zeros in E and they are all simple zeros. Let

$$G = \{\phi_1, \phi_2, \ldots, \phi_n\} \quad \text{and set} \quad X = \{\phi_i(a) \mid 1 \leq i \leq n\},$$

which consists of the zeros of $p(x)$ contained in E. Let's enumerate $X = \{a_1, a_2, \ldots, a_k\}$ where $k \leq n$. Note that $a \in X$, since the identity homomorphism is in G. Set $q(x) = \prod_{i=1}^{k}(x - a_i) \in E[x]$. Certainly, by how $q(x)$ is defined, it has only simple zeros with all its zeros in E, so in order to show that $p(x)$ has all its zeros in E and they are all simple zeros, it's enough to show that $p(x)$ and $q(x)$ differ only by a constant in F. To prove this we will show that $q(x) \in F[x]$, for if this were so, then $p|q$ (Lemma 9.2.1) and since $q|p$ and $p(x)$ is irreducible, we would then have that $p(x)$ and $q(x)$ differ only by a constant in F. Now in order to show that $q(x) \in F[x]$, we set $q(x) = b_0 + b_1 x + \cdots + b_k x^k$ and show that each $b_i \in E^G$ (which by assumption implies each $b_i \in F$ and the result is proved). To this end, take any $\phi_j \in G$. On the one hand,

$$\phi_j(q(x)) = \phi_j(b_0) + \phi_j(b_1)x + \cdots + \phi_j(b_k)x^k.$$

On the other hand,

$$\phi_j(q(x)) = \prod_{i=1}^{k}(x - \phi_j(a_i)),$$

which follows from the definition of polynomial multiplication and properties of a field homomorphism. Since, by Theorem 10.3, each $\phi_j \in G$ defines a permutation of the elements of X, we have that

$$\phi_j(q(x)) = \prod_{i=1}^{k}(x - \phi_j(a_i)) = \prod_{i=1}^{k}(x - a_i) = q(x).$$

But then

$$\phi_j(b_0) + \phi_j(b_1)x + \cdots + \phi_j(b_k)x^k = b_0 + b_1 x + \cdots + b_k x^k,$$

which implies that $\phi_j(b_i) = b_i$ for $i = 0, 1, \ldots, k$. Since ϕ_j was chosen arbitrarily in G it follows then that each $b_i \in E^G$.

Lemma 10.4 (Artin) *Let $F \subseteq E$ be fields with $[E : F] < \infty$. If $H \leq \text{Gal}(E/F)$ such that $E^H = F$, then $[E : F] \leq |H|$.*

Proof 10.13 *Set* $H = \{\phi_1, \phi_2, \ldots, \phi_n\}$ *with* ϕ_1 *being the identity homomorphism on* E. *Let* $a_1, a_2, \ldots, a_m$ *be a basis for* E *over* F *and suppose, to the contrary, that* $m > n$. *Consider the following homogeneous system of equations:*

$$\phi_1(a_1)x_1 + \phi_1(a_2)x_2 + \cdots + \phi_1(a_m)x_m = 0$$
$$\phi_2(a_1)x_1 + \phi_2(a_2)x_2 + \cdots + \phi_2(a_m)x_m = 0$$
$$\vdots$$
$$\phi_n(a_1)x_1 + \phi_n(a_2)x_2 + \cdots + \phi_n(a_m)x_m = 0$$

Since $m > n$, *the system of equations has non-trivial solutions in* E^m. *Let* $(b_1, b_2, \ldots, b_m) \in E^m$ *be a non-trivial solution containing a maximal number of zeros. Without loss of generality we can assume that* $b_1 \neq 0$ *(by simply reordering the* ϕ_i's*). Furthermore, without loss of generality, we can assume that* $b_1 = 1$, *since* $(b_1^{-1}b_1, b_1^{-1}b_2, \ldots, b_1^{-1}b_m)$ *is also a solution to the system.*

Claim 10.2 *Each* $b_i \in E^H$.

Suppose, to the contrary, the Claim were false. Without loss of generality, we can assume $b_2 \notin E^H$ *(again, by reordering) so that* $\phi_j(b_2) \neq b_2$ *for some* j. *Since* $(1, b_2, \ldots, b_m)$ *is a solution to the system, we have*

$$\phi_1(a_1) + \phi_1(a_2)b_2 + \cdots + \phi_1(a_m)b_m = 0$$
$$\phi_2(a_1) + \phi_2(a_2)b_2 + \cdots + \phi_2(a_m)b_m = 0$$
$$\vdots$$
$$\phi_n(a_1) + \phi_n(a_2)b_2 + \cdots + \phi_n(a_m)b_m = 0$$

Now apply that particular ϕ_j *to each equation in the system to get*

$$\phi_j(\phi_1(a_1)) + \phi_j(\phi_1(a_2))\phi_j(b_2) + \cdots + \phi_j(\phi_1(a_m))\phi_j(b_m) = 0$$
$$\phi_j(\phi_2(a_1)) + \phi_j(\phi_2(a_2))\phi_j(b_2) + \cdots + \phi_j(\phi_2(a_m))\phi_j(b_m) = 0$$
$$\vdots$$
$$\phi_j(\phi_n(a_1)) + \phi_j(\phi_n(a_2))\phi_j(b_2) + \cdots + \phi_j(\phi_n(a_m))\phi_j(b_m) = 0$$

Since, by Cayley's Theorem, ϕ_j *permutes the elements of* H, *by reordering the equations we can obtain*

$$\phi_1(a_1) + \phi_1(a_2)\phi_j(b_2) + \cdots + \phi_1(a_m)\phi_j(b_m) = 0$$
$$\phi_2(a_1) + \phi_2(a_2)\phi_j(b_2) + \cdots + \phi_2(a_m)\phi_j(b_m) = 0$$
$$\vdots$$
$$\phi_n(a_1) + \phi_n(a_2)\phi_j(b_2) + \cdots + \phi_n(a_m)\phi_j(b_m) = 0$$

This implies that $(1, \phi_j(b_2), \ldots, \phi_j(b_m))$ *is another solution to the original system of equations. But then the difference,*

$$(1, b_2, \ldots, b_m) - (1, \phi_j(b_2), \ldots, \phi_j(b_m)) = (0, b_2 - \phi_j(b_2), \ldots, b_m - \phi_j(b_m))$$

is yet another solution to the system. Notice, however, that this last solution is a non-trivial solution (since $\phi_j(b_2) \neq b_2$) with a greater number of zeros than $(1, b_2, \ldots, b_m)$, a contradiction. Hence, the Claim is proved.

The Claim together with the assumption that $E^H = F$ implies that the b_i's are, in fact, in F. Now observe, in particular, that $(b_1, b_2, \ldots, b_m)$ satisfying the first equation of the system (recall that ϕ_1 is the identity homomorphism on E) yields the equation $b_1 a_1 + b_2 a_2 + \cdots + b_m a_m = 0$. This is a non-trivial F-linear combination of the basis $a_1, a_2, \ldots, a_m$ of E over F equaling zero, a glaring contradiction.

EXERCISES

1 Verify in the proof of Theorem 10.7 that since E is separable over F it is separable over the intermediate field K as well.

2 Consider Exercise 1 in Section 10.6.

    a. Determine a subgroup $H \leq Gal(E/F)$ for which $E^H = F$.

    b. Confirm the conclusion of Artin's Lemma 10.4.

10.8   THE FUNDAMENTAL THEOREM OF GALOIS THEORY

We have already seen a link between groups and fields in regards to the Galois group $Gal(E/F)$ and the degree $[E : F]$. However, the link goes further and is more profound – fundamental even.

Theorem 10.8 (The Fundamental Theorem of Galois Theory) *Let $F \subseteq E$ be fields with $[E : F] < \infty$ and E Galois over F. The following are true:*

1. *There is a one-to-one inclusion reversing correspondence between intermediate subfields of E over F and subgroups of $Gal(E/F)$ given by the map $K \mapsto Gal(E/K)$ and its inverse $H \mapsto E^H$ in the sense that*

    (a) If $F \subseteq K_1 \subseteq K_2 \subseteq E$, then $Gal(E/K_2) \leq Gal(E/K_1)$

    (b) If $H_1 \leq H_2 \leq Gal(E/F)$, then $F \subseteq E^{H_2} \subseteq E^{H_1} \subseteq E$.

2. *If K is an intermediate subfield of E over F, then*

$$[E : K] = |Gal(E/K)| \qquad and \qquad [K : F] = [Gal(E/F) : Gal(E/K)].$$

3. *If $H \leq Gal(E/F)$, then*

$$[E : E^H] = |H| \qquad and \qquad [E^H : F] = [Gal(E/F) : H].$$

4. *For K an intermediate subfield of E over F, we have that K is normal over F iff $Gal(E/K) \triangleleft Gal(E/F)$ and in this case*

$$Gal(K/F) \cong Gal(E/F)/Gal(E/K).$$

5. For $H \leq Gal(E/F)$, we have $H \lhd Gal(E/F)$ iff E^H is normal over F and in this case

$$Gal(E^H/F) \cong Gal(E/F)/Gal(E/E^H).$$

Before we give the proof of this result, let's first illustrate the theorem with an example.

Example 10.13 Set $E = \mathbb{Q}(\sqrt[4]{2}, i)$ and $F = \mathbb{Q}$. First note that any element of $Gal(E/F)$ is completely determined by where it sends the generators $\sqrt[4]{2}$ and i. Second, note that any element of $Gal(E/F)$ must send $\sqrt[4]{2}$ to another zero of $x^4 - 2$, namely $\pm\sqrt[4]{2}$ and $\pm i\sqrt[4]{2}$. Third, note that any element of $Gal(E/F)$ must send i to another zero of $x^2 + 1$, namely $\pm i$. Therefore, there is an element of $Gal(E/F)$ corresponding to each of the following permutations:

$$\begin{pmatrix} \sqrt[4]{2} & -\sqrt[4]{2} & i\sqrt[4]{2} & -i\sqrt[4]{2} & i & -i \\ \sqrt[4]{2} & -\sqrt[4]{2} & i\sqrt[4]{2} & -i\sqrt[4]{2} & i & -i \end{pmatrix} \quad \begin{pmatrix} \sqrt[4]{2} & -\sqrt[4]{2} & i\sqrt[4]{2} & -i\sqrt[4]{2} & i & -i \\ -\sqrt[4]{2} & \sqrt[4]{2} & -i\sqrt[4]{2} & i\sqrt[4]{2} & i & -i \end{pmatrix}$$

$$\begin{pmatrix} \sqrt[4]{2} & -\sqrt[4]{2} & i\sqrt[4]{2} & -i\sqrt[4]{2} & i & -i \\ \sqrt[4]{2} & -\sqrt[4]{2} & -i\sqrt[4]{2} & i\sqrt[4]{2} & -i & i \end{pmatrix} \quad \begin{pmatrix} \sqrt[4]{2} & -\sqrt[4]{2} & i\sqrt[4]{2} & -i\sqrt[4]{2} & i & -i \\ -\sqrt[4]{2} & \sqrt[4]{2} & i\sqrt[4]{2} & -i\sqrt[4]{2} & -i & i \end{pmatrix}$$

$$\begin{pmatrix} \sqrt[4]{2} & -\sqrt[4]{2} & i\sqrt[4]{2} & -i\sqrt[4]{2} & i & -i \\ i\sqrt[4]{2} & -i\sqrt[4]{2} & -\sqrt[4]{2} & \sqrt[4]{2} & i & -i \end{pmatrix} \quad \begin{pmatrix} \sqrt[4]{2} & -\sqrt[4]{2} & i\sqrt[4]{2} & -i\sqrt[4]{2} & i & -i \\ i\sqrt[4]{2} & -i\sqrt[4]{2} & \sqrt[4]{2} & -\sqrt[4]{2} & -i & i \end{pmatrix}$$

$$\begin{pmatrix} \sqrt[4]{2} & -\sqrt[4]{2} & i\sqrt[4]{2} & -i\sqrt[4]{2} & i & -i \\ -i\sqrt[4]{2} & i\sqrt[4]{2} & \sqrt[4]{2} & -\sqrt[4]{2} & i & -i \end{pmatrix} \quad \begin{pmatrix} \sqrt[4]{2} & -\sqrt[4]{2} & i\sqrt[4]{2} & -i\sqrt[4]{2} & i & -i \\ -i\sqrt[4]{2} & i\sqrt[4]{2} & -\sqrt[4]{2} & \sqrt[4]{2} & -i & i \end{pmatrix}$$

So at this point we know that $|Gal(E/F)| \geq 8$. Now we compute the value of $[E : F]$. Consider the following chain of fields: $F \subseteq \mathbb{Q}(\sqrt[4]{2}) \subseteq E$. Since $x^4 - 2$ is irreducible over $\mathbb{Q}$ (Eisenstein) and has $\sqrt[4]{2}$ as a zero, we get that $[\mathbb{Q}(\sqrt[4]{2}) : F] = 4$. Since $x^2 + 1$ is irreducible over $\mathbb{Q}(\sqrt[4]{2})$ (else $i \in \mathbb{Q}(\sqrt[4]{2}) \subset \mathbb{R}$, a contradiction) and has i as a zero, we get that $[E : \mathbb{Q}(\sqrt[4]{2})] = 2$ and so $[E : F] = (4)(2) = 8$. Now since $|Gal(E/F)| \leq [E : F] = 8$ and by the above work we get that $|Gal(E/F)| = 8 = [E : F]$. Hence, we know all the elements of $Gal(E/F)$ and we also know that E is Galois over F which puts us in the context of the Fundamental Theorem of Galois Theory.

Let's now decide which group of order eight the Galois group is. There are five groups of order eight (up to isomorphism): there are the abelian ones $\mathbb{Z}_8$, $\mathbb{Z}_4 \oplus \mathbb{Z}_2$, $\mathbb{Z}_2 \oplus \mathbb{Z}_2 \oplus \mathbb{Z}_2$ and there are the non-abelian dihedral group and the quaternions. Consider the eight permutations listed earlier. The fifth such, call it ϕ, has order four and the third such, call it ψ, has order two. One can check (exercise) that $\phi\psi \neq \psi\phi$ and so the Galois group is non-abelian. In fact, one can check that ϕ and ψ generate

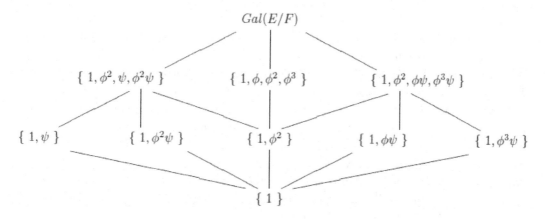

Figure 10.1 The lattice of subgroups for $Gal(E/F)$, where $E = \mathbb{Q}(\sqrt[4]{2}, i)$ and $F = \mathbb{Q}$.

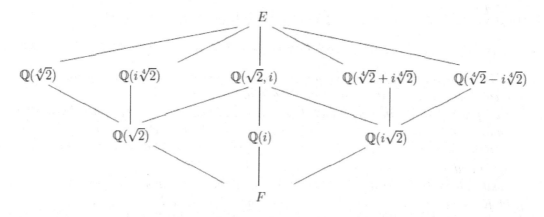

Figure 10.2 The lattice of intermediate fields for $F \subseteq E$, where $E = \mathbb{Q}(\sqrt[4]{2}, i)$ and $F = \mathbb{Q}$.

the Galois group and satisfy the relation $\psi\phi = \phi^3\psi$ which is precisely the presentation for the dihedral group. In Figure 10.1, we have the lattice of subgroups for $Gal(E/F)$.

A basis for E over F is 1, $\sqrt[4]{2}$, $\sqrt{2}$, $(\sqrt[4]{2})^3$, i, $i\sqrt[4]{2}$, $i\sqrt{2}$, $i(\sqrt[4]{2})^3$ so that every element of E can be expressed as

$$c_1 + c_2\sqrt[4]{2} + c_3\sqrt{2} + c_4(\sqrt[4]{2})^3 + c_5 i + c_6 i\sqrt[4]{2} + c_7 i\sqrt{2} + c_8 i(\sqrt[4]{2})^3,$$

where $c_1, c_2, c_3, c_4, c_5, c_6, c_7, c_8 \in F$. Using this representation one can proceed to compute the invariant subfields of E corresponding to each subgroup in the lattice of $Gal(E/F)$ presented above. Such work would yield the lattice of intermediate subfields of E over F presented in Figure 10.2.

Notice the inverted symmetry of the two lattices as predicted in the Fundamental Theorem of Galois Theory.

We now give the proof of the Fundamental Theorem of Galois Theory.

Proof 10.14 *To show the first statement, it is enough to show the composition of the two maps $K \mapsto Gal(E/K)$ and $H \mapsto E^H$ (and vice-versa) is the identity. First, for K an intermediate field of E over F, to show that $E^{Gal(E/K)} = K$ it is enough to show that E is Galois over K. To prove this, by Theorem 10.7, we show that E is both normal and separable over K. To see that E is normal over K, consider a K-homomorphism $\phi : E \to \overline{E}$. Since ϕ is also an F-homomorphism and E is normal over F, we have that $\phi(E) \subseteq E$ and this shows that E is normal over K. Just as in the proof of Theorem 10.7, E being separable over F implies E is separable over any intermediate field K. Second, we show that $Gal(E/E^H) = H$. Certainly, $Gal(E/E^H) \supseteq H$ (by the definition of $Gal(E/E^H)$). For the reverse inclusion, by Artin's Lemma, $[E : E^H] \leq |H|$ and by Corollary 10.2, $|Gal(E/E^H)| \leq [E : E^H]$. Hence, $|Gal(E/E^H)| \leq |H|$ and so $Gal(E/E^H) = H$.*

To prove second statement, as we saw in the first statement, E is both normal and separable over K so that by Corollary 10.3 it follows that $|Gal(E/K)| = [E : K]$. To show that $[K : F] = [Gal(E/F) : Gal(E/K)]$, using Lagrange's Theorem,

$$[K : F] = [E : F]/[E : K] = |Gal(E/F)|/|Gal(E/K)| = [Gal(E/F) : Gal(E/K)].$$

The third statement follows immediately from second statement and the one-to-one correspondence established in first statement.

To prove the fourth statement, first assume that some intermediate field K is normal over F. Take $\phi \in Gal(E/F)$ and $\psi \in Gal(E/K)$. By assumption, for any $a \in K$ we have that $\phi(a) \in K$ and so $\psi(\phi(a)) = \phi(a)$. But then $\phi^{-1}\psi\phi(a) = a$ which implies that $\phi^{-1}\psi\phi \in Gal(E/K)$ and so $Gal(E/K) \lhd Gal(E/F)$. For the opposite direction we prove the contrapositive statement. Suppose some intermediate field K is not normal over F. Then for some $\phi \in Gal(E/F)$ and some $a \in K$ we have that $\phi(a) \notin K$. By the first statement, $K = E^{Gal(E/K)}$, so there exists a $\psi \in Gal(E/K)$ such that $\psi(\phi(a)) \neq \phi(a)$ which implies that $\phi^{-1}\psi\phi(a) \neq a$ and so $\phi^{-1}\psi\phi \notin Gal(E/K)$. Therefore, it is not the case that $Gal(E/K) \lhd Gal(E/F)$.

To establish the isomorphism, consider the map $\Psi : Gal(E/F) \to Gal(K/F)$ by $\Psi(\phi) = \phi|K$, the restriction of ϕ to K. Since K is normal over F this function Φ maps into $Gal(K/F)$, because $\phi|K$ will map back into K. It is easy to check (exercise) that Φ is an epimorphism with kernel equaling $Gal(E/K)$, hence by the Fundamental Theorem of Homomorphisms the result follows.

The fifth statement follows immediately from the fourth statement and the one-to-one correspondence established in first statement.

Example 10.14 *Here are some additional examples illustrating the Fundamental Theorem of Galois Theory.*

1. *We have already investigated the case when $E = \mathbb{Q}(\sqrt{2}, \sqrt{3})$ and $F = \mathbb{Q}$ in Example 10.10.2. We found that E was Galois over F. We computed the Galois group of E over F to be the Klein-4 group. The intermediate subfields of E over F are $\mathbb{Q}(\sqrt{2})$, $\mathbb{Q}(\sqrt{3})$ and $\mathbb{Q}(\sqrt{6})$ and these were each found to be invariant subfields of E via the three subgroups of order 2 in $Gal(E/F)$. When you write out the lattice of subgroups and the lattice of intermediate subfields you obtain identically shaped lattices.*

2. We have also looked at $E = \mathbb{Q}(\sqrt[4]{2})$ over $F = \mathbb{Q}$. The Galois group of E over F was found to be $\mathbb{Z}_2$ and so its lattice is of the simplest type having no proper non-trivial subgroups. However, the lattice of intermediate subfields of E over F has a strictly intermediary subfield, namely $\mathbb{Q}(\sqrt{2})$, so the two lattices do not have the inverted symmetry. This is because E is not normal over F. Note that $\mathbb{Q}(\sqrt{2}) = E^G$ where $G = Gal(E/F)$.

3. Recall the example of $F = \mathbb{Z}_p(t)$ where t is transcendental over F and $E = F(a)$ where a is a root of $x^p - a = (x - a)^p$. Set $p = 2$, $L = F(a, i)$ and consider the Galois group of L over F. Any F-homomorphism in $G = Gal(L/F)$ must fix a and send i to $\pm i$. Therefore, $G \cong \mathbb{Z}_2$ and its lattice is again of the simplest type. However, the lattice of intermediate subfields of L over F includes three strictly intermediate subfields: $E = F(a)$, $F(i)$ and $F(ai)$ and once again we do not have the inverted symmetry of the two lattices. This is because E is not separable over F. Notice that of the three subfields only E is an invariant subfield, since $E = L^G$. Indeed, since

$$L = \{c_0 + c_1 a + c_2 i + c_3 ai \ : \ c_i \in F\},$$

and the only non-trivial map in G is the one sending i to $-i$, then

$$L^G = \{c_0 + c_1 a \ : \ c_i \in F\} = F(a) = E.$$

EXERCISES

1 Write out the lattice of subgroups and the lattice of intermediate subfields for Exercise 1 in Section 10.6,

2 In Example 10.13, verify that $\phi\psi \neq \psi\phi$.

3 In Theorem 10.8.4, verify that Φ is an epimorphism with kernel equaling $Gal(E/K)$.

4 Write out the lattice of subgroups and the lattice of intermediate subfields for Example 10.14.1.

10.9   SOLVABLE GROUP ESSENTIALS

We spent a whole chapter on solvable groups, but for the reader who prefers to cover only as much of the topic as needed to see the proof that there is no quintic formula, we present this material here in a self-contained manner. For those who have covered the chapter on solvable and nilpotent groups, you may wish to simply skim this section since most of it you have already seen.

Definition 10.12 *A group G is called* **solvable** *if it has a series of subgroups $1 = G_0 \triangleleft G_1 \triangleleft \cdots \triangleleft G_n = G$ with each G_{i+1}/G_i abelian.*

Example 10.15 *Here, we list several examples of solvable groups.*

1. *Every abelian group G is solvable with series $1 \lhd G$.*

2. *S_3 is solvable with series $1 \lhd A_3 \lhd S_3$, since $A_3 \cong \mathbb{Z}_3$ and $S_3 \cong \mathbb{Z}_2$.*

3. *S_4 is solvable with series $1 \lhd N \lhd A_4 \lhd S_4$ where*

$$N = \{1, (1\ 2)(3\ 4), (1\ 3)(2\ 4), (1\ 4)(2\ 3)\}.$$

Note that $|N| = 4$ and all groups of order 4 are abelian.

Our primary goal in this section is to show that S_n is not solvable for $n \geq 5$.

Proposition 10.1 *In the case that G is a finite group, G is solvable iff each G_{i+1}/G_i in the definition of solvable has prime order.*

Proof 10.15 *One direction is immediate, since a group of prime order is isomorphic to $\mathbb{Z}_p$ which is abelian. For the other direction, suppose that G is solvable. For each factor group G_{i+1}/G_i which does not have prime order we refine the series as follows: Suppose that p is a prime dividing the order of G_{i+1}/G_i. By Cauchy's Lemma, G_{i+1}/G_i has an element of order p, say gG_i. Since G_{i+1}/G_i is abelian, the subgroup $\langle gG_i \rangle$ of order p is normal in G_{i+1}/G_i. Now, by the Correspondence Theorem, there exists a subgroup $H \lhd G_{i+1}$ containing G_i such that $\langle gG_i \rangle = H/G_i$. Note that $G_i \lhd H$, since $G_i \lhd G_{i+1}$. Repeat this process on the smaller abelian factor group G_{i+1}/H. In this way and in a finite number of steps we can replace $G_i \lhd G_{i+1}$ in the series by a subseries where each factor group is of prime order (note that we are implicitly using the Third Isomorphism Theorem for groups).*

Definition 10.13 *Let X be a non-empty subset of a group G. The **subgroup generated by** X, written $\langle X \rangle$ is the collection of all finite products of elements of X and their inverses. The set X is called the **generating set** of $\langle X \rangle$.*

If $X = \{g_1, g_2, \ldots, g_n\}$ a finite set, then we write $\langle g_1, g_2, \ldots, g_n \rangle$ for $\langle X \rangle$. Note that if $X = \{g\}$, then $\langle X \rangle$ is simply the cyclic subgroup generated by g. One needs to check, of course, that $\langle X \rangle$ is indeed a subgroup of G. Furthermore, one can show that $\langle X \rangle$ is the smallest subgroup of G containing the set X.

Example 10.16 *Here are some examples of groups and their generating sets.*

1. *The Klein-4 group $V = \{e, a, b, c\}$ is generated by $X = \{a, b\}$.*

2. *The quaternions are generated by the set $X = \{i, j\}$ or $\{i, k\}$ or $\{j, k\}$.*

3. *The dihedral group D_4 (rotations and reflections of a square) is generated by any single (non-trivial) rotation and any single reflection.*

Definition 10.14 *Let G be a group.*

1. *For $g, h \in G$, the **commutator of g and h**, written*

$$[g, h] = g^{-1}h^{-1}gh.$$

2. Let X and Y be two non-empty subsets of G. The **commutator subgroup of X and Y**, written

$$[X, Y] = \langle\ [x, y]\ :\ x \in X,\ y \in Y\ \rangle,$$

the subgroup generated by the commutators $[x, y]$.

3. For a group G, the **derived subgroup of G** (or sometimes called the **commutator subgroup of G**), written $G' = [G, G]$.

Remark 10.5 Since for all $g, h \in G$ we have $[g, h]^{-1} = [h, g]$, it follows that G' is, in fact, the collection of all finite products of commutators in G (no need for their inverses). Another easy fact to verify is that G' is normal in G. We prove some additional properties of G' in the theorem which follows.

Theorem 10.9 Let G and K be groups and H be a subgroup of G.

1. If $\phi : G \to K$ a homomorphism, then $\phi(G') \leq K'$

2. $G' \leq H$ iff $H \lhd G$ and G/H is abelian.

3. G' is the smallest normal subgroup of G which will form an abelian factor group.

Proof 10.16 For the first statement, take any $[g_1, h_1] \cdots [g_n, h_n] \in G'$. By properties of a homomorphism,

$$\phi([g_1, h_1] \cdots [g_n, h_n]) = \phi([g_1, h_1]) \cdots \phi([g_n, h_n]) = [\phi(g_1), \phi(h_1)] \cdots [\phi(g_n), \phi(h_n)] \in K'.$$

For the second statement, first assume that $G' \leq H$. Then for all $g \in G$ and $h \in H$, we have $g^{-1}hg = h(h^{-1}g^{-1}hg) = h[h, g] \in H$ and so $H \lhd G$. To see that G/H is abelian, notice that for $g_1, g_2 \in G$,

$$g_1 H g_2 H = g_2 H g_1 H g_1^{-1} H g_2^{-1} H g_1 H g_2 H = g_2 H g_1 H [g_1, g_2] H = g_2 H g_1 H.$$

Now assume that $H \lhd G$ and G/H is abelian. Then for all $g_1, g_2 \in G$, we have

$$g_1^{-1} H g_2^{-1} H g_1 H g_2 H = g_1^{-1} H g_1 H g_2^{-1} H g_2 H = H,$$

and so $[g_1, g_2]H = H$. Hence, $[g_1, g_2] \in H$ and so by Remark 10.5, it follows that $G' \leq H$.

The third statement is simply a summary of the first two statements and requires no additional proof.

Definition 10.15 For any natural number n, the **nth derived subgroup**, written $G^{(n)}$, is defined recursively as follows: $G^{(0)} = G$, $G^{(1)} = G'$ and $G^{(n+1)} = (G^{(n)})'$ for $n \geq 1$.

The series, $G = G^{(0)} \geq G^{(1)} \geq G^{(2)} \geq \cdots$ is called the **derived series** of G.

The following facts about derived series will be left as exercises:

1. If the derived series of G terminates at 1 at some point, i.e. there exists a natural number n such that $G^{(n)} = 1$, then G is solvable.

2. If $H \leq G$, then $H^{(n)} \leq G^{(n)}$ for any natural number n.

Proposition 10.2 *G is solvable iff there exists a natural number n such that $G^{(n)} = 1$.*

Proof 10.17 *One direction was given as an exercise above. For the other direction assume that G is solvable. Then G has a series $1 = G_0 \lhd G_1 \lhd \cdots \lhd G_n = G$ with each G_{i+1}/G_i abelian. We show by induction on i that $G^{(i)} \subseteq G_{n-i}$ for $i = 0, 1, \ldots, n$. For $i = 0$, certainly $G = G^{(0)} \subseteq G_{n-0} = G$. For $0 < i \leq n$, Since G_{n-i+1}/G_{n-i} is abelian, by Theorem 10.9, $(G_{n-i+1})' \subseteq G_{n-i}$. Now, by induction, $G^{(i-1)} \subseteq G_{n-(i-1)}$. Therefore, by the exercise above*

$$G^{(i)} = (G^{(i-1)})' \subseteq (G_{n-i+1})' \subseteq G_{n-i}.$$

In particular, $G^{(n)} \subseteq G_{n-n} = G_0 = 1$ and the result follows.

Theorem 10.10 *If G is a solvable group, then*

1. *For each $H \leq G$, we have that H is also solvable.*

2. *For each $N \lhd G$, we have that G/N is also solvable.*

Proof 10.18 *Since G is solvable it has an abelian series, i.e. there is a subnormal series $1 = G_0 \leq G_1 \leq \cdots \leq G_n = G$ with each G_{i+1}/G_i abelian. To prove the first statement, consider the series*

$$1 = H \cap G_0 \leq H \cap G_1 \leq \cdots \leq H \cap G_n = H.$$

One can easily verify that the series is subnormal and by the Second Isomorphism Theorem, each

$$H \cap G_{i+1}/H \cap G_i \cong (H \cap G_{i+1})G_i/G_i \leq G_{i+1}/G_i.$$

In other words, $H \cap G_{i+1}/H \cap G_1$ is isomorphic to a subgroup of an abelian group and is therefore abelian as well. Hence, $1 = H \cap G_0 \leq H \cap G_1 \leq \cdots \leq H \cap G_n = H$ is an abelian series and so H is solvable.

To prove the second statement, consider the series

$$\{N\} = G_0 \leq G_1 N/N \leq \cdots \leq G_n N/N = G/N.$$

By the Third Isomorphism Theorem (part 1), each $G_i N/N \lhd G_{i+1} N/N$, since $G_i N$ and N are both normal in $G_{i+1} N$ (check). By the Third Isomorphism Theorem (part 2),

$$(G_{i+1} N/N)/(G_i N/N) \cong G_{i+1} N/G_i N \cong G_{i+1}/G_i,$$

which we know to be abelian. Hence, $\{N\} = G_0 \leq G_1 N/N \leq \cdots \leq G_n N/N = G/N$ is an abelian series for G/N and so G/N is solvable.

In preparation for showing that S_n is not solvable for $n \geq 5$ we need the following lemma:

Lemma 10.5 *If $n \geq 3$ and $\sigma \in A_n$, then σ can be expressed as a product of 3-cycles.*

Proof 10.19 *It's enough to show any pair of transpositions, $\tau\tau'$, can be written as a product of 3-cycles. There are three cases to consider. If $\tau\tau' = (a\ b)(a\ b)$, then $\tau\tau' = (a\ b\ c)(a\ b\ c)(a\ b\ c)$, where c is different from a and b (note that $n \geq 3$). If $\tau\tau' = (a\ b)(a\ c)$, then $\tau\tau' = (a\ c\ b)$. If $\tau\tau' = (a\ b)(c\ d)$, then $\tau\tau' = (a\ b\ c)(b\ c\ d)$.*

Corollary 10.4 *If $n \geq 5$, then S_n is not solvable.*

Proof 10.20 *First note that $A'_n = A_n$. Indeed, by Lemma 10.5, it is enough to show any 3-cycle is a commutator. Take any 3-cycle $(a\ b\ c)$ and d and e distinct from a, b, c (note that $n \geq 5$). Then*

$$(a\ b\ c) = (a\ b\ d)(a\ c\ f)(a\ d\ b)(a\ f\ c) = [(a\ b\ d), (a\ c\ f)].$$

Second note that $S'_n = A_n$. Indeed, $S'_n \subseteq A_n$, since $A_n \triangleleft S_n$ and $S_n/A_n \cong \mathbb{Z}_2$ abelian (see Theorem 10.9). Then $A_n = A'_n \subseteq S'_n$ yields the equality. But then for any k, $S_n^{(k)} = A_n \neq 1$ and so by Proposition 10.2, S_n is not solvable.

EXERCISES

1 Check that the subgroup N in Example 10.15.3 is the Klein-4 group.

2 Verify that the factor group G_{i+1}/H in the proof of Theorem 10.1 is abelian.

3 Prove that for a group G we have $G' \triangleleft G$.

4 If the derived series of G terminates at 1 at some point i.e. there exists a natural number n such that $G^{(n)} = 1$, then G is solvable.

5 If $H \leq G$, then $H^{(n)} \leq G^{(n)}$ for any natural number n.

10.10   SOLVABILITY BY RADICALS

Any quadratic polynomial $ax^2 + bx + c$ has roots which are solvable by radicals (formal definition to come) via the quadratic formula

$$\frac{-b \pm \sqrt{b^2 - 4ac}}{2a}.$$

In the 1500's italian mathematicians demonstrated that first any cubic polynomial and later any quartic polynomial is solvable by radicals. The quartic is rather messy and long to present here, but we will now show the solution for the cubic. Consider any cubic polynomial $ax^3 + bx^2 + cx + d$. Without loss of generality we will find the roots of the monic cubic $x^3 + bx^2 + cx + d$ (simply multiply both sides of $ax^3 + bx^2 + cx + d = 0$

by a^{-1}). Again, without loss of generality, we consider cubic polynomials of the form $x^3 + cx + d$ (replace x by $x - \frac{b}{3}$ in the monic cubic above). Then the roots of $x^3 + cx + d$ can be shown to be of the form

$$\frac{1}{3}(\sqrt[3]{z_1} + \sqrt[3]{z_2}), \qquad \zeta\sqrt[3]{z_1} + \zeta^2\sqrt[3]{z_2}, \qquad \zeta^2\sqrt[3]{z_1} + \zeta\sqrt[3]{z_2},$$

where

$$\zeta = -\frac{1}{2} + \frac{\sqrt{3}}{2}i, \qquad z_1 = \frac{1}{2}(-27d + 3\sqrt{3}\Delta i), \qquad z_2 = \frac{1}{2}(-27d - 3\sqrt{3}\Delta i),$$

and

$$\Delta = \sqrt{-4c^3 - 27d^3}.$$

Many years passed with people attempting to find a formula for the quintic. In the 1800's it was suggested by LaGrange that perhaps there were no such formula, but it was Abel who at the age of 19 proved once and for all that there was no general formula for the quintic and thus put to rest the search for a general formula. Later in the century Galois gave a criterion for when a polynomial will have a solution by radicals. It was then shown that polynomials of a degree $n \geq 5$ have no general formula for solution by radicals.

Our first goal is to obtain this criterion set out by Galois, which is aptly called the *Galois Criterion for Solution by Radicals*.

We now present the formal definition of what it means for a polynomial to be solvable by radicals.

Definition 10.16 *Let F be a field. A simple extension $F(a)$ is called a **radical extension** if $a^n \in F$. The element a is called an **nth root** of a^n.*

Note that a radical extension is an algebraic one, since the nth root a is a root of $x^n - a^n \in F[x]$.

Example 10.17 *Here, we give some examples of radical extensions.*

1. $\mathbb{Q}(\sqrt{2})$ *is a radical (quadratic) extension of $\mathbb{Q}$, since $(\sqrt{2})^2 = 2 \in \mathbb{Q}$.*

2. *Set $\zeta = -\frac{1}{2} + \frac{\sqrt{3}}{2}i$. Then $\mathbb{Q}(\zeta)$ is a radical (cubic) extension of $\mathbb{Q}$, since $\zeta^3 = 1 \in \mathbb{Q}$.*

Definition 10.17 *Let F be a field. A polynomial $f(x) \in F[x]$ is **solvable by radicals** if there exists a chain of radical extensions $F = F_0 \subseteq F_1 \subseteq \cdots \subseteq F_n$ (i.e. $F_{i+1} = F_i(a_i)$ where $a_i^{n_i} \in F_i$ for $i = 0, 1, \ldots, n - 1$) such that all the roots of $f(x)$ lie in F_n. The chain of subfields is then called a **root tower** for $f(x)$.*

Remark 10.6 *A chain of fields $F = F_0 \subseteq F_1 \subseteq \cdots \subseteq F_n$ such that F_n is a root tower over F is equivalent to the property that $a_1^{n_1} \in F$ and $a_i^{n_i} \in F(a_1, a_2, \ldots, a_{i-1})$ for positive integers $i = 2, \ldots, n$. In particular, $F_n = F(a_1, a_2, \ldots, a_n)$.*

Example 10.18 *Here, we give some examples of root towers.*

1. *Consider the polynomial $f(x) = x^2 + 4x - 1 \in \mathbb{Q}[x]$ whose roots are $-2 \pm \sqrt{3}$. Then $f(x)$ is solvable by radicals via the root tower $\mathbb{Q} = F_0 \subseteq F_1 = \mathbb{Q}(\sqrt{3})$.*

2. *In general, any quadratic polynomial $f(x) = ax^2 + bx + c \in \mathbb{Q}[x]$ is solvable by radicals via the root tower $\mathbb{Q} = F_0 \subseteq \mathbb{Q}(\sqrt{b^2 - 4ac})$.*

3. *Consider the polynomial $f(x) = (x^2 + 4x - 1)(x^2 + x + 1)$ which has roots $-2 \pm \sqrt{3}$, $-\frac{1}{2} \pm \frac{\sqrt{3}}{2}i$. Then $f(x)$ is solvable by radicals via the root tower*

$$\mathbb{Q} = F_0 \subseteq F_1 = \mathbb{Q}(\sqrt{3}) \subseteq \mathbb{Q}(\sqrt{3}, i).$$

Note that root towers are not unique either in length or in intermediate fields, for a given polynomial solvable by radicals. For instance, in this example, we could have made $F_1 = \mathbb{Q}(i)$.

4. *Consider the cubic polynomial $f(x) = x^3 + cx + d \in \mathbb{Q}[x]$. Using the cubic formula, we see that $f(x)$ is solvable by radicals via the root tower*

$$\mathbb{Q} \subseteq \mathbb{Q}(\sqrt{3}) \subseteq \mathbb{Q}(\sqrt{3}, i) \subseteq \mathbb{Q}(\sqrt{3}, i, \Delta) \subseteq \mathbb{Q}(\sqrt{3}, i, \Delta, \sqrt[3]{z_1})$$
$$\subseteq \mathbb{Q}(\sqrt{3}, i, \Delta, \sqrt[3]{z_1}, \sqrt[3]{z_2}).$$

Lemma 10.6 *Let E be the splitting field of $f(x) = x^n - 1 \in F[x]$ where $\operatorname{char} F = 0$. Then $\operatorname{Gal}(E/F)$ is abelian.*

Proof 10.21 *Since $f'(x) = nx^{n-1} \neq 0$ and $\gcd(f, f') = 1$ we know that $f(x)$ has distinct roots. Set H equal to the roots of $f(x)$. Note that $H \leq E^*$. Indeed, if $a, b \in H$, then $a^n = 1$ and $b^n = 1$ and so $(ab^{-1})^n = a^n (b^n)^{-1} = 1$ which implies $ab^{-1} \in H$. Since any finite subgroup of the multiplicative group of a field is cyclic it follows that H is cyclic. Enumerate the subgroup $H = \{h_1, h_2, \ldots, h_n\}$ and consider the map $\Psi : \operatorname{Gal}(E/F) \to \operatorname{Aut}(H)$ where $\Psi(\phi)(h_i) = \phi(h_i)$. First note that Ψ maps into $\operatorname{Aut}(H)$, since ϕ permutes the roots of H and is a field homomorphism. It's easy to check that Ψ is a homomorphism with trivial kernel (since ϕ is completely determined by where it sends the roots of $f(x)$). Now $\operatorname{Aut}(H) \cong \operatorname{Aut}(\mathbb{Z}_n) \cong U(\mathbb{Z}_n)$ which is abelian. Hence, $\operatorname{Gal}(E/F)$ embeds in an abelian group and therefore must be abelian.*

Definition 10.18 *The roots of $x^n - 1 \in F[x]$ with $\operatorname{char} F = 0$ are called the **nth roots of unity**. Any generator of the cyclic group of nth roots of unity is called a **primitive** nth root of unity.*

Lemma 10.7 *Let F be a field of characteristic 0 containing all the nth roots of unity. Let $c \in F$ and set $f(x) = x^n - c \in F[x]$ and let E be the splitting field of $f(x)$. Then $\operatorname{Gal}(E/F)$ is cyclic of order a divisor of n.*

Proof 10.22 *Fix $a \in E$ a root of $f(x)$ and $\zeta \in F$ a primitive nth root of unity. Observe that the n roots of $f(x)$ are of the form $\zeta^i a$ for $i = 0, 1, \ldots, n-1$, since*

$$f(\zeta^i a) = (\zeta^i a)^n - c = (\zeta^n)^i a^n - c = 1 \cdot c - c = 0.$$

Therefore, $E = F(a, \zeta a, \zeta^2 a, \ldots, \zeta^{n-1}a) = F(a)$, because F contains all the nth roots of unity. Hence, any element of $Gal(E/F)$ is completely determined by where it sends a and we know it must send a to another root of $f(x)$, i.e. $\phi(a) = \zeta^i a$ for some $i \in \{0, 1, \ldots, n-1\}$. Therefore, we can define the map $\Psi : Gal(E/F) \to \mathbb{Z}_n$ by $\Psi(\phi) = i$ where $\phi(a) = \zeta^i a$. It's easy to check that Ψ is a monomorphism and so $Gal(E/F)$ embeds in a cyclic group of order n and so must be cyclic of order a divisor of n.

Lemma 10.8 *Let p be a prime and F a field containing all the pth roots of unity. If E is a field containing F such that $[E : F] = |Gal(E/F)| = p$, then E is a radical extension of F where $E = F(u)$ and $u^p \in F$.*

Proof 10.23 *Set $G = Gal(E/F)$ and fix an $a \in E - F$. Since $[E : F]$ is prime it must be that $E = F(a)$. Set $H = \{\zeta_1, \zeta_2, \ldots, \zeta_p\}$ the pth roots of unity in F. Since $Gal(E/F)$ has prime order, it must be cyclic. Set $Gal(E/F) = \langle \phi \rangle$. Define*

$$a_1 = a, \quad a_2 = \phi(a_1), \quad a_3 = \phi(a_2), \quad \ldots, \quad a_p = \phi(a_{p-1}).$$

Set $u_i = a_1 + \zeta_i a_2 + \cdots + \zeta_i^{p-1} a_p$. Notice that

$$\phi(u_i) = \phi(a_1) + \zeta_i \phi(a_2) + \cdots + \zeta_i^{p-1} \phi(a_p) = a_2 + \zeta_i a_3 + \cdots + \zeta_i^{p-2} a_p + \zeta_i^{p-1} a_1$$

$$= \zeta^{-1}(a_1 + \zeta_i a_2 + \cdots + \zeta_i^{p-1} a_p) = \zeta_i^{-1} u_i.$$

Therefore, $\phi(u_i^p) = \zeta_i^{-p} u_i^p = u_i^p$. Hence, $u_i^p \in E^G = F$, since E is Galois over F.

Claim 10.3 *$E = F(u_i)$ for some $i \in \{1, 2, \ldots, p\}$ (which proves the lemma)*

Let's express in matrix form the identities we have derived thus far concerning the u_i's:

$$\begin{bmatrix} u_1 \\ u_2 \\ \vdots \\ u_p \end{bmatrix} = \begin{bmatrix} 1 & \zeta_1 & \zeta_1^2 & \cdots & \zeta_1^{p-1} \\ 1 & \zeta_2 & \zeta_2^2 & \cdots & \zeta_2^{p-1} \\ & & \vdots & & \\ 1 & \zeta_p & \zeta_p^2 & \cdots & \zeta_p^{p-1} \end{bmatrix} \begin{bmatrix} a_1 \\ a_2 \\ \vdots \\ a_p \end{bmatrix}$$

*The coefficient matrix is called the **Vandermonde** matrix and one can show it is invertible by computing the determinant to be $\prod_{i<j}(\zeta_j - \zeta_i) \neq 0$, since the ζ_i's are distinct and $char F = p$. So the linear system has a solution in $a_1, a_2, \ldots, a_p$ in terms of $u_1, u_2, \ldots, u_p$. In particular, $a = a_1 \in F(u_1, u_2, \ldots, u_p)$. Thus, not all the u_i are in F for otherwise $a \in F$, a contradiction. Let i be such that $u_i \notin F$. Then $E = F(u_i)$ (again, since $[E : F]$ is prime).*

Lemma 10.9 *If K has a root tower over F of some polynomial in F, then there exists a field $E \supseteq K$ such that E is normal over F and E has a root tower over F of some polynomial over F.*

Proof 10.24 *By Assumption, $K = F(a_1, \ldots, a_n)$ where $a_1^{n_1} \in F$ and $a_i^{n_i} \in F(a_1, a_2, \ldots, a_{i-1})$ for $i = 2, 3, \ldots, n$. Let $f(x)$ be the product of all the minimal polynomials of the a_i over F and set E to be the splitting field of $f(x)$. Since E is a splitting field of a polynomial over F we know that E is normal over F. Since the elements of $Gal(E/F)$ are determined by where they send the roots of $f(x)$, each root of $f(x)$ has the form $\phi(a_i)$ for some $i \in \{1, 2, \ldots, n\}$ and some $\phi \in Gal(E/F)$. Notice also that for a given $\phi \in Gal(E/F)$ we have $\phi(a_1)^{n_1} = \phi(a_1^{n_1}) = a_1^{n_1} \in F$ and for $i = 2, 3, \ldots, n$ we have*

$$\phi(a_i)^{n_i} = \phi(a_i^{n_i}) \in \phi(F(a_1, a_2, \ldots, a_{i-1})) = F(\phi(a_1), \phi(a_2), \ldots, \phi(a_{i-1})),$$

since the image of an element in $F(a_1, a_2, \ldots, a_n)$ by ϕ is completely determined by where it sends $a_1, a_2, \ldots, a_n$. Enumerate the elements in $Gal(E/F) = \{\phi_1, \phi_2, \ldots, \phi_k\}$. Using the elements $\{\phi_j(a_i) : 1 \le i \le n, 1 \le j \le k\}$ we can construct a root tower for E over F.

We will need the Fundamental Theorem of Algebra for our final result which will illustrate the insolvability of the quintic. Our goal for the moment will be this important result. But first we need some additional results about separability.

Lemma 10.10 *If E is algebraic over F and $char(F) = 0$, then E is separable over F.*

Proof 10.25 *Take any $a \in E$ and set $p(x) \in F[x]$ to be the irreducible polynomial of a over F. Since $deg(p') < deg(p)$ and $p(x)$ is irreducible, it follows that $gcd(p, p') = 1$, and since $char(F) = 0$ it must also be the case that $p'(x)$ is not the zero polynomial. Hence, by an earlier result, $p(x)$ has no multiple roots.*

Lemma 10.11 *If E is finite and separable extension of F, then E is a simple extension.*

Proof 10.26 *We may assume that F is infinite, for it is always the case that a finite extension of a finite field is a simple extension. Since E is a finite and algebraic extension of F, we know that E is finitely generated over F. Set $E = F(a_1, \ldots, a_n)$ and the proof will be by induction on n, but to avoid a messy presentation we will simply prove the case of two generators and this proof easily generalizes. So set $E = F(a, b)$, let $p(x)$ be the minimal polynomial of a over F of degree k and $q(x)$ the minimal polynomial of b over F of degree m. Let L be a field in which both $p(x)$ and $q(x)$ split. Since a and b are both separable over F the distinct roots of $p(x)$ and $q(x)$ are $a = a_1, a_2, \ldots, a_k$ and $b = b_1, b_2, \ldots, b_m$ (respectively). Choose $c \in F$ such that $a_i + b_j c \ne a + bc$ for all $i = 1, 2, \ldots, k$ and $j = 2, 3, \ldots, m$ (such a c exists since $a_i + b_j x = a + bx$ has a unique solution, namely $(a - a_i)(b_j - b)^{-1}$, and F is infinite). Set $t = a + cb$.*

Claim 10.4 $E = F(t)$.

Certainly, $F(t) \subseteq E$, *since* $t \in E$. *For the reverse inclusion, its enough to show that* $b \in F(t)$, *for then* $a = t - cb \in F(t)$ *and so* $F(a,b) \subseteq F(t)$. *In order to show* $b \in F(t)$, *we show that the irreducible polynomial of* b *over* $F(t)$ *has degree 1 from which it follows that* $[F(b,t) : F(t)] = 1$ *and so* $F(b,t) = F(t)$ *and* $b \in F(t)$.

Let $r(x)$ *be the irreducible polynomial of* b *over* $F(t)$ *and set* $f(x) = p(t - cx) \in F(t)[x]$. *Since* b *is a root of both* $f(x)$ *and* $q(x)$ *it follows that* r *is a common divisor of* f *and* q *in* $F(t)[x]$. *Now* $t - cb_j \neq a_i$ *for* $i = 1, 2, \ldots, k$ *and* $j = 2, 3, \ldots, m$ *by how* c *was defined and so* b_j *is not a root of* $f(x)$ *for* $j = 2, 3, \ldots, m$. *Hence,* $x - b_j$ *does not divide* $f(x)$ *in* $F(t)[x]$ *for* $j = 2, 3, \ldots, m$. *Now* $q(x) = (x-b_1)(x-b_2)\cdots(x-b_m)$ *in* $F(t)[x]$ *and so* $\gcd(f, q) = x - b_1 = r(x)$.

We now prove the Fundamental Theorem of Algebra. We make use of four facts which we do not expressly prove here, but are easily attainable.

1. Any polynomial $f(x) \in \mathbb{R}[x]$ of odd degree has a real root.

2. Every complex number has square roots which are also complex numbers. Hence, there are no irreducible quadratics over $\mathbb{C}$ and so there are no quadratic extensions of $\mathbb{C}$.

3. If $f(x) \in \mathbb{C}[x]$, then $f(x)\overline{f}(x) \in \mathbb{R}$, where $\overline{f}(x)$ is the polynomial obtained by replacing all the coefficients in $f(x)$ by their complex conjugates.

Theorem 10.11 (Fundamental Theorem of Algebra) $\mathbb{C}$ *is algebraically closed.*

Proof 10.27 *We use the Fundamental Theorem of Galois Theory so many times in this proof that for brevity we shall call it FTG. Let* $f(x) \in \mathbb{C}[x]$ *and we show* $f(x)$ *has a root in* $\mathbb{C}$. *Let* E *be the splitting field of* $f(x)$ *over* $\mathbb{C}$ *and set* $g(x) = (x^2+1)f(x)\overline{f}(x) \in \mathbb{R}[x]$. *Let* L *be the splitting field of* $g(x)$ *over* $\mathbb{R}$. *Note that* $\mathbb{C} \subseteq E \subseteq L$, *since* L *contains* $\mathbb{R}$, i, *and the roots of* $f(x)$. *We will show that* $L = \mathbb{C}$ *and so* $E = \mathbb{C}$ *and the result is proved. Since* $char(\mathbb{R}) = 0$ *we know by Lemma 10.10 that* L *is Galois over* $\mathbb{R}$. *Set* $G = Gal(L/\mathbb{R})$, *let* H *be a 2-Sylow of* G *(which may be trivial) and* $K = L^H$. *Certainly,* $[G : H]$ *is odd and so by FTG, so is* $[K : \mathbb{R}]$. *Since* $char(\mathbb{R}) = 0$, *by Lemma 10.10 and Lemma 10.11, we know there is an* $a \in K$ *such that* $K = \mathbb{R}(a)$. *Therefore,* $[\mathbb{R}(a) : \mathbb{R}]$ *is odd and so the irreducible polynomial of* a *over* $\mathbb{R}$ *is odd. However, we know every polynomial of odd degree over* $\mathbb{R}$ *has a real root, thus it must be the case that the minimal polynomial of* a *over* $\mathbb{R}$ *is linear. But then* $[K : \mathbb{R}] = [\mathbb{R}(a) : \mathbb{R}] = 1$ *and so* $K = \mathbb{R}$ *which in turn implies* $G = H$ *is a 2-group. Set* $G_1 = Gal(L/\mathbb{C})$ *which is a 2-subgroup of* G.

Claim 10.5 $G_1 = 1$.

Suppose not. Set $|G_1| = 2^k$. *By the First Sylow Theorem, there exists a subgroup of* G_1 *of order* 2^{k-1} *and since it has index 2 in* G_1 *it must be normal in* G_1. *Call it* $N \lhd G_1$ *and set* $K = L^N$. *By FTG,* $[K : \mathbb{C}] = [G_1 : N] = 2$. *Since* $char(\mathbb{C}) = 0$, *by Lemma 10.10 and Lemma 10.11, $K = \mathbb{C}(a)$ for some $a \in K$. But this contradicts that $\mathbb{C}$ has no quadratic extension and thus the claim is proved.*

Having proved the claim we see now that $Gal(L/\mathbb{C}) = G_1 = 1$ and so $[L : \mathbb{C}] = |Gal(L/\mathbb{C})| = 1$, which implies that $L = \mathbb{C}$, and so $E = \mathbb{C}$.

Theorem 10.12 (Galois Criterion for Solvability by Radicals) *Let $f(x) \in F[x]$ and $char(F) = 0$. Let E be the splitting field of $f(x)$ over F. Then $f(x)$ is solvable by radicals iff $Gal(E/F)$ is a solvable group.*

Proof 10.28 *First assume that $f(x)$ is solvable by radicals. Therefore, there exists $F = F_0 \subseteq F_1 \subseteq \cdots \subseteq F_m = L$ a root tower for $f(x)$ over F. Let E be the splitting field for $f(x)$ over F. Since E is the smallest field containing F and the roots of $f(x)$ we know that $E \subseteq L$. By Lemma 10.9, we may assume that L is normal over F and so L is the splitting field of some polynomial $g(x) \in F[x]$ over F. L being a root tower of $f(x)$ over F we know that each $F_{i+1} = F_i(a_i)$ with $a_i^{n_i} \in F_i$ for some positive integer n_i (i=0,1,...,m-1). Set n equal to the least common multiple of all the n_i's and ζ a primitive nth root of unity. Note that $F(\zeta) = F_0(\zeta) \subseteq F_1(\zeta) \subseteq \cdots \subseteq F_m(\zeta) = L(\zeta)$ is a chain of radical extensions as well as the splitting field of $g(x)(x^n - 1)$. Thus, $L(\zeta)$ is normal over $F(\zeta)$ and since $char(F) = 0$, by Lemma 10.10, $L(\zeta)$ is Galois over $F(\zeta)$. Since $F \subseteq E \subseteq L(\zeta)$ and E is normal over F, by FTG, we know that*

$$Gal(E/F) \cong Gal(L(\zeta)/F)/Gal(L(\zeta)/E).$$

Thus, it's enough to show $Gal(L(\zeta)/F)$ is a solvable group, by Theorem 10.10.

Claim 10.6 *$Gal(L(\zeta)/F)$ is a solvable group.*

Set $K_i = F(\zeta, a_1, \ldots, a_i)$ and $N_i = Gal(L(\zeta)/K_i)$ for $i = 1, 2, \ldots, m$, $G = Gal(L(\zeta)/F)$ and $N = Gal(L(\zeta)/F(\zeta))$. We will show that $1 \lhd N_m \lhd \cdots \lhd N_1 \lhd N \lhd G$ is a series of subgroups of G such that the quotient group of adjacent subgroups is abelian and so G is solvable. Note that by how ζ was defined, $\zeta^{(n/n_i)}$ is a primitive n_ith root of unity and so K_i is the splitting field of $x^{n_i} - \zeta^{(n/n_i)}a_i$ over F_{i-1} for $i = 2, 3, \ldots, m$. Hence, F_i is normal over F_{i-1} and by FTG, $N_{i-1} \lhd N_i$ for $i = 2, 3, \ldots, m$. Furthermore, by Lemma 10.7, $N_{i-1}/N_i \cong Gal(F_i/F_{i-1})$ is cyclic and therefore abelian. Finally, $N \lhd G$ and by Lemma 10.6, $G/N \cong Gal(F(\zeta)/F)$ is abelian.

Therefore, to prove the insolvability of the quintic, it is enough to produce a polynomial of degree five such that the Galois group of its splitting field over its field of coefficients is not solvable. We will do exactly this by constructing a polynomial whose Galois group is S_5 which we know is not solvable. This construction can be generalized to any prime $p \geq 5$ and so this construction can be used to show the insolvability of polynomials of degree any prime ≥ 5. For simplicity we will just show $p = 5$. We decided not to show the insolvability of polynomials of arbitrary degree ≥ 5, since this involves more preliminaries regarding what are called *symmetric* functions.

Consider the function $g(x) = (x^2 + k)(x - l)(x - m)(x - n)$ where k, l, m, n are even integers, $k > 0$ and $l < m < n$. Certainly $g(x)$ has exactly three real roots,

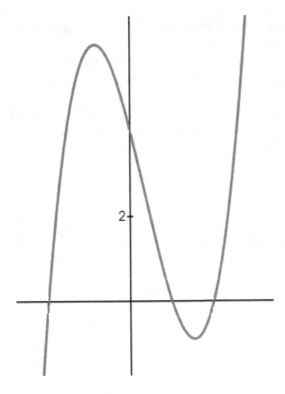

Figure 10.3 The graph of a fifth degree polynomial.

namely l, m, n and by consider sign changes for outputs of $g(x)$, it is clear the graph has the general shape depicted in Figure 10.3.

Therefore, $g(x)$ has exactly one local maximizer c and one local minimizer d. Now the local maximum $g(c)$ has value larger than 2, since at any odd integer r, we have

$$|g(r)| = |r^2 + k||r - l||r - m||r - n| \geq |r^2 + k| = r^2 + k > 2.$$

Hence, the polynomial $f(x) = g(x) - 2$ has the same local minimizer and maximizer with $f(c) > 0$ and so $f(x)$ also has exactly three real roots. The reason we shifted from $g(x)$ to $f(x)$ is because (as we shall see) $f(x)$ is irreducible, a property we will need. To see why $f(x)$ is irreducible, first note that all but the leading coefficient of $g(x)$ is divisible by 2 and its constant coefficient is divisible by 4. Therefore, the same holds true for $f(x)$ except that now 4 does not divide its constant coefficient. Therefore, by Eisenstein's Criterion using $p = 2$ it follows that $f(x)$ is irreducible over $\mathbb{Q}$.

Lemma 10.12 *If a subgroup H of S_5 contains a transposition and a 5-cycle, then $H = S_5$.*

Proof 10.29 *By renumbering we may assume that $(1\ 2) \in H$. For the 5-cycle $\sigma \in H$ there exists a positive integers k, m, n such that $(1\ 2\ k\ m\ n) \in H$. Again, by renumbering we may assume that $(1\ 2\ 3\ 4\ 5) \in H$. But then $(1\ 2)(1\ 2\ 3\ 4\ 5) =$*

$(2\ 3\ 4\ 5) \in H$. *Therefore,*

$$(2\ 3\ 4\ 5)^2 = (2\ 4)(3\ 5) \in H \qquad and \qquad (2\ 3\ 4\ 5)^3 = (2\ 5\ 4\ 3) \in H.$$

$$(2\ 3\ 4\ 5)(1\ 2)(2\ 5\ 4\ 3) = (1\ 3) \in H,$$

$$(2\ 4)(3\ 5)(1\ 2)(2\ 4)(3\ 5) = (1\ 4) \in H, \ and$$

$$(2\ 5\ 4\ 3)(1\ 2)(2\ 3\ 4\ 5) = (1\ 5) \in H.$$

Therefore, for any transposition $(m\ n) \in S_5$, *we have* $(m\ n) = (1\ n)(1\ m)(1\ n) \in H$. *Since any permutation is a product of transpositions, it follows that* $S_5 = H$.

Lemma 10.13 *Let* $f(x) = (x^2 + k)(x - l)(x - m)(x - n) - 2$ *where* k, l, m, n *are even integers,* $k > 0$ *and* $l < m < n$ *and* E *the splitting field of* $f(x)$ *over* $\mathbb{Q}$. *Then* $Gal(E/\mathbb{Q}) \cong S_5$.

Proof 10.30 *Since* $\mathbb{C}$ *is algebraically closed we know that* $E \subseteq \mathbb{C}$. *Set* $G = Gal(E/\mathbb{Q})$ *and since each element of* G *permutes the 5 roots of* $f(x)$ *we know* $G \leq S_5$. *Let* $a \in E$ *be one of the five roots of* $f(x)$ *and set* $K = \mathbb{Q}(a)$. *Since* $f(x)$ *is irreducible over* $\mathbb{Q}$ *we know* $[K : \mathbb{Q}] = 5$. *Since* $char(\mathbb{Q}) = 0$ *we know that* E *is Galois over* $\mathbb{Q}$ *and so by FTG,* $[G : Gal(E/K)] = [K : \mathbb{Q}] = 5$ *which implies* 5 *divides* $|G|$. *By Cauchy's Lemma,* G *has an element of order* 5 *which corresponds to a 5-cycle in* S_5. *Since* $f(x)$ *has exactly 3 real roots and 2 complex conjugate roots, it follows that the map in* G *which sends one complex root to its conjugate has order two and thus corresponds to a transposition in* S_5. *Therefore, by Lemma 10.12, it follows that* $G \cong S_5$.

Corollary 10.5 *Not every polynomial of degree 5 is solvable by radicals.*

Proof 10.31 *Apply the Galois Criterion for Solvability by Radicals to the polynomial defined in Lemma 10.13 which has Galois group* S_5 *which we know is not a solvable group.*

EXERCISES

1 In the proof of Lemma 10.6, check that Ψ is a homomorphism with trivial kernel.

2 In the proof of Lemma 10.7, check that Ψ is a monomorphism.

3 Verify that the Vandermonde matrix is invertible.

4 Prove the following three facts that were used in the Fundamental Theorem of Algebra:

a. Any polynomial $f(x) \in \mathbb{R}[x]$ of odd degree has a real root.

b. Every complex number has square roots which are also complex numbers. Hence, there are no irreducible quadratics over $\mathbb{C}$ and so there are no quadratic extensions of $\mathbb{C}$.

c. If $f(x) \in \mathbb{C}[x]$, then $f(x)\overline{f}(x) \in \mathbb{R}$, where $\overline{f}(x)$ is the polynomial obtained by replacing all the coefficients in $f(x)$ by their complex conjugates.

5 In the proof of Theorem 10.12, verify that all but the leading coefficient of $g(x)$ is divisible by 2 and its constant coefficient is divisible by 4.

6 In the proof of Lemma 10.12, explain why for the 5-cycle $\sigma \in H$ there exists a positive integers k, m, n such that $(1\ 2\ k\ m\ n) \in H$.

References

[1] M. Artin. *Algebra*. Prentice Hall, NY, USA, 2nd edition, 2011.

[2] J.A. Beachy and W.D. Blair. *Abstract Algebra*. Waveland Press, Inc., IL, USA, 2nd edition, 1996.

[3] J.B. Fraleigh. *A First Course in Abstract Algebra*. Addison Wesley, NY, USA, 7th edition, 2002.

[4] N. Jacobson. *Basic Algebra I*. W.H. Freeman and Co., NY, USA, 2nd edition, 1985.

[5] D.J.S. Robinson. *A Course in the Theory of Groups*. Springer-Verlag, NY, USA, 1st edition, 1993.

Index

Printed in the United States
by Baker & Taylor Publisher Services